High Performance Functional Bio-based Polymers for Skin-contact Products

High Performance Functional Bio-based Polymers for Skin-contact Products

Editors

Maria Beatrice Coltelli
Serena Danti

MDPI • Basel • Beijing • Wuhan • Barcelona • Belgrade • Manchester • Tokyo • Cluj • Tianjin

Editors
Maria Beatrice Coltelli
University of Pisa
Italy

Serena Danti
University of Pisa
Italy

Editorial Office
MDPI
St. Alban-Anlage 66
4052 Basel, Switzerland

This is a reprint of articles from the Special Issue published online in the open access journal *Journal of Functional Biomaterials* (ISSN 2079-4983) (available at: https://www.mdpi.com/journal/jfb/special_issues/Skin_contact_Products).

For citation purposes, cite each article independently as indicated on the article page online and as indicated below:

LastName, A.A.; LastName, B.B.; LastName, C.C. Article Title. *Journal Name* **Year**, *Volume Number*, Page Range.

ISBN 978-3-03943-861-7 (Hbk)
ISBN 978-3-03943-862-4 (PDF)

Cover image courtesy of Maria-Beatrice Coltelli.

© 2020 by the authors. Articles in this book are Open Access and distributed under the Creative Commons Attribution (CC BY) license, which allows users to download, copy and build upon published articles, as long as the author and publisher are properly credited, which ensures maximum dissemination and a wider impact of our publications.

The book as a whole is distributed by MDPI under the terms and conditions of the Creative Commons license CC BY-NC-ND.

Contents

About the Editors vii

Preface to "High Performance Functional Bio-based Polymers for Skin-contact Products" . . . ix

Maria-Beatrice Coltelli and Serena Danti
Biobased Materials for Skin-Contact Products Promoted by POLYBIOSKIN Project
Reprinted from: J. Funct. Biomater. 2020, 11, 77, doi:10.3390/jfb11040077 1

Maria-Beatrice Coltelli, Laura Aliotta, Alessandro Vannozzi, Pierfrancesco Morganti,
Luca Panariello, Serena Danti, Simona Neri, Cristina Fernandez-Avila, Alessandra Fusco,
Giovanna Donnarumma and Andrea Lazzeri
Properties and Skin Compatibility of Films Based on Poly(Lactic Acid) (PLA)
Bionanocomposites Incorporating Chitin Nanofibrils (CN)
Reprinted from: J. Funct. Biomater. 2020, 11, 21, doi:10.3390/jfb11020021 5

Maria-Beatrice Coltelli, Luca Panariello, Pierfrancesco Morganti, Serena Danti,
Adone Baroni, Andrea Lazzeri, Alessandra Fusco and Giovanna Donnarumma
Skin-Compatible Biobased Beauty Masks Prepared by Extrusion
Reprinted from: J. Funct. Biomater. 2020, 11, 23, doi:10.3390/jfb11020023 29

Pooja Basnett, Elena Marcello, Barbara Lukasiewicz, Rinat Nigmatullin, Alexandra Paxinou,
Muhammad Haseeb Ahmad, Bhavana Gurumayum and Ipsita Roy
Antimicrobial Materials with Lime Oil and a Poly(3-hydroxyalkanoate) Produced via
Valorisation of Sugar Cane Molasses
Reprinted from: J. Funct. Biomater. 2020, 11, 24, doi:10.3390/jfb11020024 49

Aleksandra Miletić, Ivan Ristić, Maria-Beatrice Coltelli and Branka Pilić
Modification of PLA-Based Films by Grafting or Coating
Reprinted from: J. Funct. Biomater. 2020, 11, 30, doi:10.3390/jfb11020030 63

Kudirat A. Obisesan, Simona Neri, Elodie Bugnicourt, Inmaculada Campos
and Laura Rodriguez-Turienzo
Determination and Quantification of the Distribution of CN-NL Nanoparticles Encapsulating
Glycyrrhetic Acid on Novel Textile Surfaces with Hyperspectral Imaging
Reprinted from: J. Funct. Biomater. 2020, 11, 32, doi:10.3390/jfb11020032 79

Bahareh Azimi, Lily Thomas, Alessandra Fusco, Ozlem Ipek Kalaoglu-Altan, Pooja Basnett,
Patrizia Cinelli, Karen De Clerck, Ipsita Roy, Giovanna Donnarumma,
Maria-Beatrice Coltelli, Serena Danti and Andrea Lazzeri
Electrosprayed Chitin Nanofibril/Electrospun Polyhydroxyalkanoate Fiber Mesh as Functional
Nonwoven for Skin Application
Reprinted from: J. Funct. Biomater. 2020, .11, 62, doi:10.3390/jfb11030062. 95

Clément Lacoste, Benjamin Gallard, José-Marie Lopez-Cuesta, Olzem Ipek Kalaoglu-Altan
and Karen De Clerck
Development of Bionanocomposites Based on Poly(3-Hydroxybutyrate-co-3-Hydroxyvalerate)
/PolylActide Blends Reinforced with Cloisite 30B
Reprinted from: J. Funct. Biomater. 2020, 11, 64, doi:10.3390/jfb11030064 113

Maria-Beatrice Coltelli, Serena Danti, Karen De Clerk, Andrea Lazzeri and Pierfrancesco Morganti
Pullulan for Advanced Sustainable Body- and Skin-Contact Applications
Reprinted from: *J. Funct. Biomater.* **2020**, *11*, 20, doi:10.3390/jfb11010020 . . . . **125**

Chayane Karla Lucena de Carvalho, Beatriz Luci Fernandes and Mauren Abreu de Souza
Autologous Matrix of Platelet-Rich Fibrin in Wound Care Settings: A Systematic Review of Randomized Clinical Trials
Reprinted from: *J. Funct. Biomater.* **2020**, *11*, 31, doi:10.3390/jfb11020031 - **143**

Mario Milazzo, Giuseppe Gallone, Elena Marcello, Maria Donatella Mariniello, Luca Bruschini, Ipsita Roy and Serena Danti
Biodegradable Polymeric Micro/Nano-Structures with Intrinsic Antifouling/Antimicrobial Properties: Relevance in Damaged Skin and Other Biomedical Applications
Reprinted from: *J. Funct. Biomater.* **2020**, *11*, 60, doi:10.3390/jfb11030060 . . . . **157**

Bahareh Azimi, Homa Maleki, Lorenzo Zavagna, Jose Gustavo De la Ossa, Stefano Linari, Andrea Lazzeri and Serena Danti
Bio-Based Electrospun Fibers for Wound Healing
Reprinted from: *J. Funct. Biomater.* **2020**, *11*, 67, doi:10.3390/jfb11030067 . . . . **177**

About the Editors

Maria Beatrice Coltelli, Associate Professor of Materials Science and Technology of the University of Pisa, Department of Civil and Industrial Engineering, degree in Chemistry and Ph.D. in Chemical Science at the University of Pisa, has 20 years of experience on material science and technology, in particular the characterization and study of the correlations between molecular structure, morphology and properties of polymers, biopolymers, blends and composites, especially those related to their relationship with the environment. She co-founded the SPIN-PET company aimed at facilitating material recycling and considering circularity aspects linked to the production and use of polymers and materials. She is the author and co-author of more than 70 publications in international journals, a book related to the reuse of polymers and three polymers. She followed the scientific activity of European projects on Material Science, Advanced Materials and Nanotechnology. Currently, she is participating in three European BioBased Industry Joint Undertaking (BBI JU) funded projects promoting the use of renewable and biodegradable materials in packaging, personal care, cosmetic and sanitary applications. She is pro-tempore president of European Network of MATerials Research Centers (ENMAT). She has supported, for many years, the didactical activity of degree courses for Chemical Engineers (courses: Materials Science and Engineering and Science and Tchnology of Polymeric Materials) and Materials and Nanotechnology master courses (course: Reactive processing and recycling of polymers).

Serena Danti, Senior Researcher, is Professor of Biomaterials for the Master of Science degree in Materials and Nanotechnology (University of Pisa). She received an M.S. in Chemical Engineering from the University of Pisa in 2003, with a major in Materials Science. She obtained a Ph.D. in "Health Technologies: evaluation and management of Innovations in Biomedical field" from the Center for Excellence in Computer-Assisted Surgery (ENDOCAS), Faculty of Medicine and Surgery, University of Pisa, in 2007. In 2006, she was a visiting scholar at Rice University, Houston, TX, USA. Her postdoctoral training was performed at the Center for the Clinical use of Stem Cells, Faculty of Medicine and Surgery, University of Pisa and at the Otorhinolaryngology Unit, Cisanello Hospital, Pisa (Italy). She is affiliate researcher at Massachusetts Institute of Technology (MIT), Cambridge, MA, USA since 2017, and at the BioRobotics Institute, Scuola Superiore Sant'Anna, Pontedera (Pisa) since 2014. She is a member of 3R's center (Italy). Dr. Danti is involved in international research projects and has published about 90 papers indexed in Scopus, including 12 book chapters. She is an author of more than 120 presentations for national and international conferences, including presentations on biomaterials, tissue engineering and regenerative medicine, ear biology, among others. Her research interests are in smart biomaterials and technologies for tissue regeneration and pathology resolutions, including skin, bone, ear and cancer.

Preface to "High Performance Functional Bio-based Polymers for Skin-contact Products"

The personal care, cosmetic and biomedical industries deal with high-value and/or large-volume consumption of polymer-based products that are often derived from fossil sources. Although several bio-based alternatives have been the subject of recent research, more effort is still needed to increase their specific functionalities and performances in order to proceed with their true translation into the market. Recently, many researchers, such as those involved in the European POLYBIOSKIN project funded by BBI JU, are working in the field of biomaterials with anti-microbial, anti-inflammatory and anti-oxidant properties, as well as biobased materials, which are renewable and biodegradable in the environment. Both types can be transformed into final products by innovative technologies, allowing for the control of bulk and surface properties down to the nanostructure. By merging such biomedical functionalities and environmental aspects, new research could have a great impact on skin-contact biomaterials. The present issue aims to gather research and review papers where the attention paid to health and environmental impact is efficiently integrated, thus considering both source and final waste management.

Maria Beatrice Coltelli, Serena Danti
Editors

Editorial

Biobased Materials for Skin-Contact Products Promoted by POLYBIOSKIN Project

Maria-Beatrice Coltelli * and Serena Danti

Department of Civil and Industrial Engineering, University of Pisa, 56122 Pisa, Italy; serena.danti@unipi.it
* Correspondence: maria.beatrice.coltelli@unipi.it

Received: 26 October 2020; Accepted: 26 October 2020; Published: 29 October 2020

The skin is the body outermost tissue and acts as a barrier and defense line to protect our organs. As such, it is the first one to be in contact with stressing external agents and pathogens. Skin structure and function naturally reflect its physiology, and for this reason skin is provided with high capacity for self-repair [1]. Recently, due to the many (also new) substances that we are exposed to, skin shows hyperreactivity, often leading to dermatitis from irritations and allergies. More than 3000 substances are known to cause dermatitis and the sensitization time after exposure to synthetic or chemically treated materials is decreasing. Moreover, once deeply or largely injured, skin repairing and regenerative capacity cannot be completely sufficient, leading to the need of adjuvant materials. The abovementioned examples suggest why the biomaterial–skin interaction may be crucial in our daily life, when healthy skin is in touch with products for sanitary and cosmetical use, as well as for wound healing (Figure 1).

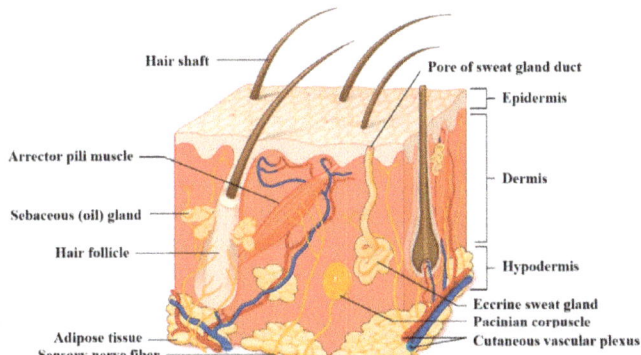

Figure 1. Structure of the skin: a superficial thin layer called epidermis is the first barrier towards the external, provided with cells apt for protection and sensation. Beneath is the dermis layer, which provides both mechanical consistency and nourishing functions; it also hosts sensory receptors and vasculature. The underneath subcutaneous layer is mainly composed of fat for thermic insulation and energy storage. Reprinted with permission from [2] (© 16 January 2020 OpenStax. Textbook content produced by OpenStax is licensed under a Creative Commons Attribution License 4.0 license).

The European project POLYBIOSKIN [3] is aimed at replacing fossil-based materials with natural origin materials for three relevant skin contact products (sanitary pads/diapers, beauty masks and wound dressings), to enable better biocompatible and eco-friendly options. The project was funded by the Bio-Based Industry Joint Undertaking (BBI JU), which is a public–private partnership between the EU and the Bio-based Industries Consortium. Operating under Horizon 2020, this EU body is driven by the Vision and Strategic Innovation and Research Agenda (SIRA) developed by the industry. The BBI

JU is dedicated to realizing the European bioeconomy potential, turning biological residues and wastes into greener everyday products through innovative technologies and biorefineries, which are at the heart of the bioeconomy. BBI's scope is bridging key sectors, creating new value chains and producing a range of innovative bio-based products to ultimately form a new bio-based community and economy.

In the framework of POLYBIOSKIN activities, several research centers and academies were involved aimed at selecting biobased materials and biomolecules compatible with skin that can show high performances. International collaborations were also activated on these relevant topics. The present Special Issue is thus offering some interesting papers and reviews regarding some strategic scientific objectives that were fundamental for the development of high-performance functional bio-based skin-contact and skin-repair products. Many papers were prepared by researchers involved in this EU project or collaborating with them from a scientific point of view. Other contributes were collected from researchers actively working on this topic around the world.

Several polymers and biomolecules fulfil the criteria of being bio-based, as they are derived from renewable biological resources, e.g., produced from living organisms. These polymers are eco-friendly, biodegradable and often highly biocompatible with the human body. The way these polymers can provide benefits when used in body contact by virtue of their functional properties is a subject of recent research.

Two main classes of bio-based polymers relevant for next generation bio-based industry—biopolyesters, such as poly(lactic acid) and polyhydroxyalkanoates (PHAs), being fully renewable and biodegradable [4,5]—as well as natural polysaccharides—cellulose/starch, pullulan and chitin/chitosan, highly available from biomass or waste food—were mainly investigated for their peculiar properties, such as absorbency and anti-infectivity [6]. In addition, proteins derived from animals (gelatin) and humans (plasma) are also considered for their important biological properties.

Functional biomolecules with specific properties were then considered for the modification of the bio-based substrates. Both incorporation in bio-based composites [7,8] and surface modifications [9] are methodologies that can be applied to bio-based and biocompatible substrates to make them anti-microbial, anti-oxidant or anti-inflammatory. These methodologies, enhancing the high performance of bio-based materials, greatly enhance the competitiveness of bio-based towards fossil-based products.

Antimicrobial and antioxidant treatments used to date commonly rely on the use of chemicals and vehicles that may result in toxic or inflammatory side effects, allergies, sensitization and irritations. Diaper dermatitis is one of the most common dermatological problems in infancy and in the elderly, but it is also becoming a gynecological problem [10,11]. Such eruptions can be subdivided into primary diaper dermatitis, an acute inflammation of the skin in the diaper area with an ill-defined and multifactorial etiology, and secondary diaper dermatitis, which refers to eruptions in the diaper area with other etiologies. The most important factors in the development of primary diaper dermatitis are due to: (i) water/moisture contact, (ii) friction, (iii) urine, (iv) feces, and (v) microorganisms. Possible treatments so far include the use of disposable diapers, barrier creams, mild topical cortisones, and antifungal agents, which can be insufficient and ineffective. In diapers and sanitary pads, the use of biobased materials on the topsheet and the modification of the surface with functional molecules can be thus fundamental for matching both the production of more healthy and environmentally friendly products.

Beauty masks are cosmetic products aimed at revitalizing the skin by releasing active substances through the stratum corneum that is active in the enzymatic synthesis of lipids. Indeed, barrier recovery and skin homeostasis are the result of restoration of these lipids, which also involves control and normalization of keratinocyte turnover. Non-woven tissues have been obtained by the gelation method and electrospinning technology, using hyaluronic acid and chitin–lignin microparticles by MAVI [12]. In particular, several possibilities were investigated in POLYBIOSKIN, considering both electrospun bio-based tissues and films obtained by the use of biopolymers like starch and PHAs.

Nonetheless, deep, wide, chronic and acute lesions can challenge skin healing and adjuvant therapies may be needed to restore its function. The injured skin needs to be immediately covered with a dressing capable of restoring tissue integrity, maintaining homeostasis, and preventing invasion of toxic substances and pathogens [13]. Therefore, a medical dressing has to establish a barrier to environmental irritants, impede microbial growth, maintain a moist environment, and allow an exchange of gaseous and nutritive ingredients. Moreover, it should not adhere to the wound, in order to allow new tissue growth, and it must be easily removable. For this purpose, special non-wovens made from engineered biomaterials are used and are biocompatible, non-allergenic, and non-toxic. They also promote wound healing by modulating extracellular matrix (ECM) synthesis and regulating microbial growth. To stimulate self-repair of skin, naturally occurring biomaterials, such as decellularized tissues or tissue-derived biomolecules are usually preferred, often in combination with specific chemical factors. The reasons underlying the use of tissue-derivatives rely on the outstanding biointegration, bioresorption and healing capacity exerted by natural biomaterials in lesion repair. However, important shortcomings are observed, like inter-batch variability and biohazard issues [5,6]. In this context, bioresorbable biomaterials offer their most valuable opportunities, not only as dressings enabling in vivo cell migration, but also, and most interestingly, as scaffolds replicating the tissue morphology with controlled chemical and physical properties.

Novel tissue engineering approaches exploiting scaffolds made of biopolymers and combined with natural antimicrobial agents offer great potential for skin regeneration, as they provide enhanced safety and reliability, owing to their controllable synthesis and fabrication processes. PHAs have been proven to have very high skin regenerative properties and will be the leading biopolymer used for the novel wound dressing developing POLYBIOSKIN. Finally, advanced analytic methods are important to detect the presence and distribution of these particles and molecules on a biomaterial surface.

Ethically sustainable tests for materials represent one priority of the European Community. Indeed, advanced cellular tools, including specific primary cell models and three dimensional (3D) tissue analogues (first of all, the "skin equivalent", which has become a widely used standard test in cosmetics) have been revealed to be highly predictive, more consistent and less expensive than animal testing. Although the in vitro tests based on human immortalized keratinocytes, representing the epidermis layer, seem the best option for studying the skin compatibility of products, novel cellular tests based on human mesenchymal stromal cells (hMSCs) can be used to validate the materials for skin-contact, because they can differentiate into fibroblasts to represent the dermis layer. hMSCs play a central role in tissue regeneration and repair, maintenance and turnover as they can differentiate into specific cell phenotypes, and/or mediate other cells' functions by secreting bioactive factors, known as the trophic function of hMSCs. For this reason, hMSCs are a powerful cellular model to evaluate the influence of new materials for skin that have still yet to be fully exploited [14]. This test, together with cytocompatibility and immunomodulation tests performed on dermal keratinocytes and fibroblasts, can improve and refine the biocompatibility test panel for skin and will serve as a spill-over for fostering the biocompatibility testing in biomedical devices.

By collecting valuable manuscripts for this Special Issue, we aim to focus the attention of the biomaterial and biomedical researchers on the great advantages that bio-based polymers can offer in the wide set of applications dealing with human hygiene, wellbeing and health, which POLYBIOSKIN started to explore, but that still require research and innovation to impact our daily lives.

Funding: This research was supported by the Bio-Based Industries Joint Undertaking under the European Union Horizon 2020 research program (BBI-H2020), PolyBioSkin project, grant number G.A 745839.

Conflicts of Interest: The authors declare no conflict of interest.

References

1. Danti, S.; D'Alessandro, D.; Mota, C.; Bruschini, L.; Berrettini, S. Applications of bioresorbable polymers in skin and eardrum. In *Bioresorbable Polymers for Biomedical Applications*; Elsevier: Amsterdam, The Netherlands, 2017; pp. 423–444.

2. Betts, J.G.; Young, K.A.; Wise, J.A.; Johnson, E.; Poe, B.; Dean, H.K.; Korol, O.; Jody, E.J.; Womble, M.; DeSaix, P. *Anatomy and Physiology*; OpenStax: Houston, TX, USA, 2013.
3. High Performance Functional Bio-Based Polymers for Skin-Contact Products in Biomedical, Cosmetic and Sanitary Industry (POLYBIOSKIN), H2020-EU.3.2.6., 745839. Available online: http://polybioskin.eu/ (accessed on 29 September 2020).
4. Misra, S.K.; Valappil, S.P.; Roy, I.; Boccaccini, A.R. Polyhydroxyalkanoate (PHA)/Inorganic Phase Composites for Tissue Engineering Applications. *Biomacromolecules* **2006**, *7*, 2249–2258. [PubMed]
5. Gigante, V.; Coltelli, M.-B.; Vannozzi, A.; Panariello, L.; Fusco, A.; Trombi, L.; Donnarumma, G.; Danti, S.; Lazzeri, A. Flat Die Extruded Biocompatible Poly(Lactic Acid) (PLA)/Poly(Butylene Succinate) (PBS) based Films. *Polymers* **2019**, *11*, 1857. [CrossRef] [PubMed]
6. Li, Q.; Wang, S.; Jin, X.; Huang, C.; Xiang, Z. The Application of Polysaccharides and Their Derivatives in Pigment, Barrier, and Functional Paper Coatings. *Polymers* **2020**, *12*, 1837. [CrossRef] [PubMed]
7. Coltelli, M.-B.; Cinelli, P.; Gigante, V.; Aliotta, L.; Morganti, P.; Panariello, L.; Lazzeri, A. Chitin Nanofibrils in Poly(Lactic Acid) (PLA) Nanocomposites: Dispersion and Thermo-Mechanical Properties. *Int. J. Mol. Sci.* **2019**, *20*, 504. [CrossRef]
8. Aliotta, L.; Vannozzi, A.; Panariello, L.; Gigante, V.; Coltelli, M.-B.; Lazzeri, A. Sustainable Micro and Nano Additives for Controlling the Migration of a Biobased Plasticizer from PLA-Based Flexible Films. *Polymers* **2020**, *12*, 1366. [CrossRef]
9. Baran, E.H.; Erbil, H.Y. Surface Modification of 3D Printed PLA Objects by Fused Deposition Modeling: A Review. *Colloids Interfaces* **2019**, *3*, 43. [CrossRef]
10. Scheinfeld, N. Diaper Dermatitis. *Am. J. Clin. Dermatol.* **2005**, *6*, 273–281. [PubMed]
11. Margesson, L.J. Contact dermatitis of the vulva. *Dermathol. Ther.* **2004**, *17*, 20–27. [CrossRef] [PubMed]
12. Morganti, P.; Febo, P.; Cardillo, M.; Donnarumma, G.; Baroni, A. Chitin Nanofibril and Nanolignin: Natural Polymers of Biomedical Interest. *J. Clin. Cosmet. Dermatol.* **2017**, *1*, 1–7. [CrossRef]
13. Cho, M.; Hunt, T.K. The Overall Approach to Wounds. In *Cutaneous Wound Healing*; Falanga, V., Ed.; Martin Dunitz: London, UK, 2001; pp. 141–154.
14. Sharma, S.; Venkatesan, V.; Prakhya, B.M.; Bhonde, R. Human mesenchymal stem cells as a novel platform for simultaneous evaluation of cytotoxicity and genotoxicity of pharmaceuticals. *Mutagenesis* **2015**, *30*, 391–399. [CrossRef] [PubMed]

Publisher's Note: MDPI stays neutral with regard to jurisdictional claims in published maps and institutional affiliations.

© 2020 by the authors. Licensee MDPI, Basel, Switzerland. This article is an open access article distributed under the terms and conditions of the Creative Commons Attribution (CC BY) license (http://creativecommons.org/licenses/by/4.0/).

Article

Properties and Skin Compatibility of Films Based on Poly(Lactic Acid) (PLA) Bionanocomposites Incorporating Chitin Nanofibrils (CN)

Maria-Beatrice Coltelli [1,2,*], Laura Aliotta [1,2], Alessandro Vannozzi [1,2], Pierfrancesco Morganti [3], Luca Panariello [1,2], Serena Danti [1], Simona Neri [4], Cristina Fernandez-Avila [4], Alessandra Fusco [1,5], Giovanna Donnarumma [1,5] and Andrea Lazzeri [1,2]

1. Department of Civil and Industrial Engineering, University of Pisa, 56122 Pisa, Italy; laura.aliotta@dici.unipi.it (L.A.); alessandrovannozzi91@hotmail.it (A.V.); luca.panariello@ing.unipi.it (L.P.); serena.danti@unipi.it (S.D.); alessandra.fusco@unicampania.it (A.F.); giovanna.donnarumma@unicampania.it (G.D.); andrea.lazzeri@unipi.it (A.L.)
2. Consorzio Interuniversitario Nazionale per la Scienza e Tecnologia dei Materiali (INSTM), 50121 Florence, Italy
3. Academy of History of Health Care Art, 00193 Rome, Italy; pierfrancesco.morganti@iscd.it
4. IRIS Technology Solutions S.L, 08860 Castelldefels, Barcelona, Spain; sneri@iris.cat (S.N.); cfernandez@iris.cat (C.F.-A.)
5. Department of Experimental Medicine, University of Campania "Luigi Vanvitelli", 80138 Naples, Italy
* Correspondence: maria.beatrice.coltelli@unipi.it; Tel.: +39-050-2217856

Received: 29 January 2020; Accepted: 18 March 2020; Published: 1 April 2020

Abstract: Nanobiocomposites suitable for preparing skin compatible films by flat die extrusion were prepared by using plasticized poly(lactic acid) (PLA), poly(butylene succinate-co-adipate) (PBSA), and Chitin nanofibrils as functional filler. Chitin nanofibrils (CNs) were dispersed in the blends thanks to the preparation of pre-nanocomposites containing poly(ethylene glycol). Thanks to the use of a melt strength enhancer (Plastistrength) and calcium carbonate, the processability and thermal properties of bionanocomposites films containing CNs could be tuned in a wide range. Moreover, the resultant films were flexible and highly resistant. The addition of CNs in the presence of starch proved not advantageous because of an extensive chain scission resulting in low values of melt viscosity. The films containing CNs or CNs and calcium carbonate resulted biocompatible and enabled the production of cells defensins, acting as indirect anti-microbial. Nevertheless, tests made with *Staphylococcus aureus* and *Enterobacter* spp. (Gram positive and negative respectively) by the qualitative agar diffusion test did not show any direct anti-microbial activity of the films. The results are explained considering the morphology of the film and the different mechanisms of direct and indirect anti-microbial action generated by the nanobiocomposite based films.

Keywords: poly(lactic acid); poly(butylene succinate); chitin nanofibrils; starch; skin compatibility; anti-microbial

1. Introduction

In the last decades, extensive academic and industrial research has been focused on the development of bio-based and biodegradable polymers with specific functional properties for decreasing the environmental impact of many general products, especially those with a very short life, and thus responsible for much of the concern related to environmental issues [1–3]. The production of biobased films with anti-microbial and anti-inflammatory properties that can be easily integrated in the current industrial processes for producing diapers or other pads can be thus extremely useful for decreasing the environmental impact of sanitary products. As an example, disposable baby diapers contribute globally about 77 million tons of solid waste to landfills, with a degradation period of at least 500 years [4]. Regarding the parts of the diaper in contact with skin, biobased and biodegradable versions with improved compatibility with skin would be a very good alternative for health reasons also. In fact, by combining biobased materials and functional biopolymers, it seems possible to limit cases of rash, dermatitis, and inflammation phenomena that are still present in the population using sanitary products. For instance, adding to films anti-microbial and/or anti-inflammatory substances can be a good strategy [5]. It is known that topical exposure to a variety of xenobiotics may result in irritant as well as allergic contact dermatitis in humans and that the cytokines play a pivotal role in immune and inflammatory reactions [6]. In this article, cytokine levels were used as a marker for skin sensitizer prediction. On the other hand, a direct anti-microbial activity can be attributed to substances that are toxic for bacteria cells, but sometimes these substances are toxic also for human cells. An indirect anti-microbial action is, on the contrary, due to substances stimulating skin cells vitality and inducing them to produce antimicrobial peptides (defensins) and can be anyway very desirable in sanitary applications. Moreover, thanks to this indirect approach, the problem of cytotoxicity is not relevant.

Chitin nanofibrils (CNs) possess good antimicrobial properties [7], but they also act as indirect anti-microbial, stimulating the immunomodulatory activity of skin cells [8,9]. Chitin nano-fibrils have an average size of 240 nm × 7 nm × 5 nm and their shape is like thin needles. They can be easily metabolized by the human body and promote the proliferation and adhesion of cells [9]. Consequently, they can be used in cosmetic and biomedical sectors [10], but their potential can also be exploited in sanitary and food packaging sectors [11].

For the production of sanitary films, aliphatic polyesters are the most promising materials. In fact, they are environmentally friendly and at the same time they offer good compromise of mechanical properties; furthermore, thanks to their biocompatibility, they were intensively investigated for applications in the biomedical sector [12,13]. Among the biodegradable polyesters, one of the most studied is the poly(lactic acid) (PLA). PLA is completely biobased, as it is obtained from the fermentation of corn starch, sugar beets or other renewable resources [14,15]. Regarding its properties, at room temperature it exhibits a Young's modulus of about 3 GPa and a tensile strength between 50 and 70 MPa [16,17]. However, the low deformation at break (around 4%) and low impact properties limits its applications. To improve the flexibility of PLA as well its the processability, different research activities were focused on the addition of biocompatible and biodegradable plasticizers like acetyl tributyl citrate (ATBC), triethylene glycol, lactic acid oligomer (OLA) etc. [18–20].

Polymer blending is also a successful strategy to improve the drawbacks of PLA. Blending PLA with poly(butylene adipate-co-terephthalate) (PBAT) revealed significant improvements in both mechanical properties (in particular, flexibility and toughness) and processability [21–23]. Nevertheless, PBAT is currently not completely derived from renewable resources.

A very promising biodegradable and biobased polymer that can be blended to PLA with good results is poly(butylene succinate) (PBS), which is a semicrystalline polymer synthetized from butanediol and succinic acid (both are available from biobased renewable resources). The starting mechanical properties of PBS are quite promising. In fact, it has excellent biodegradability, good thermal properties and it is very easily processable [24]. Nevertheless, PLA/PBS blends have some limitations related to an insufficient ductility and stiffness. To overcome these limits, many attempts were done to improve the PLA/PBS blends properties via plasticization. To this purpose, in order to have a completely biobased product, the use of a suitable biobased plasticizer is of fundamental importance [25]. PLA/PBS blends plasticized with poly(ethylene) glycol (PEG) have been investigated and an increment in elongation at break and softness thanks to plasticization were reported [1,26,27]. The use of citrate derivatives such as tributyl citrate, acetyl triethyl citrate, and acetyl tributyl citrate (ATBC), have also been investigated in literature [28,29]. It was found that the addition of ATBC leads to flexibility and processability improvements in PLA blends. Furthermore, PLA/PBS/ATBC films, produced via flat-die extrusion, have shown an immunomodulatory behavior in tests performed with keratinocytes, suggesting a slight indirect anti-microbial effect [24]. The use of chain extenders has been revealed as fundamental in order to modulate the melt viscosity, thus improving the processability [30].

As a starting point, and on the basis of previous studies [25,31–34], in this work, a PLA/PBSA blend plasticized with ATBC is considered the matrix for producing nanobiocomposites containing CNs as functional filler.

The addition of CNs to PLA based blends can lead to a bioplastic nanocomposite material having improved mechanical and functional properties [35,36]. However, in order to reach good mechanical properties, an efficient dispersion of CNs is fundamental. In the literature, different research methodologies are reported for CN dispersion. A preliminary chemical modification of chitin nanofibrils was investigated by several authors [37–40], however the final mechanical properties of the material were not significantly improved. Furthermore, the preliminary chemical modification complicates the processing of the material and increases the final cost. It is essential to use an efficient process that can be easily transferable to an industrial scale considering that films are produced by the melt extrusion technique. In this sense, the production of pre-composite by using a suitable plasticizer can represent a good solution. The plasticizer has a synergistic effect: it improves the processability and flexibility and at the same time it can facilitate the dispersion of the CNs. Citrates plasticizers have been investigated in literature; thanks to their ester groups, they have a very good affinity with the PLA matrix and favor the CNs dispersion [41–43]. However, it has been demonstrated that another very effective plasticizer for the dispersion of CNs is the poly(ethylene glycol) (PEG) [32,34,44,45].

For its low cost, renewability, and high compatibility with CNs, starch can be another good filler that can be added with CNs in order to reduce the final material cost [45]. However, native starch is not chemically compatible with the PLA matrix, and if there is not a good dispersion of the starch granules [46,47], the resulting PLA composite material will be more brittle. A good technique that was investigated in the literature is starch granule plasticization, allowing a better dispersion of the starch granules. Different types of plasticizer have been widely studied such as water [48], urea [49], citric acid [50], glycerol [51], and polyethylene glycol (PEG) [52]. These plasticizers are capable of breaking the hydrogen bonds within the starch granules. However, a drawback that can occur using these plasticizers, is that they can accelerate the degradation of PLA by hydrolysis when they are processed in the molten state [46].

In this work, starting from a PLA/PBS/ATBC matrix, where the ratio between the various components was selected based on a previous work [25], bionanocomposites including CNs were prepared. The effect of a melt strength enhancer and contemporarily of CNs alone or mixed with starch was analyzed. In order to have a better dispersion of CNs, a master-batch preparation with PEG was carried out before the blend extrusion. PEG 6000, solid, was used for composites containing only CNs, while for composites also containing the native starch, the liquid PEG 400 was used for an easy pre-mixing with starch at room temperature.

The blends were characterized in terms of their processability through torque and melt fluidity measurements. Moreover, they were compared in terms of mechanical and thermal properties discussing the results of tensile tests and differential scanning calorimetry (DSC) investigations respectively. The blends showing the best processability and thermo-mechanical performances were characterized in terms of compatibility with keratinocytes and immunomodulatory behavior as well as anti-microbial activity based on agar diffusion method with the aim of investigating biocompatibility and direct and indirect anti-microbial properties. All the results were discussed to identify materials and methodologies to enlarge the use of functional bionanocomposites in specific widely diffused skin contact applications.

2. Materials and Methods

2.1. Materials

Poly(lactic) acid (PLA) was purchased from NatureWorks LLC (Minnetonka, MN, USA), trade name PLA2003D. This is a special grade of PLA for extrusion process. According to data sheet it contains about 4% of D-lactic acid units, melt flow index (MFI) of 6 g/10 min (210 °C, 2.16 kg), nominal average molar mass 200,000 g/mol and density of 1.24 g/cm^3.

Poly(butylene succinate) (PBS) was purchased from Mitsubishi Chemical Corporation (Tokyo, Japan), trade name BioPBS FD92PM. It is a copolymer of succinic acid, adipic acid and 1,4-butandiol. It is a soft and flexible semi-crystalline polyester suitable for both blown and cast film extrusion having a melt flow index (MFI) of 4 g/10 min (190 °C, 2.16 kg) and a density of 1.24 g/cm^3.

Acetyl Tributyl Citrate (ATBC) from Tecnosintesi S.p.A (Bergamo, Italy) was used as plasticizer. It is a biobased and biodegradable plasticizer obtained from the acetylation of tributylcitrate. It appears as a colorless liquid [density 1.05 g/cm^3, molecular weight: 402.5 g/mol].

Two typologies of Poly(ethylene glycol) (PEG) purchased from Sigma-Aldrich (St. Louis, MO, USA) were used without any further purification. In particular, a liquid PEG, having a low molecular weight of 400 g/mol (PEG 400), and solid PEG, with a high molecular weight of 6000 g/mol (PEG 6000), were used. PEG 6000 is a colorless solid with a solubility in water of 50 mg/mL at 20 °C. PEG 400 is a liquid soluble in Toluene, Acetone and Ethanol, dispersible in water and with a density of 0.985 g/mL.

Plastistrength 550 (named PS for brevity) was purchased from Arkema (Paris, France). It is a medium molecular-weight acrylic copolymer that appears as a white powder (density: 1.17 g/cm^3). It is a commercial processing aid added to improve the melt processability [53].

Wheat native starch is a white odorless powder, insoluble in water, commercialized by Sacchetto SPA (Cuneo, Italy). It is a dry product with a very low protein content of vegetable origin, with a humidity content <13% and a pH (sol. 20% water dist.) between 5.0 and 7.0. It consists mainly of amylose, amylopectin and water with an ash content <0.25%.

Chitin nano-fibrils (CN) water suspension (2 wt.% of concentration) were provided by MAVI SUD (Latina, Italy). These CNs are produced by means of a patented process starting from chitin coming from seafood waste [54]. Thanks to this process a stable water suspension, containing 300 billion of chitin nano-crystals for each millimeter, can be obtained. Furthermore, MAVI optimized the methodology necessary to produce these chitin nano-crystals industrially maintaining the specific biologic activity [31]. Consequently, CNs are perfect for those applications (like cosmetic) in which the skin-compatibility is an essential requirement. The CN water suspension was concentrated at 20% by weight for the pre-composites preparation [33].

Calcium carbonate was received from Omya SpA (Massa Carrara, Italy), trade name Omyacarb 2-AV. It is a white powder with a specific weight of 2.7 g/cm^3 and a refractive index of 1.59. The product has a micrometric particle size, a diameter value relative to the maximum distribution curve (d 98%) of 15 μm, and 38% of particles having a diameter lower than 2 μm. The average statistical diameter (d 50%) is 2.6 μm.

2.2. Methods

2.2.1. Brunauer–Emmett–Teller (BET) Characterization of Chitin Nano-Fibrils

A BET analysis of the chitin nano-fibrils was carried out, with a Micromeritics instrument -Gemini V analyzer (Micromeritics, Atlanta, GA, USA), in order to determine the surface area and the total porosity of chitin. The measurement was performed with nitrogen as measuring gas and helium for the calibration phase. Samples were conditioned at 120 °C for 6 h to remove the humidity. The degasification time was 30 min. The range of the ratio P/P_0 (where P_0 is the room pressure) for the analysis was between 0.05 and 0.3. The results of the measurements are the BET values that represent the surface area of the sample.

The morphology of chitin nanofibrils was investigated by a field emission scanning electron microscope (FESEM) FEI Quanta 450 FEG (FEI, Hillsboro, OR, USA). The water suspension, at 2 wt.% of chitin nanofibrils, was diluted 1:1000 and then one drop of the diluted suspension was deposited onto a glass window in order to make the FESEM analysis. At this point, from the SEM analysis on the chitin diluted sample, it was possible to obtain a theoretical area per gram of sample using the Image J software (NIH, Bethesda, MD, USA) (to calculate the average width and length of the nano-fibrils). Considering a chitin density value of 1.425 g/cm^3 [55], the average mass of each fibril was obtained and the number of particles per gram was calculated by dividing a gram for the mass of a single fibril. Considering the fibrils as parallelepipeds, it was possible to evaluate the area and then a BET theoretical value by multiplying the area for the number of particles per gram.

2.2.2. Chitin Master-Batch Preparation

To better disperse and to avoid the agglomeration of chitin nanofibrils, a dispersion with PEG was prepared. On the basis of a previous study it was observed that PEG can be intercalated between the CNs and it avoids the formation of compact CNs agglomerates [32].

Two pre-composite master batches were prepared with a high pre-dispersed chitin nano-fibrils content. The first one (named MB1) contained 50% of chitin nano-fibrils and 50% of PEG6000. The second one (named MB2) was prepared at 80 °C with chitin nano-fibrils, PEG400 and plasticized starch at 2:2:3 ratio. In both cases, the additives were added to the stirred chitin suspension and the drying was done in a rotavapor (Buchi R-210, buchi, Flawil, Switzerland) equipment at 50 °C and 100 mbar to avoid chitin degradation.

PEG 6000 in the MB1 was chosen in order to obtain a solid powder that can be easily fed during the extrusion process; at this purpose the use of PEG 6000 is fundamental instead of PEG 400 (that have a lower molecular weight). On the other hand, the use of liquid PEG 400 is essential for the preparation of MB2 due to the use of starch in powder form.

2.2.3. Blends Preparation

Composites containing CNs and starch were obtained using an HAAKE Minilab II twin-screw mini-compounder (HAAKE, Vreden, Germany). This equipment is able not only to compound the molten material, but at the same time it is able to make torque measurements of the molten material.

For selecting the processing conditions, the thermal stability of the different additives were considered [9,25,51,56]. Moreover, it was taken into account that the thermal degradation of PLA/PBS blends occurs above 340 °C [57].

Before the extrusion, the polymers granules were dried in an oven at 60 °C for 24 h. In order to record the torque values for all the formulations, for each extrusion compounding 6 g of PLA/PBS pellets were manually mixed together with the other additives. The mixture was fed into the co-rotating mini-extruder with the help of a suitable little hopper. The processing temperature was set at 190 °C and the screws rotating speed was 110 rpm. After the introduction of the material, the molten material, pushed by the screws, was flushed in a back-flow channel (with the exit valve of the die closed) for 1 min. During this period, the torque value was recorded as a function of time. The extruded material was recovered (opening the die valve) after one minute of rotation inside the mini-extruder chamber to ensure a correct mixing.

At least ten experimental torque measurements were carried out for each blend to assure the reliability and consistency of the test and also to recover enough material (about 60 g per blend) for further tests. The final torque value represents the most significant value for the sample as the melt stabilizes. All the extrusion process, including the feeding operations, had a duration of 120 s to avoid degradation phenomena.

2.2.4. Melt Flow Rate

The melt flow behavior of the blends was investigated with a Melt Flow Tester M20 (CEAST, Torino, Italy) equipped with an encoder. The instrument is able to measure the melt volume rate (MVR) of the polymer blend acquired by the encoder that follows the movement of the piston. Before the test, granules of the blends obtained from the mini-extruder were dried in an oven (set at 60 °C) for one day.

The melt flow rate (MFR) is defined as the weight of the polymer, at the molten state, that passes through a capillary (having a specific length and diameter) in 10 min under pressure applied by a specific weight (according the ISO 1133:2005). For this work the standard used was the ISO 1133D custom TTT where the following procedure was adopted: the sample was preheated without the weight for 40 s at 190 °C, then the weight of 2.160 kg was released on the piston and after 5 s a blade cut the spindle starting the real test. At this point the MVR was recorded, every 3 s, by the encoder.

All the MVR data were reported with their standard deviation thanks to the CEAST Visuamelt software of the equipment. The MFR values standard deviations were calculated by considering the results obtained by the measurements.

2.2.5. Mechanical Testing

The mechanical characterization of the blends was carried out on films prepared by compression molding. At this purpose, the recovered material coming out from the mini-extruder (having the die rectangular die section) was manually pelletized. Before the compression molding the pellets were kept in a circulating air oven at 60 °C for 24 h to avoid water uptake. Different film for each formulation were prepared in a NOSELAB ATS manual laboratory heat press (Noselab, Milano, Italy). About 5 g of granules were put between two Teflon sheets and then pressed at 180 °C for 1 min with a pressure of 3 tons. The samples for tensile tests were obtained from the film using an Elastocon cutting die (Elastocon, Brämhult, Sweden), into dump-bell shaped tensile specimens (ISO 527-2 type A).

Tensile tests were carried out on an INSTRON universal testing machine 5500R (INSTRON, Buckinghamshire, UK), interfaced with a MERLIN software (version 4.42, INSTRON, Buckinghamshire, UK) and equipped with a 100 N load cell. Compressed air grips were used and the initial grip separation was 25 mm while the crosshead speed was set at 100 mm/min. At least ten specimens were tested for each blend and the average values were reported.

2.2.6. Differential Scanning Calorimetry

To perform the DSC characterization of selected formulations, a Q200-TA Instrument differential scanning calorimeter (DSC, TA Instruments, New Castle, DE, USA) equipped with a RSC cooling system was used with nitrogen flow set at 50 mL/min, as purge gas. Indium was adopted as a standard for temperature and enthalpy calibration of the instrument.

The thermal program was set according to the following procedure: the sample was heated, at 10 °C/min, to 220 °C where was held for 2 min (in order to delete the thermal history of the sample). Subsequently, the sample was cooled again to −40 °C at 10 °C/min and then reheated again up to 200 °C.

Melting temperature (T_m) and cold crystallization temperature (T_{cc}) of the blend were determined by considering the maximum of the melting peaks and at the minimum of the cold crystallization peak respectively. As a consequence, the enthalpies of melting and of the cold crystallization were determined from the corresponding peak areas in the thermograms. The crystallinity percentage of PLA was evaluated according the following equation:

$$X_c = \frac{\Delta H_{m,PLA} - \Delta H_{cc,PLA}}{\Delta H°_{m,PLA} \cdot X_{PLA}}, \tag{1}$$

where $\Delta H_{m,PLA}$ and $\Delta H_{c,PLA}$ are the melting enthalpy and the enthalpy of cold crystallization of PLA, while X_{PLA} is the weight fraction of PLA in the selected formulation. $\Delta H°_{m,PLA}$ is the melting enthalpy of the 100% crystalline PLA (considered 93 J/g [58]).

2.2.7. Skin Compatibility Tests

HaCaT cells are adherent cells often utilized in skin test for their high capacity to differentiate and proliferate in vitro [59]. Immortalized human keratinocyte HaCaT cell line (purchased from CLS–Cell Lines Service, Eppelheim, Germany), were cultured in Dulbecco's Modified Eagle Medium (DMEM) supplemented with 1% Penstrep, 1% glutamine and 10% fetal calf serum (Invitrogen, Carlsbad, CA, USA) at 37 °C in air and 5% CO_2. The HaCaT cells, seeded in 12-well plates until 80% of confluence, were incubated for 24 h with the films F4 and F7. F4 and F7 films were sterilized by washing with ethanol before the test. At the end of this time, resazurine was added to the concentration of 0.5 mg/mL and incubated for 4 h.

2.2.8. Evaluation of Inflammatory and Indirect Antimicrobial Properties

The cells, cultured as described above, were seeded inside 12-well TC plates until 80% of confluence, and incubated for 24 h with the F4 and for 6 h and 24 h with the F7 (n = 3). F4 and F7 films were sterilized by washing with ethanol before the test. At these endpoints, total RNA was isolated with TRIzol and 1 µm of RNA was reverse-transcribed into complementary DNA (cDNA) using random hexamer primers, at 42 °C for 45 min, according to the manufacturer's instructions. Real time polymer chain reaction (PCR) was carried out with the LC Fast Start DNA Master SYBR Green kit using 2 µL of cDNA, corresponding to 10 ng of total RNA in a 20 µL final volume, 3 mM $MgCl_2$ and 0.5 µM sense and antisense primers (Table 1). Real-Time PCR was used to evaluate the expression of interleukins TNF-α, TGF-β, IL-6, IL-8, IL-1α, IL 1β and antimicrobial peptide HBD-2.

Table 1. Primers sequences and Real-Time conditions.

Gene	Primer Sequence	Conditions	Size (bp)
IL-1 α	5'-CATGTCAAATTTCACTGCTTCATCC-3' 5'-GTCTCTGAATCAGAAATCCTTCTATC-3'	5 s at 95 °C, 8 s at 55 °C, 17 s at 72 °C for 45 cycles	421
IL-1 β	5'-GCATCCAGCTACGAATCTCC-3' 5'-CCACATTCAGCACAGGACTC-3'	5 s at 95 °C, 14 s at 58 °C, 28 s at 72 °C for 40 cycles	708
TNF-α	5'-CAGAGGGAAGAGTTCCCCAG-3' 5'-CCTTGGTCTGGTAGGAGACG-3'	5 s at 95 °C, 6 s at 57 °C, 13 s at 72 °C for 40 cycles	324
IL-6	5'-ATGAACTCCTTCTCCACAAGCGC-3' 5'-GAAGAGCCCTCAGGCTGGACTG-3'	5 s at 95 °C, 13 s at 56 °C, 25' s at 72 °C for 40 cycles	628
IL-8	5-ATGACTTCCAAGCTGGCCGTG-3' 5-TGAATTCTCAGCCCTCTTCAAAAACTTCTC-3'	5 s at 94 °C, 6 s at 55 °C, 12 s at 72 °C for 40 cycles	297
TGF-β	5'-CCGACTACTACGCCAAGGAGGTCAC-3' 5'-AGGCCGGTTCATGCCATGAATGGTG-3'	5 s at 94 °C, 9 s at 60 °C, 18 s at 72 °C for 40 cycles	439
HBD-2	5'-GGATCCATGGGTATAGGCGATCCTGTTA-3' 5'-AAGCTTCTCTGATGAGGGAGCCCTTTCT-3'	5 s at 94 °C, 6 s at 63 °C, 10 s at 72 °C for 50 cycles 5 s at 94 °C, 6 s at 63 °C, 10 s at 72 °C for 50 cycles	198

2.2.9. Anti-Microbial Tests

The antimicrobial capacity of the samples was assayed in vitro by the qualitative agar diffusion test. This method consists in the determination of the spectrum of action of the sample, according to resistance of the selected microorganisms. In this case, as an exploratory approximation, the test was performed against two typical bacterial colonizers of the normal human skin: *Staphylococcus aureus*

(ATCC9144) and *Enterobacter* spp. (ATCC13047) (Gram positive and negative respectively). All test strains were obtained from American Type Culture Collection (ATCC).

Briefly, aliquots of 100 µL microbial suspensions adjusted to 5×10^5 CFU/mL were added to a melted soft agar solution which was poured onto the surface of solid agar plate (soft-agar overlay method). Microbial suspension densities were previously quantified by the common plate count method in the appropriate culture conditions.

Samples were cut into disks with a diameter of 6 mm and sterilized by UV irradiation for 15 min. Also, sterile filter disks of same size were used as controls. Afterwards, disks were placed on the previously inoculated agar plates (4–5 disks per plate) and incubated at 37 °C for 24 h. After the incubation time, the presence of clear zones of growth inhibition was determined visually and the diameter of the inhibition halos was measured with a caliper. The assay was performed in triplicate for each microorganism.

The microorganisms were declared sensitive in the agar diffusion test against bacteria, when an inhibition zone of ≥10 mm was observed, indicating a positive result of the test, while as negative when no inhibition zone was present.

3. Results

3.1. Comparison between Experimental and Theoretical Chitin BET Values

In Figure 1, the morphology of chitin nanofibrils is shown. By using image J software and following the procedure described in Section 4, the theoretical BET value, indicating the specific surface of the sample, was calculated and compared with the experimental one, measured for a sample obtained by drying up to constant weight the suspension of chitin nanofibrils. For the calculation of the average fibrils volume, a thickness of 5 nm was supposed based on what is reported in the literature [60]. The results are shown in Table 2.

Figure 1. SEM image of a diluted 1:1000 sample of chitin suspension.

Table 2. Results of SEM and BET analysis related to chitin nanofibrils.

Chitin Average Length (nm)	Chitin Average Width (nm)	Average Fibrils Volume (cm^3)	Average Mass of Each Fibril (g)	Number of Chitin Fibrils per Gram	Experimental BET (m^2/g)	Theoretical BET (m^2/g)
11,300	300	1.695×10^{-14}	2.4×10^{-14}	4.15×10^{13}	39.14	286.18

From the data reported in Table 2, it is possible to observe a consistent discrepancy between the theoretical and the experimental BET values. This discrepancy can be ascribed to the formation of agglomerates because of the inter-fibrils interactions during the drying.

A sort of agglomeration factor (that will be named "weldability index" (WI)), capable to consider the percentage of agglomerated fibrils, was evaluated according to the following equation:

$$WI = \frac{\text{theoretical BET} - \text{experimental BET}}{\text{theoretical BET}} \times 100 \qquad (2)$$

A WI of 80% was obtained. The higher the weldability (WI) is, the more the particles are agglomerated providing a low experimental BET value. The result obtained supports that it is necessary to avoid this agglomeration by preparing a pre-nanocomposite with a plasticizer (PEG in this case) used as dispersant to maintain separated the nanofibrils in accordance with other literature results [32,42]. Moreover, based on the work of Pereira et al. [54], in the slightly acidic water suspension of CNs, a positive zeta potential is present. PEG, being a non-ionic surfactant, does not alter the full positive charge of CNs, but decreases its density, allowing a better dispersion in the more hydrophobic polyester blend. In good agreement it was demonstrated that this solid pre-nanocomposite, containing 50% by weight of chitin nanofibrils, can be then dispersed in the PLA based material [31,32].

3.2. Melt Properties

Different PLA/PBS formulations (Table 3) were prepared by extrusion to obtain biobased, biocompatible and flexible films. The ATBC was selected as plasticizer to increase the mechanical flexibility of the final material [7,48]. The melt strength enhancer Plastistrength (PS), based on acrylic copolymers, was also added up to 2 wt.% to modulate the viscosity during the processing of the blends containing chitin nanofibrils [25]. On the basis of the literature in this field, the CNs, added preparing pre-composites, and starch quantities were fixed to 2 wt.% and 3 wt.% respectively [52]. The F1 formulation, considered as a reference, consisted of only PLA, PBS, and ATBC. Their composition was selected considering the promising results of a previous work [25]. To better understand the effect of the Plastistrength (PS) addition, the F2 blend was prepared. Subsequently, the F3–F6 formulations were obtained replacing 2 wt.% of ATBC with PEG, having a similar plasticizing effect. Chitin nanofibrils were added only at 2 wt.% because, as they are dispersed at nanometric scale, low percentages are enough to obtain a homogenous distribution [32]. The addition of starch in the F5 and F6 blend was also evaluated as it is an additive acting as chitin nanofibrils carrier, because of its good compatibility with chitin, renewability, biodegradability and low cost. The F7 blend has the same composition of F4 but with the addition of micrometric calcium carbonate. The $CaCO_3$ is used both as a slip agent and as a viscosity regulator during the processing.

Table 3. PLA/PBS blends name and compositions.

Blends	PLA (wt.%)	PBS (wt.%)	ATBC (wt.%)	PS (wt.%)	PEG (%wt)	NC (wt.%)	Starch (wt.%)	Calcium Carbonate (wt.%)
F1	63	17	20	-	-	-	-	-
F2	62	16	20	2	-	-	-	-
F3	62	16	18	-	2 (PEG 6000)	2	-	-
F4	61	15	18	2	2 (PEG 6000)	2	-	-
F5	59	16	18	-	2 (PEG 400)	2	3	-
F6	58	15	18	2	2 (PEG 400)	2	3	-
F7	57.5	14.5	15	2	2 (PEG 6000)	2	-	7

The results of MFR and Torque for each blend are reported in Table 4. It can be observed that the addition of the melt strength enhancer (PS) to the starting formulation (F1), provokes an increment of torque (indirectly connected to the increase in melt viscosity) and consequently, a decrease in the MFR value. This result is in accordance with previous results [24] indicating that the viscosity increase due to PS addition is attributed to a combination of high miscibility and interactions between PLA and the acrylic copolymer-based products.

Table 4. Torque, MVR and MFR values of the prepared blends.

Blends	Torque (N·cm)	MVR (cm^3/10 min)	MFR (g/10 min)
F1	67.8 ± 5.4	22.5 ± 2.0	23.6 ± 2.1
F2	72.8 ± 6.0	11.8 ± 0.9	12.4 ± 0.9
F3	60.0 ± 4.1	16.9 ± 1.5	18.4 ± 1.3
F4	63.3 ± 3.6	11.8 ± 1.1	13.1 ± 1.3
F5	42.0 ± 4.1	40.2 ± 10.5	43.5 ± 11.4
F6	38.6 ± 2.2	50.7 ± 6.0	55.0 ± 6.5
F7	73.5 ± 5.2	10.0 ± 1.3	11.4 ± 1.5

3.3. Tensile Properties

The results of the main mechanical properties (stress at break, elongation at break, and stress at yielding) are reported in Table 5.

Table 5. Tensile properties of the films obtained from each blend.

Blends	Stress at Break σ_b (MPa)	Elongation at Break ε_b (%)	Stress at Yielding σ_y (MPa)
F1	31.8 ± 1.4	572.7 ± 20.7	-
F2	33.0 ± 1.2	554.2 ± 12.3	10.2 ± 0.7
F3	25.5 ± 1.2	455.2 ± 16.4	8.3 ± 1.5
F4	25.5 ± 1.3	421.9 ± 25.1	11.6 ± 0.7
F5	21.3 ± 3.1	398.7 ± 49.2	-
F6	19.7 ± 1.5	381.2 ± 27.5	-
F7	25.5 ± 1.0	400.1 ± 21.9	10.8 ± 1.8

It is evident that the addition of PS does not alter significantly the elongation at break and the stress at break of the starting blend based on PLA/PBS/ATBC. A slight increment of the stress at break followed by a decrement of the elongation at break is registered for the F2 blend in which the PS was added. However the addition of PS led to an increase in the yield stress compared to F1 that exhibits a behavior quite similar to an elastomer [25]. The addition of chitin alone (F3) or combined with PS (F4) induced a slight decrease in the elongation and stress at break. Both stress and elongation at break are lower than the F1 and F2 mixtures. Nevertheless, the yield stress of the F4 mixture is improved if compared to F2 mixture. The addition of chitin combined with PS led to an improvement of the final mechanical properties. The partial decrement of elongation at break and tensile stress at break is balanced by the increment of yield stress which provides to the final material an improved resistance in the elastic field. The final mechanical properties of the F4 blend are typical of flexible and ductile films and the CNs presence makes the use of this blend interesting for biomedical or cosmetic applications. The potential anti-microbicity and biocompatibility of these films can thus allow several new applications.

Contrary to what is reported in the literature [60], a decrement in mechanical performances was observed with the starch addition. Probably, during the processing, the starch carries a quantity of water higher than chitin and is not crystalline, so its hydroxyl groups have a stronger nucleophilic action. Consequently, the blends with starch, having a major quantity of humidity, lead to an inevitable matrix degradation (confirmed also by the MFR and Torque values) that worsens the final mechanical properties of the material. The degradation leads to shorter polymeric chains. Therefore, the addition of PS becomes useless because the interactions between macromolecules generated by PS are not sufficient for determining an effect similar to a physical cross-linking. This explains why no significant differences can be observed between F5 and F6 mixtures where the only difference is the PS addition.

The addition of CaCO$_3$ to F4 formulation (F7 blend) led to a good balance of mechanical properties and at the same time it is possible to obtain a formulation with a good processability. The CaCO$_3$ micro-particles are rigid fillers and decrease the elongation at break, but increasing the stress at yielding. However, it can be observed from Table 5 that the final mechanical properties of both CNs and Calcium Carbonate particles, probably due to their aggregation [61], causes a decrement of both elongation at break and yield stress.

3.4. DSC Characterization

The thermal characterization, by DSC analysis, was carried out on the best formulations selected by considering the mechanical tests. The F4 formulation showed the best performances, thus its thermal characteristics were compared with the starting blends (F1 and F2) and with the blend containing CaCO$_3$ (F7).

The results of the first heating scan, are reported in Table 6 and the relative thermograms in Figure 2. The first heating scan reveals the thermal history of the samples that are rapidly cooled after compression molding. It can be observed that there is not a significant change in the melting temperature for all the analyzed mixtures. Despite the rapid cooling, the F1 and F2 mixtures possess a significant crystalline fraction. Nevertheless, it can be observed that a net decrement of crystalline fraction is registered when PS is added (F2 blend). The result is in accordance to what is found in literature; in fact, the addition of acrylic copolymer-based product like PS, depresses the PLA crystal formation [25,62]. The addition of CNs provokes a significant further decrement of PLA crystallinity, thus the polymer remains almost amorphous.

In the presence of PS and also CNs (F2 and F4 blends), the exothermic crystallization peak is shifted to lower temperature. Thus, the presence of these additives favors the cold crystallization of PLA during the heating process. For the F1 mixture, the cold crystallization peak is almost negligible as it has a very small area coincident with a low value of cold crystallization enthalpy (3.05 J/g).

Table 6. Results of differential scanning calorimetry analysis (first heating).

Blends	T_g (°C)	T_C (°C)	ΔH_C (J/g)	T_m (°C)	ΔH_m (J/g)	X_C %
F1	43.64	~96	3.05	143.74	18.84	27
F2	36.29	86.71	12.97	145.58	19.94	12
F4	43.90	87.52	19.31	145.91	20.57	2
F7	43.55	85.72	11.97	145.46	18.82	13

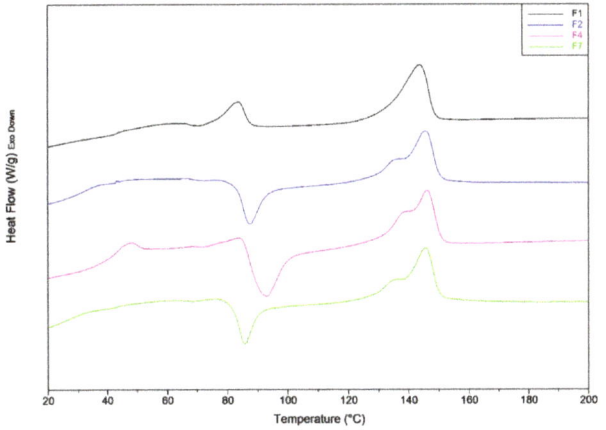

Figure 2. DSC first heating thermograms of F1, F2, F3 and F4 mixtures.

An enthalpic relaxation peak above the glass transition temperature due to the aging [63] is present markedly for F4 blend. Consequently, it can be deducted that the addition of CNs favors the PLA aging.

The CaCO$_3$ addition, causes a marked increment of PLA crystallinity that passes from 2% for F4 blend to 13%. This result is in agreement with what can be found in the literature. In fact, it is known that CaCO$_3$ acts as a heterogeneous nucleation enhancer for PLA, favoring and accelerating the crystallization process [64,65].

The results of the second heating scan are reported in Table 7 and Figure 3. Thanks to the plasticization of PLA, the glass transition temperatures of all mixtures are moved towards lower values (if compared with the glass transition temperatures of the first heating).

Table 7. Results of differential scanning calorimetry analysis (second heating).

Blends	T_g (°C)	$T_{m,PBS}$ (°C)	$\Delta H_{m,PBS}$ (J/g)	$T_{C,PLA}$ (°C)	$\Delta H_{C,PLA}$ (J/g)	$T_{m,PLA}$ (°C)	$\Delta H_{m,PLA}$ (J/g)	X_C %
F1	26.96	79.63	3.74	88.21	11.49	144.27	22.62	19
F2	33.54	83.40	6.83	96.72	16.90	145.40	20.54	6
F4	34.75	84.87	5.81	96.53	20.06	147.24	22.87	5
F7	30.24	83.72	6.68	94.87	15.00	146.01	20.07	9

The PLA plasticization coupled with the annulment of thermal history, led to mixtures with a lower crystallinity content. In particular, the addition of the PS melt enhancer, allows to the samples to crystallize only during the heating step (not during the cooling), in agreement with a lower value of the crystallinity content in the sample. In fact, it can be observed that, for the second heating scan, the cold crystallization peak temperatures for F2 and F4 blends, as well as the crystalline content are similar. Probably, the crystal growth during cooling is blocked by intermolecular interactions that occur between PS and PLA.

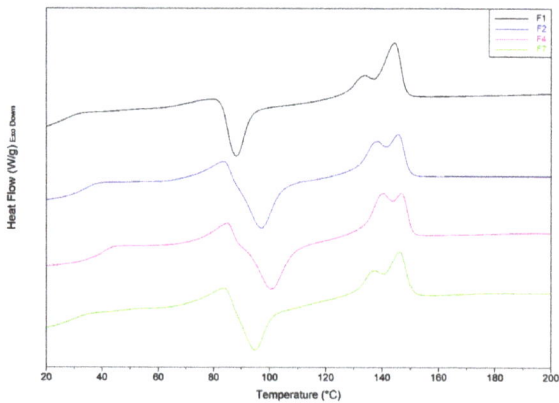

Figure 3. DSC second heating thermograms of F1, F2, F3 and F4 mixtures.

3.5. Morphology of the Composites

The plasticized PLA/PBSA blend containing 2% of PS showed the morphology represented by the micrograph of Figure 4a, where the spherical plasticized PBSA domains can be easily seen in the plasticized PLA matrix. By adding the CNs in the PEG masterbatch (Blends F4), the morphology of the bionanocomposites also results as in Figure 4b, but some additional CNs bundles can be observed.

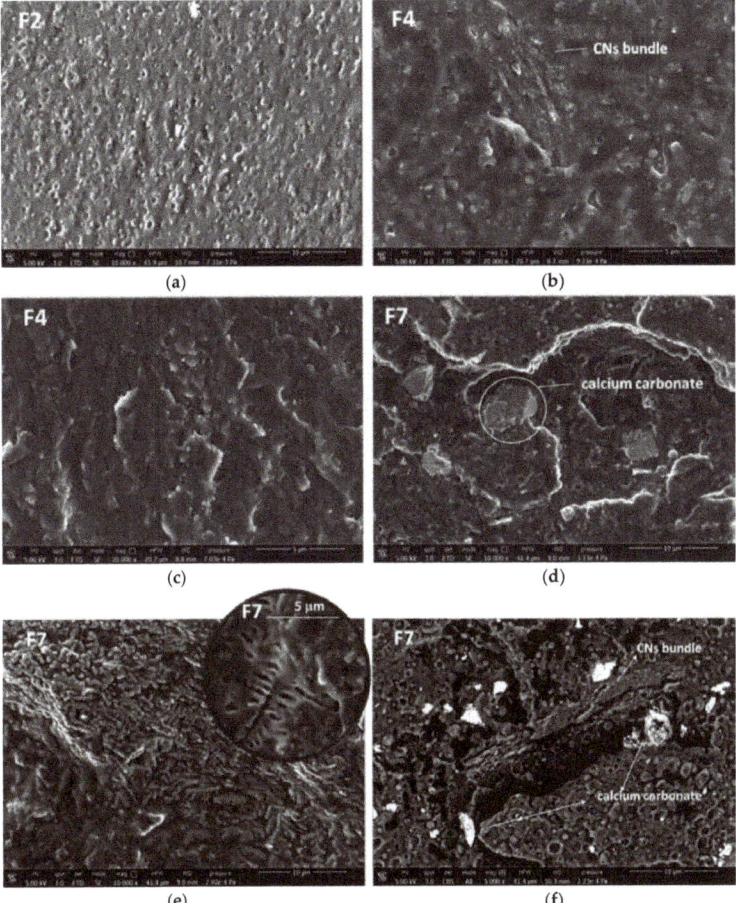

Figure 4. SEM micrograph related to: (**a**) F2; (**b**,**c**) F4; (**d**–**f**) F7. The micrographs (**a**–**e**) were obtained by the signal of secondary electrons (SEI modality). Micrograph (**f**) was obtained by signal due to backscattered electrons (CBS modality).

The general morphology was homogenous, with CNs present inside the matrix as extended and well compatible bundles in all the nanocomposite (Figure 4c). When the calcium carbonate was also added the micrometric $CaCO_3$ particles can be observed (Figure 4d) in the PLA/PBSA blend. Interestingly, in this blend some very extended regular structures are present with the shape typical of CNs bundles. Inside these bundles it can be observed a regular structure with a branched geometry (Figure 4e) having a shape and dimensions similar to the one shown by crystals of chitin nanofibrils in a previous work [7]. Hence these peculiar shapes can allow to observe the quite extended structure of CNs in the nanobiocomposite. By obtaining micrographs by backscattered electrons, it was possible to identify better the calcium carbonate as lighter particles, because of the higher atomic number of Calcium with respect to the other atoms in the blend. This kind of analysis evidenced that calcium carbonate was dispersed quite homogeneously in the F7 bionanocomposite, with some particles resulting close to CNs bundles (Figure 4f).

3.6. Skin Compatibility Results

The assessment of cell viability was performed by Alamar Blue assay in order to confirm the biocompatibility of the films. As shown in Table 8, F4 and F7 were not cytotoxic, indeed they promote, after 24 h of incubation, keratinocytes proliferation.

Table 8. % of ABred in keratinocytes treated for 24 h with F4 and F7.

Sample	% AB$_{RED}$
F4	150
F14	137

3.7. Evaluation of Antinflammatory and Indirect Antimicrobial Properties

Being inflammatory response a key factor in inducing skin sensitization, after 6 and 24 h of incubation Real-Time PCR was performed to evaluate the expression levels of pro- and anti-inflammatory cytokines produced by HaCaT cells treated with F4 and F7. In addition, the indirect antimicrobial activity of the films was examined by evaluating the expression levels of antimicrobial peptide HBD-2.

As shown in Figure 5, our results indicated that both F4 and F7 were able to reduce the expression of most proinflammatory cytokines, with the exception of IL-1, and were also able to strongly upregulate the expression of HBD-2.

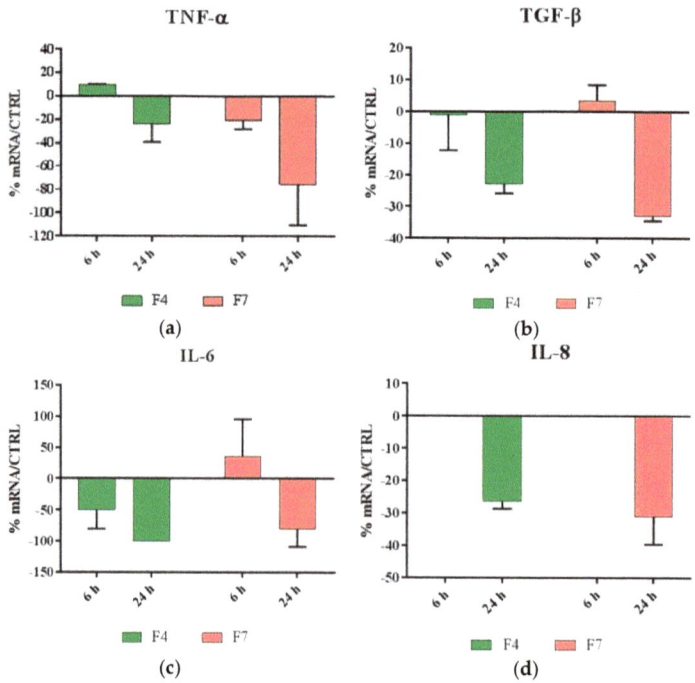

Figure 5. *Cont.*

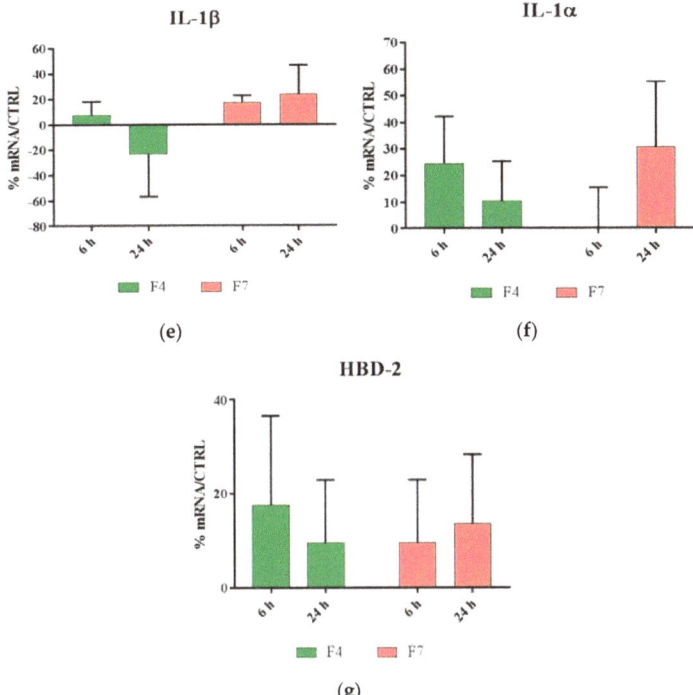

Figure 5. Relative gene expression of (**a**)TNF-α, (**b**)TGF-β, (**c**) IL-6, (**d**) IL-8, (**e**) IL-1α, (**f**) IL-1β and (**g**) HBD-2 in HaCat cells treated with F4 and F7 for 6 and 24 h. Data are mean ± SD and are expressed as percentage of increment relative to untreated cells (ctrl).

3.8. Anti-Microbial Tests

Results of the agar diffusion test of the two PLA blend films formulations (F4 and F7) towards both Gram positive and negative bacteria reveal no effect. Hence, the films did not show any anti-microbial activity against these two bacteria (Figure 6).

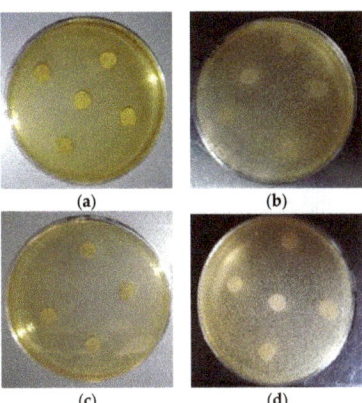

Figure 6. Plates of Agar diffusion test against (**a,c**) *S. aureus* and (**b,d**) *Enterobacter* spp., with no inhibition halos observed (inhibition zone ≤10 mm) for any sample tested: F4, F13 or F14.

F4 and F7 contained 2% (w/w) of dispersed CNs. This concentration was selected for its potential effect of reducing the growth of the bacteria strains tested, as observed in similar applications of the same CNs regarding food packaging [7]. Therefore, the lack of antimicrobial activity observed could be attributable to many factors, e.g., the low concentration of CNs distributed along the surface of the samples, since CNs are mainly embedded in the plasticized blend matrix.

4. Discussion

It can be observed that, compared to the F2 blend, the addition of chitin nanofibrils to PLA-PBS blend caused a slight decrease in melt viscosity. This behavior can be caused by the hydrolytic degradation of the polyester matrix caused by a not perfect drying of the CN/PEG pre-composite. In fact, the chitin nanofibrils are very hygroscopic, and the complete elimination of humidity is not an easy process. So, this torque decrement can be attributed to humidity traces that contribute to decrease the viscosity by shortening the polymeric PLA chains because of chain scission.

Nevertheless, the MFR and MVR, slightly decreased when the CNs were added, comparing F1 and F3. This effect, relevant in conditions of higher shear stress typical of the MFR test, also indicates the presence of good interactions between CNs, behaving as reinforcing agent, and the polymeric matrix.

The variation of MVR due to the addition of the melt strength enhancer is reported in Figure 7a, and evidences very well that, in the presence of chitin nanofibrils, the melt strength enhancer is significant, but it seems less effective due to a more extensive chain scission. The strategy of regulating the fluidity of the melt by using PS is effective for controlling and tuning the nanobiocomposite processability.

In the samples where the starch was present, the difference is not significant at all. In these conditions the interactions occurring between PLA and PS are not effective in counterbalancing the very extensive chain scission. In the blends containing starch the torque resulted strongly decreased with respect to F1 and F2 because of extensive chain scission of the biopolyesters (Table 4).

In good agreement, the addition of starch (F5 and F6 blends) influenced in a significant way the MFR values. For F5 and F6 blend that contain starch, it was necessary to modify the test procedure because the mixtures were so fluid that the chamber emptied quickly not allowing the MVR measurement. For these mixtures the measurement time was accelerated (10 s instead of 1 min). From the results obtained it can be deducted that the addition of very small amount of pre-gelatinized starch (3 wt.%), are sufficient to cause an extensive degradation of both F5 and F6 mixtures even with the PS addition.

The $CaCO_3$ addition to the F4 blend causes, as expected, a viscosity incrase that is reflected into a Torque increment and a lower MFR value.

Regarding the viscosity variation as a function of time during processing, in the case of extrusion lasting only 60 s, it was almost negligible (Figure 7b). Regarding the MVR as a function of time trends, as evidenced in Figure 7c, the reduction of slope observed between the F1 and F2 blend, suggesting a better stability of the F2 melt after the addition of PS, was not obtained comparing F4 and F3. The presence of CNs is thus making more complex the general processing behavior of the composites as a function of time. However, these variations, being slight, can be tuned and well controlled without compromising the general processability of the blends.

In Figure 7b,c, the MVR and Torque trends are reported as a function of time.

Regarding the properties of blends films, In Figure 8a the first part of the stress-strain curves (up to 100% of elongation) for F1 up to F6 blends are reported. The addition of PS provoked an increase of yielding stress in both F2 and in F4 blends.

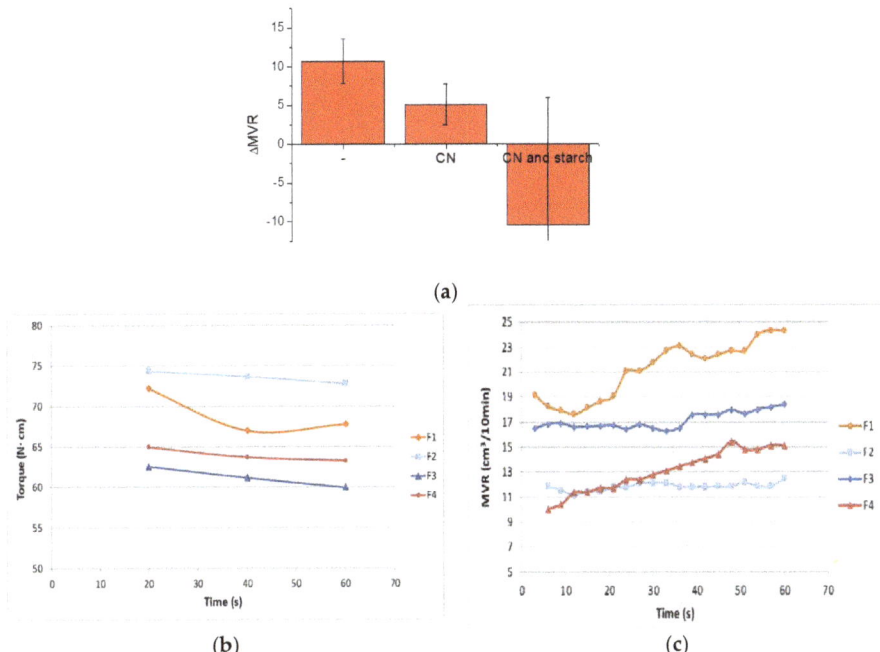

Figure 7. (a) MVR variation due to PS addition; (b) MVR trends as a function of time for F1, F2, F3 and F4 formulations; (c) Torque trends as a function of time for F1, F2, F3 and F4 formulations.

It can be noticed the improvement of the yield stress for the F4 mixture containing PS and CNs. However, also the CNs determined a change on the stress-strain curve shape. The CNs act like fillers increasing the yielding stress value due to significant interactions with the polymer matrix also thanks to orientation along the solicitation axis. The yielding is the point corresponding to the formation of neck in the specimens. More energy is required for necking if the interactions between macromolecules are more intense. Interestingly, the starch allowed to obtain stress-strain curves with a behavior more similar to an elastomer. The action of starch is thus different. In the first part of the stress-strain curve, its action completely deletes the reinforcing effect of CNs and seems to enhance the plasticization action of ATBC. In a similar way, but with a minor effect, the $CaCO_3$ determines a decrease in stress at yield with respect to F4 (Figure 8b). These results evidence different mechanisms of interaction between the different fillers and the biopolyester chains during tensile tests.

Regarding thermal properties, CNs tend to hinder the PLA crystallization and the combination of CNs and PS showed a similar effect. This can be ascribed to the inhibition of crystal growth during cooling due to intermolecular interactions between PS and PLA.

The enthalpic relaxation peak above the glass transition temperature due to aging observed for F4 blend indicated that the addition of CNs favors the PLA aging. This result can be attributed to the orientation and alignment of macromolecules along the CNs surfaces, that favor the slow formation of ordered regions in the material. This effect is present in F4 but not in F7. This is due to the less homogeneous distribution obtained for CNs in the presence of calcium carbonate. In fact, the presence of fillers with different shapes, makes more difficult the macromolecular orientation. The complexity is also in agreement with the morphology analysis carried out by SEM. The PBSA formed a spherical dispersed phase in the plasticized PLA matrix and the CNs dispersion resulted homogeneous despite of the presence of complex bundles highly compatible with the matrix could be revealed in the blends. The addition of $CaCO_3$ showed that both the fillers were well dispersed in the matrix where CNs bundles could crystallize, and interactions in between the two fillers could be also reasonably hypothesized.

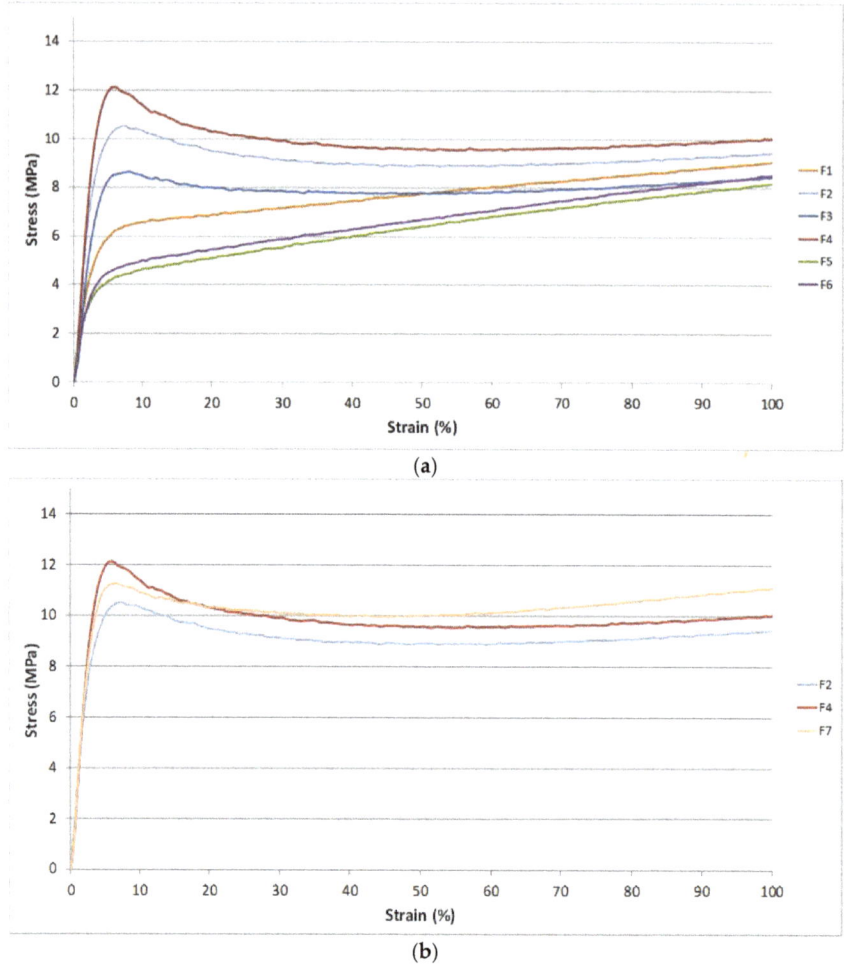

Figure 8. (a) First part of the stress-strain curves for the formulations without calcium carbonate; (b) Figure 2. F4 and F7.

The biocompatibility tests evidenced that both F4 and F7 films promoted cell proliferation. Moreover, they possess a significant immunomodulatory activity, in fact they are capable to downregulate the expression of IL, 6, IL-8, TGF-beta and TNF-alpha, and to upregulate the expression of IL-1 alpha and beta and HBD-2.

One of the functions of IL-1 in the epidermis is to promote keratinocyte differentiation, so we can hypothesize that F4 and F7 can stimulate cell differentiation while preserving an anti-inflammatory and antibacterial indirect activity.

It should be noted that the antibacterial activity occurs, in our experimental model, following the contact of the films with keratinocytes; this contact causes the induction of the production of defensins, antimicrobial peptides produced by epithelial cells. Hence, only when the film will be in contact with skin, will this indirect anti-microbial action be present.

The F4 and F7 films resulted not anti-microbial towards *S. aureus* and *Enterobacter* spp. by the agar diffusion test. The lack of antimicrobial activity observed could be attributable to a low density of CNs distributed along the surface of the samples. Li et al. [64] highlighted the importance of the CNs surface chemistry in their antimicrobial performance when loaded to a film making solution, confirming that partially deacetylated CNs exhibited superior antimicrobial activity despite the amount loaded when compared to non-deacetylated CNs. This could be explained by the increase in the number of amino groups being exposed on the chitin backbone, when a deacetylation process is involved [65].

On the other hand, Foster and Butt [66] reported an apparent lack of antimicrobial activity of thin chitosan films, with a 2% of CNs loaded, when tested against Gram positive and negative bacteria. The authors affirmed that even though chitosan solutions have demonstrated strong bactericidal activity against a wide range of bacteria, there is a loss of this beneficial property in thin films cast from the same solutions. Other authors reported antimicrobial activity of CNs when loaded to film making solutions in concentration range between 0.5% and 10% w/w [64,67], where agar diffusion tests had been performed on the solution, not on the films obtained from those solutions.

A relevant anti-microbial activity of CNs was observed for boards coated with CNs water suspension and this effect was demonstrated on fresh pasta [7]. In this case, the humidity and the hygroscopic behavior of cellulose granted the necessary condition for CNs to be active towards bacteria, discouraging their proliferation.

In agreement with the previous literature survey, the anti-microbial effects of CN and Chitosan are currently still under debate because of the different methodologies of applications. In our case, as the CNs are dispersed in the film by extrusion, the CNs are not reasonably exposed on the surface but embedded in the polymer bulk. Hence, the direct anti-microbial activity of the film is not significant, unless the concentration of CNs on the film surface is increased.

5. Conclusions

Nanobiocomposites suitable for preparing films by flat die extrusion were prepared based on plasticized poly(lactic acid) (PLA), poly(butylene succinate-co-adipate) (PBSA) and Chitin nanofibrils as functional filler. Chitin nanofibrils (CNs) were well dispersed thanks to the preparation of a pre-nanocomposites containing poly(ethylene glycol).

Thanks to the use of a melt strength enhancer (Plastistrength) and calcium carbonate the processability and thermal properties of bionanocomposites films containing CNs could be tuned in a wide range. The films resulted flexible and highly resistant.

The addition of CNs in the presence of starch proved not advantageous because of an extensive chain scission resulting in very low values of melt viscosity, making the material not suitable to be processed by flat die extrusion.

The films containing CNs or CNs and calcium carbonate were proven biocompatible and able to stimulate the production of cells defensins, acting as indirect anti-microbials.

Tests made with *Staphylococcus aureus* and *Enterobacter* spp. (Gram positive and negative respectively) by the qualitative agar diffusion test did not show any anti-microbial activity. The results can be explained considering that the CNs are present in the material bulk and not on the surface. Hence the direct anti-microbial activity, linked to the direct contact of the CNs with the bacteria cells, should be optimized.

The work suggests that for a compatible film with a mild indirect anti-microbial activity due to the defensins produced by skin cells, the incorporation of CNs in the film can be a good strategy. However, if a direct anti-microbial activity is required, it is necessary to increase the concentration of CNs on the film surface. Hence, developing coatings containing chitin nanofibrils can be an interesting perspective for future research regarding bioplastic films with a direct anti-microbial activity.

Author Contributions: Conceptualization, M.-B.C., S.D. and G.D.; methodology, A.V., A.F., L.P.; writing—original draft preparation, L.A., M.-B.C., A.V.; writing—review and editing, G.D., P.M., S.N., C.F.-A.; supervision, P.M., A.L.; project administration, S.N. All authors have read and agreed to the published version of the manuscript.

Funding: This research was funded by the Bio-Based Industries Joint Undertaking under the European Union Horizon 2020 research program (BBI-H2020), PolyBioSkin project, grant number G.A. 745839.

Acknowledgments: Alessandro Gagliardini and Pietro Febo of ATERTEK company (Pescara, Italy) are thanked for their suggestions regarding the sanitary market and its technological objectives; Tony Virdis and Fabio Di Berardino of TEXOL (Pescara, Italy) are thanked for kind discussion regarding the industrial production of sanitary films. Omya and Sacchetto companies are thanked for providing calcium carbonate and wheat starch respectively. Adone Baroni is thanked for helpful discussion. The CISUP—Centre for Instrumentation Sharing—University of Pisa is thanked for its support.

Conflicts of Interest: The authors declare no conflict of interest.

References

1. Moustafa, H.; El Kissi, N.; Abou-Kandil, A.I.; Abdel-Aziz, M.S.; Dufresne, A. PLA/PBAT Bionanocomposites with Antimicrobial Natural Rosin for Green Packaging. *ACS Appl. Mater. Interf.* **2017**, *9*, 20132–20141. [CrossRef] [PubMed]
2. Sharma, R.; Jafari, S.M.; Sharma, S. Antimicrobial bio-nanocomposites and their potential applications in food packaging. *Food Control* **2020**, *112*, 107086. [CrossRef]
3. Liao, C.; Li, Y.; Tjong, S.C. Antibacterial Activities of Aliphatic Polyester Nanocomposites with Silver Nanoparticles and/or Graphene Oxide Sheets. *Nanomaterials* **2019**, *9*, 1102. [CrossRef] [PubMed]
4. Febo, P.; Gagliardini, A. Baby Diapers Past and Present: A Critical Review. In *Bionanotechnology to Save the Environment: Plant and Fishery's Biomass as Alternative to Petrol*; Morganti, P., Ed.; MDPI: Basel, Switzerland, 2018; pp. 227–238.
5. Huang, K.S.; Yang, C.H.; Huang, S.L.; Chen, C.Y.; Lu, Y.Y.; Lin, Y.S. Recent Advances in Antimicrobial Polymers: A Mini-Review. *Int. J. Mol. Sci.* **2016**, *17*, 1578. [CrossRef] [PubMed]
6. Corsini, E.; Galli, C.L. Epidermal cytokines in experimental contact dermatitis. *Toxicology* **2000**, *142*, 203–211. [CrossRef]
7. Panariello, L.; Coltelli, M.B.; Buchignani, M.; Lazzeri, A. Chitosan and nano-structured chitin for biobased anti-microbial treatments onto cellulose based materials. *Eur. Polym. J.* **2019**, *113*, 328–339. [CrossRef]
8. Donnarumma, G.; Fusco, A.; Morganti, P.; Palombo, M.; Anniboletti, T.; Del Ciotto, P.; Baroni, A. Advanced medications made by green nanocomposites. *Int. J. Res. Pharmaceut. Nano Sci.* **2016**, *5*, 261–270.
9. Danti, S.; Trombi, L.; Fusco, A.; Azimi, B.; Lazzeri, A.; Morganti, P.; Coltelli, M.-B.; Donnarumma, G. Chitin Nanofibrils and Nanolignin as Functional Agents in Skin Regeneration. *Int. J. Mol. Sci.* **2019**, *20*, 2669. [CrossRef]
10. Morganti, P.; Danti, S.; Coltelli, M.B. Chitin and lignin to produce biocompatible tissues. *Res. Clin. Dermatol.* **2018**, *1*, 5–11. [CrossRef]
11. Morganti, P.; Morganti, G.; Nunziata, M.L. Chitin Nanofibrils, a Natural Polymer from Fishery Waste: Nanoparticle and Nanocomposite Characteristics. In *Bionanotechnology to Save the Environment: Plant and Fishery's Biomass as Alternative to Petrol*; MDPI: Basel, Switzerland, 2018; pp. 60–81.
12. Pivsa-Art, W.; Fujii, K.; Nomura, K.; Aso, Y.; Ohara, H.; Yamane, H. The effect of poly(ethylene glycol) as plasticizer in blends of poly(lactic acid) and poly(butylene succinate). *J. Appl. Polym. Sci.* **2016**, *133*, 1–10. [CrossRef]
13. Lim, S.T.; Hyun, Y.H.; Choi, H.J.; Jhon, M.S. Synthetic biodegradable aliphatic polyester/montmorillonite nanocomposites. *Chem. Mater.* **2002**, *14*, 1839–1844. [CrossRef]
14. Bogaert, J.; Coszach, P. Poly(lactic acids): A potential solution to plastic waste dilemma. *Macromol. Symp.* **2000**, *153*, 287–303. [CrossRef]
15. Helmes, R.J.K.; López-Contreras, A.M.; Benoit, M.; Abreu, H.; Maguire, J.; Moejes, F.; van den Burg, S.W.K. Environmental impacts of experimental production of lactic acid for bioplastics from Ulva spp. *Sustainaiblity* **2018**, *10*, 2462. [CrossRef]
16. Aliotta, L.; Cinelli, P.; Coltelli, M.B.; Righetti, M.C.; Gazzano, M.; Lazzeri, A. Effect of nucleating agents on crystallinity and properties of poly (lactic acid) (PLA). *Eur. Polym. J.* **2017**, *93*, 822–832. [CrossRef]
17. Raquez, J.M.; Habibi, Y.; Murariu, M.; Dubois, P. Polylactide (PLA)-based nanocomposites. *Prog. Polym. Sci.* **2013**, *38*, 1504–1542. [CrossRef]
18. Martin, O.; Avérous, L. Poly(lactic acid): Plasticization and properties of biodegradable multiphase systems. *Polymers* **2001**, *42*, 6209–6219. [CrossRef]

19. Ljungberg, N.; Wesslén, B. Preparation and properties of plasticized poly(lactic acid) films. *Biomacromolecules* **2005**, *6*, 1789–1796. [CrossRef]
20. Fehri, S.; Cinelli, P.; Coltelli, M.-B.; Anguillesi, I.; Lazzeri, A. Thermal Properties of Plasticized Poly (Lactic Acid) (PLA) Containing Nucleating Agent. *Int. J. Chem. Eng. Appl.* **2016**, *7*, 85–88. [CrossRef]
21. Gigante, V.; Canesi, I.; Cinelli, P.; Coltelli, M.B.; Lazzeri, A. Rubber Toughening of Polylactic Acid (PLA) with Poly(butylene adipate-co-terephthalate) (PBAT): Mechanical Properties, Fracture Mechanics and Analysis of Ductile-to-Brittle Behavior while Varying Temperature and Test Speed. *Eur. Polym. J.* **2019**, *115*, 125–137. [CrossRef]
22. Al-Itry, R.; Lamnawar, K.; Maazouz, A. Improvement of thermal stability, rheological and mechanical properties of PLA, PBAT and their blends by reactive extrusion with functionalized epoxy. *Polym. Degrad. Stab.* **2012**, *97*, 1898–1914. [CrossRef]
23. Teamsinsungvon, A.; Ruksakulpiwat, Y.; Jarukumjorn, K. Preparation and Characterization of Poly(lactic acid)/Poly(butylene adipate-co-terepthalate) Blends and Their Composite. *Polym. Plast. Technol. Eng.* **2013**, *52*, 1362–1367. [CrossRef]
24. Liu, G.; Zheng, L.; Zhang, X.; Li, C.; Jiang, S.; Wang, D. Reversible lamellar thickening induced by crystal transition in poly(butylene succinate). *Macromolecules* **2012**, *45*, 5487–5493. [CrossRef]
25. Gigante, V.; Coltelli, M.; Vannozzi, A.; Panariello, L.; Fusco, A.; Trombi, L.; Donnarumma, G.; Danti, S.; Lazzeri, A. Flat Die Extruded Biocompatible Poly(Lactic Acid). *Polymers* **2019**, *11*, 1857. [CrossRef] [PubMed]
26. Hassouna, F.; Raquez, J.M.; Addiego, F.; Dubois, P.; Toniazzo, V.; Ruch, D. New approach on the development of plasticized polylactide (PLA): Grafting of poly(ethylene glycol) (PEG) via reactive extrusion. *Eur. Polym. J.* **2011**, *47*, 2134–2144. [CrossRef]
27. Dorez, G.; Taguet, A.; Ferry, L.; Lopez-Cuesta, J.M. Thermal and fire behavior of natural fibers/PBS biocomposites. *Polym. Degrad. Stab.* **2013**, *98*, 87–95. [CrossRef]
28. Coltelli, M.-B.; Della Maggiore, I.; Bertoldo, M.; Signori, F.; Bronco, S.; Ciardelli, F. Poly(lcatic acid) Properties as a Consequnce of Poly(butylene adipate-co-terephtalate) Blending and Acetyl trybutyl Citrate Plasticization. *J. Appl. Polym. Sci.* **2008**, *110*, 1250–1262. [CrossRef]
29. Fortunati, E.; Puglia, D.; Iannoni, A.; Terenzi, A.; Kenny, J.M.; Torre, L. Processing conditions, thermal and mechanical responses of stretchable poly (lactic acid)/poly (butylene succinate) films. *Materials* **2017**, *10*, 809. [CrossRef]
30. Coltelli, M.B.; Bronco, S.; Chinea, C. The effect of free radical reactions on structure and properties of poly(lactic acid) (PLA) based blends. *Polym. Degrad. Stab.* **2010**, *95*, 332–341. [CrossRef]
31. Coltelli, M.B.; Gigante, V.; Panariello, L.; Morganti, P.; Cinelli, P.; Danti, S.; Lazzeri, A. Chitin nanofibrils in renewable materials for packaging and personal care applications. *Adv. Mater. Lett.* **2018**, *10*, 425–430. [CrossRef]
32. Coltelli, M.; Cinelli, P.; Gigante, V.; Aliotta, L.; Morganti, P.; Panariello, L.; Lazzeri, A. Chitin Nanofibrils in Poly(Lactic Acid)(PLA) Nanocomposites: Dispersion and Thermo-Mechanical Properties. *Int. J. Mol. Sci.* **2019**, *20*, 504. [CrossRef]
33. Morganti, P. Chitin Nanofibrils in Skin Treatment. *J. Appl. Cosmetol.* **2014**, *27*, 251–270.
34. Cinelli, P.; Coltelli, M.B.; Mallegni, N.; Morganti, P.; Lazzeri, A. Degradability and sustainability of nanocomposites based on polylactic acid and chitin nano fibrils. *Chem. Eng. Trans.* **2017**, *60*, 115–120. [CrossRef]
35. Liu, Y.; Liu, M.; Yang, S.; Luo, B.; Zhou, C. Liquid Crystalline Behaviors of Chitin Nanocrystals and Their Reinforcing Effect on Natural Rubber. *ACS Sustain. Chem. Eng.* **2018**, *6*, 325–336. [CrossRef]
36. Liu, M.; Zheng, H.; Chen, J.; Li, S.; Huang, J.C.; Zhou, C. Chitosan-chitin nanocrystal composite scaffolds for tissue engineering. *Carbohydr. Polym.* **2016**, *152*, 832–840. [CrossRef]
37. Rizvi, R.; Cochrane, B.; Naguib, H.; Lee, P.C. Fabrication and characterization of melt-blended polylactide-chitin composites and their foams. *J. Cell. Plast.* **2011**, *47*, 283–300. [CrossRef]
38. Zhang, Q.; Wei, S.; Huang, J.; Feng, J.; Chang, P.R. Effect of surface acetylated-chitin nanocrystals on structure and mechanical properties of poly(lactic acid). *J. Appl. Polym. Sci.* **2014**, *131*, 2–9. [CrossRef]
39. Araki, J.; Kurihara, M. Preparation of sterically stabilized chitin nanowhisker dispersions by grafting of poly(ethylene glycol) and evaluation of their dispersion stability. *Biomacromolecules* **2015**, *16*, 379–388. [CrossRef]

40. Guan, Q.; Naguib, H.E. Fabrication and Characterization of PLA/PHBV-Chitin Nanocomposites and Their Foams. *J. Polym. Environ.* **2014**, *22*, 119–130. [CrossRef]
41. Herrera, N.; Singh, A.A.; Salaberria, A.M.; Labidi, J.; Mathew, A.P.; Oksman, K. Triethyl citrate (TEC) as a dispersing aid in polylactic acid/chitin nanocomposites prepared via liquid-assisted extrusion. *Polymers* **2017**, *9*, 406. [CrossRef]
42. Herrera, N.; Roch, H.; Salaberria, A.M.; Pino-Orellana, M.A.; Labidi, J.; Fernandes, S.C.M.; Radic, D.; Leiva, A.; Oksman, K. Functionalized blown films of plasticized polylactic acid/chitin nanocomposite: Preparation and characterization. *Mater. Des.* **2016**, *92*, 846–852. [CrossRef]
43. Scatto, M.; Salmini, E.; Castiello, S.; Coltelli, M.B.; Conzatti, L.; Stagnaro, P.; Andreotti, L.; Bronco, S. Plasticized and nanofilled poly(lactic acid)-based cast films: Effect of plasticizer and organoclay on processability and final properties. *J. Appl. Polym. Sci.* **2013**, *127*, 4947–4956. [CrossRef]
44. Li, J.; Gao, Y.; Zhao, J.; Sun, J.; Li, D. Homogeneous dispersion of chitin nanofibers in polylactic acid with different pretreatment methods. *Cellulose* **2017**, *24*, 1705–1715. [CrossRef]
45. Coltelli, M.-B.; Cinelli, P.; Anguillesi, I.; Salvadori, S.; Lazzeri, A. Structure and properties of extruded composites based on bio-polyesters and nano-chitin. In Proceedings of the Symposium E-MRS Fall Meeting, Warsaw University of Technology, Warsaw, Poland, 15–18 September 2014; p. 192.
46. Xiong, Z.; Yang, Y.; Feng, J.; Zhang, X.; Zhang, C.; Tang, Z.; Zhu, J. Preparation and characterization of poly(lactic acid)/starch composites toughened with epoxidized soybean oil. *Carbohydr. Polym.* **2013**, *92*, 810–816. [CrossRef] [PubMed]
47. Wang, H.; Sun, X.; Seib, P. Strengthening blends of poly(lactic acid) and starch with methylenediphenyl diisocyanate. *J. Appl. Polym. Sci.* **2001**, *82*, 1761–1767. [CrossRef]
48. Teixeira, E.M.; Da Róz, A.L.; Carvalho, A.J.F.; Curvelo, A.A.S. The effect of glycerol/sugar/water and sugar/water mixtures on the plasticization of thermoplastic cassava starch. *Carbohydr. Polym.* **2007**, *69*, 619–624. [CrossRef]
49. Ma, X.F.; Yu, J.G.; Wan, J.J. Urea and ethanolamine as a mixed plasticizer for thermoplastic starch. *Carbohydr. Polym.* **2006**, *64*, 267–273. [CrossRef]
50. Shi, R.; Zhang, Z.; Liu, Q.; Han, Y.; Zhang, L.; Chen, D.; Tian, W. Characterization of citric acid/glycerol co-plasticized thermoplastic starch prepared by melt blending. *Carbohydr. Polym.* **2007**, *69*, 748–755. [CrossRef]
51. Rodriguez-Gonzalez, F.J.; Ramsay, B.A.; Favis, B.D. Rheological and thermal properties of thermoplastic starch with high glycerol content. *Carbohydr. Polym.* **2004**, *58*, 139–147. [CrossRef]
52. Coltelli, M.-B.; Danti, S.; Trombi, L.; Morganti, P.; Donnarumma, G.; Baroni, A.; Fusco, A.; Lazzeri, A. Preparation of Innovative Skin Compatible Films to Release Polysaccharides for Biobased Beauty Masks. *Cosmetics* **2018**, *5*, 70. [CrossRef]
53. ARKEMA Technical Datasheet—Plastistrength® 550. 2019. Available online: https://www.additives-arkema.com/export/shared/.content/media/downloads/products-documentations/organic-peroxides/functional-additives-acrylic-modifiers/plastistrength/plastistrength-550-tds.pdf (accessed on 24 January 2020).
54. Muzzarelli, C.; Morganti, P. Preparation of Chitin and Derivatives Thereof for Cosmetic and Therapeutic Use. U.S. Patent 8,552,164 B2, 8 October 2013.
55. Morganti, P.; Carezzi, F.; Del Ciotto, P.; Morganti, G.; Nunziata, M.L.; Gao, X.; Tishenko, G.; Yudin, V. Chitin Nanofibrils: A Natural Multifunctional Polymer. In *Nanobiotechnology*; Phoenix, D.A., Ahmed, W., Eds.; One Central Press: Altrincham, UK, 2014; pp. 1–31. ISBN 978-1-910086-03-2.
56. Signori, F.; Coltelli, M.B.; Bronco, S. Thermal degradation of poly(lactic acid) (PLA) and poly(butylene adipate-co-terephthalate) (PBAT) and their blends upon melt processing. *Polym. Degr. Stab.* **2009**, *94*, 74–82. [CrossRef]
57. Bhatia, A.; Gupta, R.K.; Bhattacharya, S.; Choi, H.J. Effect of Clay on Thermal, Mechanical and Gas Barrier Properties of Biodegradable Poly(lactic acid)/Poly(butylene succinate) (PLA/PBS) Nanocomposites. *Int. Polym. Process.* **2010**, *25*, 5–14. [CrossRef]
58. Fischer, E.W.; Sterzel, H.J.; Wegner, G. Investigation of the structure of solution grown crystals of lactide copolymers by means of chemical reactions. *Colloid Polym. Sci.* **1973**, *251*, 980–990. [CrossRef]
59. Wilson, V.G. Growth and Differentiation of HaCaT Keratinocytes. In *Epidermal Cells. Methods in Molecular Biology (Methods and Protocols)*; Turksen, K., Ed.; Springer: New York, NY, USA, 2013; Volume 1195, pp. 33–41. [CrossRef]

60. Olaiya, N.G.; Surya, I.; Oke, P.K.; Rizal, S.; Sadiku, E.R.; Ray, S.; Farayibi, P.K.; Hossain, M.S.; Abdul Khalil, H.P.S. Properties and characterization of a PLA-Chitin-Starch Biodegradble Polymer Composite. *Polymers* **2019**, *11*, 1656. [CrossRef]
61. Pereira, A.G.B.; Muniz, E.C.; Hsieh, Y.L. Chitosan-sheath and chitin-core nanowhiskers. *Carbohydr. Polym.* **2014**, *107*, 158–166. [CrossRef]
62. Kaczmarek, H.; Nowicki, M.; Vuković-Kwiatkowska, I.; Nowakowska, S. Crosslinked blends of poly(lactic acid) and polyacrylates: AFM, DSC and XRD studies. *J. Polym. Res.* **2013**, *20*. [CrossRef]
63. Mallegni, N.; Phuong, T.V.; Coltelli, M.B.; Cinelli, P.; Lazzeri, A. Poly(lactic acid) (PLA) based tear resistant and biodegradable flexible films by blown film extrusion. *Materials* **2018**, *11*, 148. [CrossRef]
64. Li, M.; Wu, Q.; Song, K.; Cheng, H.N.; Suzuki, S.; Lei, T. Chitin Nanofibers as Reinforcing and Antimicrobial Agents in Carboxymethyl Cellulose Films: Influence of Partial Deacetylation. *ACS Sustain. Chem. Eng.* **2016**, *4*, 4385–4395. [CrossRef]
65. Min, B.; Lee, S.W.; Lim, J.N.; You, Y.; Lee, T.S.; Kang, P.H. Chitin and chitosan nanofibers: Electrospinning of chitin and deacetylation of chitin nanofibers. *Polymer* **2004**, *45*, 7137–7142. [CrossRef]
66. Foster, L.J.R.; Butt, J. Chitosan films are NOT antimicrobial. *Biotechnol. Lett.* **2011**, *33*, 417–421. [CrossRef]
67. Rai, S.; Dutta, P.K.; Mehrotra, G.K. Lignin Incorporated Antimicrobial Chitosan Film for Food Packaging Application. *J. Polym. Mater.* **2017**, *34*, 171.

© 2020 by the authors. Licensee MDPI, Basel, Switzerland. This article is an open access article distributed under the terms and conditions of the Creative Commons Attribution (CC BY) license (http://creativecommons.org/licenses/by/4.0/).

Article

Skin-Compatible Biobased Beauty Masks Prepared by Extrusion

Maria-Beatrice Coltelli [1,2,*], Luca Panariello [1,2], Pierfrancesco Morganti [3,4], Serena Danti [2], Adone Baroni [1,5], Andrea Lazzeri [1,2], Alessandra Fusco [1,5] and Giovanna Donnarumma [1,5,*]

1. Consorzio Interuniversitario Nazionale per la Scienza e Tecnologia dei Materiali (INSTM), 50121 Florence, Italy; luca.panariello@ing.unipi.it (L.P.); adone.baroni@unicampania.it (A.B.); andrea.lazzeri@unipi.it (A.L.); alessandra.fusco@unicampania.it (A.F.)
2. Department of Civil and Industrial Engineering, University of Pisa, 56122 Pisa, Italy; serena.danti@unipi.it
3. Academy of History of Health Care Art, 00193 Rome, Italy; morgantipf@gmail.com
4. Dermatology Department, China Medical University, Shenyang 110001, China
5. Department of Experimental Medicine, University of Campania "Luigi Vanvitelli", 80138 Naples, Italy
* Correspondence: maria.beatrice.coltelli@unipi.it (M.-B.C.); giovanna.donnarumma@unicampania.it (G.D.); Tel.: +39-050-2217856 (M.-B.C.)

Received: 31 January 2020; Accepted: 20 March 2020; Published: 6 April 2020

Abstract: In the cosmetic sector, natural and sustainable products with a high compatibility with skin, thus conjugating wellness with a green-oriented consumerism, are required by the market. Poly(hydroxyalkanoate) (PHA)/starch blends represent a promising alternative to prepare flexible films as support for innovative beauty masks, wearable after wetting and releasing starch and other selected molecules. Nevertheless, preparing these films by extrusion is difficult due to the high viscosity of the polymer melt at the temperature suitable for processing starch. The preparation of blends including poly(butylene succinate-co-adipate) (PBSA) or poly(butylene adipate-co-terephthalate) (PBAT) was investigated as a strategy to better modulate melt viscosity in view of a possible industrial production of beauty mask films. The release properties of films in water, connected to their morphology, was also investigated by extraction trials, infrared spectroscopy and stereo and electron microscopy. Then, the biocompatibility with cells was assessed by considering both mesenchymal stromal cells and keratinocytes. All the results were discussed considering the morphology of the films. This study evidenced the possibility of modulating thanks to the selection of composition and the materials processing of the properties necessary for producing films with tailored properties and processability for beauty masks.

Keywords: starch; poly(hydroxyalkanoate); biopolyesters; beauty masks; releasing; skin compatible

1. Introduction

Over the past two decades, declining fertility and mortality rates have resulted in an increased aged population [1]. This phenomenon has created room for innovation, leading to a robust demand for anti-aging products in order to prevent wrinkles, age-spots, dry skin, uneven skin tone and even hair weakening. By 2050, in fact, the worldwide population over 60 years of age is expected to reach 2.09 billion: the life expectancy for women has been predicted to rise from 82.8 years in 2005 to 86.3 in 2050, while for men this increase is expected to be from 78.4 to 83.6 years. As a consequence, new demographic trends, technologies, and consumer insights are impacting the cosmetic sector that has shown a general resistance to macroeconomic events such as recession [2]. Thus, today the global beauty industry reached a value of $532 billion, with the largest market represented by the US (with about 20% share), followed by China (13%) and Japan (8%). With a compound annual growth rate between 5% and 7%, the beauty market is estimated to reach $800 billion by 2025, remaining impervious to the ups and down of the global economy [2,3].

However, while 70% of women aged more than 40 years old want to see more beauty products targeting anti-aging issues, more specific men's personal care products are also emerging. Moreover, consumers are increasingly demanding about the real composition of the products they are buying, and they wish to have more information regarding the use of the different cosmetics and beauty products that work both internally and topically, being less interested in the usual advertising and marketing announcements. Thus, ingredient sourcing and effectiveness have become a major concern for the cosmetic industry together with the use of recyclable and refillable packaging materials necessary to reduce their carbon footprint, especially if obtained by industrial and agro-food waste.

In the cosmetic sector, the consumers' naturally oriented demand is towards the use of biobased active molecules and materials; this so called clean and natural beauty, therefore, is not only about what is in consumers' products, but also about how products are produced and packaged. In conclusion, in moving the raw materials from waste and/or land to the lab, biotechnologies are increasingly impacting the production of future beauty ingredients as well as the final products. Biobased materials to make innovative beauty masks go in this direction because such renewable raw materials seem able to protect the skin from pollution and modulate its microbiota [4].

For all these reasons, the production of more and more commercial products made by biobased and biodegradable polymers [5] (especially those products having a brief life-cycle) can be no longer delayed.

Beauty masks are currently mainly produced by using wet, nonwoven tissues, often prepared with fossil-based fibers. After their use, these products and their packaging materials are not selectively collected, contributing to increase the nondifferentiated part of the urban waste, often incinerated or landfilled.

Biobased and biodegradable polymeric materials, with the capacity of adhering to skin after wetting and releasing active molecules that are skin-friendly and easily compostable after usage, represent highly eco-sustainable options and at the same time meet consumers' expectations. They can be commercialized at the dry state, avoiding the use of preservatives [5–7]. Moreover, they can effectively release starch [8] together with the functional molecules previously added by several techniques to the starch or to the obtained film, for instance as coating.

Among the biobased polymers already available on the market, poly(hydroxyalkanoates) (PHAs), a family of polymers obtained from bacteria [9], are suitable for this application because of their very high biocompatibility [10,11], lower greenhouse gas emissions [12] and both soil and marine compostability [13,14]. These properties make the PHA-based materials very promising to be used in applications where environmental concern and biocompatibility are both fundamental. On the other hand, starch is the major carbohydrate reserve in higher plants [15] and is also a very abundant biopolymer still much used in nonfood applications (e.g., glues, gums, thickener, soothing for skin).

Skin-compatible PHA/starch films including calcium carbonate were prepared in a previous paper [16], where a PHA elastomer (EM 5400F EM 5400F, obtained from Shenzhen Ecomann Biotechnology Co.), a specific calcium carbonate and an easy methodology to pregelatinize starch were investigated and selected for preparing films by compression molding. These films resulted as promising to make beauty masks being flexible enough to follow the skin curvature, being sticky when applied on wet skin and having a fast release kinetic in water. However, their preparation was based on compression molding of biopolymeric powder with calcium carbonate as additive. This methodology is different from the ones currently used in industry to prepare bioplastic flat films characterized by a high productivity [17–19] and therefore can limit the real feasibility of the process.

Moreover, the preliminary extrusion of the blend consisting of PHA, starch and calcium carbonate showed a high value of torque in agreement with a high viscosity of the melt. Therefore, the processing should be made easier to allow the use of these biobased films in industrial applications.

Poly(hydroxybutyrate) (PHB), the most investigated biopolyester of the PHA family, has a limited stability at the high temperatures required for the melting processes. It, in fact, undergoing thermal degradation affects its physical and mechanical properties, making the industrial process difficult [20]. The very low resistance to thermal degradation seems to be the most serious problem related to its processing. The main reaction involves chain scission, which results in a rapid decrease in molecular weight [21]. Copolymers of 3-hydrobutyrate with other hydroxyacids present a lower melting point and an increased processing window. However, the critical molecular weight of these polymers is high [22] and, when processed, it is necessary to control the temperature, keeping it slightly above the melting point to avoid the polymer degradation. Therefore, plasticization is generally adopted as the methodology to control the processing temperature [9].

The blending of PHA, especially PHB with starch, has been extensively investigated [23,24]. Godbole et al. [25] discussed the compatibility of PHB with starch to achieve the improved properties and reduce cost. The results revealed that films had a single glass transition temperature (Tg) for all proportions of PHB/starch blends. However, the necessity of increasing the compatibility of PHB and PHA with starch is evidenced in other papers, and several methodologies including the chemical modification of polymers are proposed [26,27].

The blending with copolyesters having a low viscosity in the processing temperature window of PHA seems to be a good strategy to enhance processability. Elastomeric commercial copolyesters are the most promising, while copolymers based on poly(butylenesuccinate) with adipic acid (PBSA) are commercially available. They are currently partially renewable, but they will be fully renewable in the near future [28], being biodegradable in composting plant and soil [29]. Thus, blends of PBS or PBSA with PHA were investigated for improving the compatibility [30], and their promising application in flexible packaging films was recently demonstrated [31].

Blends of PHB and PBS filled with starch were recently investigated by Zhang et al. [32]. It was evidenced that starch nanoparticles localized on the PHB/PBS phase interfaces improve phase adhesion, while those dispersed in the continuous PHB phase are able to prevent the coalescence of PBS droplets during the melt mixing, thus remarkably decreasing the droplet domain size. Although these results were achieved for specific nanocomposites, they suggest that ternary PHA/PBSA/starch blends can show an affordable compatibility. On the other hand, some authors have shown that blends of PBS with starch showed a good adhesion at the interface [33], with an improved processability [34] that enabled these blends to be used in food packaging.

Another biopolyester that can be blended with PHAs is poly(butylene adipate-co-terephthalate) (PBAT). Currently, it is obtained only partially from renewable sources, but in a short time it will be fully renewable [28], and it is biodegradable in a composting plant [35]. Consequently, it will be suitable for preparing bioplastic films, incorporating starch with mechanical properties suitable for food packaging applications [36,37].

Regarding the blending of PHA and PBAT, Larsson et al. [38] evidenced that PBAT can be a very efficient additive to improve the processability of PHA. Matos Costa et al. [39] studied the thermal behavior of the blends and evidenced that the crystallization of PBAT is very fast during cooling, whereas that of PHB is slow. Lin et al. [40] reported that these blends can be used in electrospinning to obtain antibacterial hydrophobic nanofibrous membranes. Zarrinbakhsh et al. [41] investigated the compatibilization of PHB/PBAT blends by considering the effect of a compatibilizer, polymeric methylene diphenyl diisocyanate (PMDI), and corn oil as lubricant for preparing composites, including a waste of bioethanol production. The change in melt processing force suggested the occurrence of chemical reactions during the process time. The glass transition peaks pertaining to the PBAT and PHBV matrix shifted slightly towards each other, suggesting the occurrence of crosslinking at the PBAT–PHBV interface due to the reactivity of PMDI. Belyamani et al. [42] investigated composites based on PHA/PBAT reinforced with trisilanolisobutyl polyhedral oligomeric silsesquioxanes (POSS) and calcium phosphate glass (CaP-g) under simulated physiological and human body temperature conditions. Biodegradation studies regarding PHB/PBAT blends [43] demonstrated that the amount of PBAT in the blend can impact the degradation rate, with formation of porous structure based on PBAT, but the material remains as substantially biodegradable as PBAT is.

The biocompatibility of materials for cosmetic use with the skin is a very important aspect in their production and marketing and mainly concerns the mechanisms of the innate immune response; the biological mediators of innate immunity are cytokines, which are multifunctional molecules implicated in various biological activities and endowed with pro- and anti-inflammatory activities. The best-studied members of this group are proinflammatory cytokines IL-1, tumor necrosis factor α (TNF-α), IL-6, chemokine IL-8 and anti-inflammatory cytokine transforming growth factor β (TGF-β). IL-1 promotes coagulation, increases the expression of adhesion molecules, causes the release of chemokines that recruit other leukocytes to the site of inflammation, and stimulates the growth and differentiation of B lymphocytes and of the many effector cell response [44,45]; The tumor necrosis factor α (TNF-α) is an essential mediator in inflammation [46]. The production of TNF-α and its release in the site of the inflammation involves a localized vascular endothelial activation, the release of NO and vasodilation with increased vascular permeability [44]; IL-6 is involved in synthesis of fibrinogen, which contributes to the inflammatory acute phase response [47], and IL-8 is involved in chemotaxis of basophils and has a role in angiogenesis [48]. Finally, The TGF-β or transformation and growth factor is part of the family of anti-inflammatory cytokines and is considered as the most powerful, able to negatively modulate almost all the inflammatory responses [49].

Coltelli et al. [16] described the use of a commercial elastomeric poly(hydroxyalkanoate) and starch for obtaining compression-molded bioplastic films with the necessary resistance in wet conditions, skin compatibility and capacity for a fast release of polysaccharides. Starting from these results, the preparation of blends by extrusion including poly(butylene succinate-co-adipate) (PBSA) or poly(butylene adipate-co-terephthalate) (PBAT) was successfully exploited in the present paper to better modulate melt viscosity to obtain films by extrusion. Their behavior and morphologies upon release in water were correlated using spectroscopic and microscopic evidence to understand the different releasing mechanisms. Consequentially, the biocompatibility of prepared films with cells was assessed by considering keratinocytes, and the mechanisms of the innate immune response were also investigated. The results were discussed considering the effect of extrusion and blending on the morphology of the films, their releasing capacity, their compatibility with cells and their immunomodulatory behavior. The objective is preparing a substrate suitable for adding specific functional molecules or complexes (in the starch phase or as coating) that can be released during the application of the mask onto skin.

2. Materials and Methods

2.1. Materials

PHA (Ecomann EM F5400 F) was supplied as pellet from Shenzhen Ecomann Biotechnology Co., Ltd., Shandong, China.

PBAT (Ecoflex C1200) was purchased as pellet from BASF. It is a statistical, aromatic–aliphatic copolymer based on the monomers 1.4-butanediol, adipic acid and terephthalic acid. It will biodegrade to the basic monomers 1,4-butanediol, adipic acid and terephthalic acid and eventually to carbon dioxide, water and biomass when metabolized in the soil or compost under standard conditions [50]. Specifically, PBAT C1200 has a density of 1.26 g/cm^3 and MW = 126,000 g/mol. PBSA (BioPBS FD92PM) was supplied as pellet from Mitsubishi Chemical Corporation. It consists of a copolymer of succinic acid, adipic acid and 1,4-butandiol [51]. In particular, BioPBS FD92PM has a density of 1.24 g/cm^3.

Treated calcium carbonate (further indicated as CC) (OmyaSmartfill 55—OM) was supplied from Omya SPA and is characterized by 55% of particles with a diameter less than 2 μm.

Wheat native starch was supplied by Sacchetto SPA (Lagnasco, CN, Italy). Poly(ethylene glycol) with Mn = 400 (PEG400) and Glycerol were purchased from Sigma-Aldrich (Milan, Italy). Absolute ethanol was purchased from Bio Optica S.p.A. (Milan, Italy). Immortalized human keratinocytes (HaCaT cells) were purchased from Cell Lines Service GmbH (Eppelheim, Germany). Resazurin sodium salt (used to prepare AlamarBlue test reagent), phosphate-buffered saline (PBS), fungizone and trypsin were provided by Sigma-Aldrich (Milan, Italy). Dulbecco's modified Eagle's medium (D-MEM), L-glutamine, penicillin–streptomycin (penstrep) and fetal calf serum (FCS) were obtained from Invitrogen (Carlsbad, CA, USA). Tri Reagent was purchased by Sigma-Aldrich/Merck (Darmstadt, Germany). LC Fast Start DNA Master SYBR Green kit was provided by Roche Applied Science (Euroclone S.p.A., Pero, Italy).

2.2. Preparation of Samples and Films

Preparation of PHA/starch films was accomplished in the following steps [14]: (1) Starch was mixed with glycerol and PEG 400 in an oven at 80 °C for 16 h to obtain its pregelatinization; (2) PHA powder was added to the starch in order to obtain a homogenous powder; (3) films were prepared through compression molding for 1 min at 190 °C.

Preplasticization of starch was performed by mixing wheat starch (RH 75%), PEG 400 and glycerol in a 60:10:30 ratio into a mortar. The mixture was kept overnight in a ventilated oven at 80 °C to obtain the starch gelatinization. Preplasticized starch (P-PLS) was then mixed with the powder of PHA in a 50:50 ratio into a mortar. Samples E-BM1, E-BM2, E-BM3, E-BM4 and E-BM5 were obtained by extrusion of the powder composed by PHA, pregelatinized native wheat starch (P-PLST), PBAT or PBSA and CC in a Minilab II Haake TMRheomex CTW 5 conical twin-screw extruder (Haake, Vreden, Germany). Materials were mixed for 1 min at the temperature of 140 °C and with a screw speed of 60 rpm. Torque values were recorded during all the extrusion process. The composition of the different blends is reported in Table 1. Some strands obtained from the extrusion were recovered and characterized.

Table 1. Extruded P-PLST/EM/PBAT or P-PLST/EM/PBSA blends prepared by micro-compounder.

Blends	P-PLST (wt%)	PHA (wt%)	PBSA (wt%)	PBAT (wt%)	CC (wt%)
E-BM1	46.5	46.5	-	-	7
E-BM2	46.5	-	46.5	-	7
E-BM3	46.5	23.25	23.25	-	7
E-BM4	46.5	-	-	46.5	7
E-BM5	46.5	23.25	-	23.25	7

The extruded strands (approximately 4 g for each extrusion) were minced and transferred between two Teflon square sheets for film preparation. The mixture was placed into the compression molding equipment at 190 °C, applying no pressure for the first 30 s, followed by the application of 4 metric tons for 30 s. After that pressure was removed, each film was rapidly removed and quenched with a cold air flow. Formed films were finally detached from the Teflon sheets. Several films were prepared for successive tests. Some nonextruded film samples were also prepared. They were produced only by compression molding the reference formulation named BM.

2.3. Characterization

2.3.1. Material Characterization

Melt flow index (MFI) value was measured with a CEAST Melt Flow Tester M20 (Instron, Canton, MA, USA) equipped with an encoder. Samples were conditioned for 3 h in an oven at 60 °C before the tests. MFI is defined as the weight of molten polymer passed in 10 min through a capillary of known diameter and length, applying pressure through a weight. Measure was performed according to the ISO 1133:2005 and, in particular, the ISO1133D. In this work, a weight of 2160 kg was used, and melt volume rate (MVR) data were recorded each 6 s for 60 s. Different measurements of MVR were performed, and a mean value (with their standard deviation) was reported. MFR was calculated starting from MVR value, and its standard deviation was estimated by MVR standard deviation data multiplied for the viscosity in the melt (ratio of MFR to MVR).

From the starting films, some small square specimens of about 20 mm side length were cut in two replicates for each formulation. The squares were poured at room temperature in distilled water for 30 min. After soaking with water, the specimens were dried in an oven at 60 °C until constant weight. The replicates were left in water for 16 h and then washed and dried in the same way.

Before and after release in water, the surface morphology of films was analyzed by stereomicroscopy using a Wild Heerbrugg M3 microscope equipped with a Pulnix TMC-6 camera (Heerbrugg, Switzerland) and by scanning electron microscopy using an FEI Quanta 450 FEG scanning electron microscope (SEM) (Thermo Fisher Scientific, Waltham, MA, USA). Samples for SEM analysis were cryo-fractured and covered with a tiny metallic layer of Au, in a way that the surface could be electrically conductive, to observe the sectional surface.

Furthermore, before and after tests of release in water, films were also characterized by infrared spectroscopy using a Nicolet T380 Thermo Scientific instrument equipped with a Smart ITX ATR accessory with diamond plate (Thermo Fisher Scientific, Waltham, MA, USA). The spectra were normalized in intensity with respect to the band at 1720 cm^{-1} typical of PHA polymer.

2.3.2. Epidermal Cell Culture and Viability Assay

HaCaT cells were cultured in D-MEM supplemented with 1% penstrep, 1% glutamine and 10% fetal calf serum at 37 °C in air and 5% CO_2. The HaCaT cells, seeded in 12-well plates until 80% of confluence, were incubated for 24 h with the film (BM or E-BM5). At the end of this time, resazurin solution (AB) diluted in DMEM to a final concentration of 0.5 mg/mL was added to the cells and incubated for 4 h in an incubator. Resazurin incorporates a redox indicator that changes color according to cell metabolic activity. The supernatants were read with a spectrophotometer using a double wavelength reading at 570 and 600 nm. Finally, the reduced percentage of the dye (%AB_{RED}) was calculated by correlating the absorbance values and the molar extinction coefficients of the dye at the selected wavelengths, following the protocol provided by the manufacturer.

2.3.3. Evaluation of Immunomodulatory Properties

The immunomodulatory properties of BM and E-BM5 films were analyzed using HaCaT cells.

The cells, cultured as described above, were seeded inside 12-well TC plates until 80% confluence and were incubated for 24 h with the films for 6 and 24 h ($n = 3$). At these endpoints, total RNA was isolated with TRizol and 1 µm of RNA was reverse-transcribed into complementary DNA (cDNA) using random hexamer primers, at 42 °C for 45 min, according to the manufacturer's instructions. Real-time polymerase chain reaction (PCR) was carried out with the LC Fast Start DNA Master SYBR Green kit using 2 µL of cDNA, corresponding to 10 ng of total RNA in a 20 µL final volume, 3 mM MgCl2 and 0.5 µM sense and antisense primers (Table 2). Real-time PCR was used to evaluate the expression of interleukins IL-1α, IL 1β, IL-6 and IL-8, as well as the expression of TNF-α and TGF-β.

Table 2. Real-time PCR conditions for HaCaT cells.

Gene	Primer Sequence	Conditions	Size (bp)
IL-1 α	5′-CATGTCAAATTTCACTGCTTCATCC-3′ 5′-GTCTCTGAATCAGAAATCCTTCTATC-3′	5 s at 95 °C, 8 s at 55 °C, 17 s at 72 °C for 45 cycles	421
IL-1 β	5′-GCATCCAGCTACGAATCTCC-3′ 5′-CCACATTCAGCACAGGACTC-3′	5 s at 95 °C, 14 s at 58 °C, 28 s at 72 °C for 40 cycles	708
TNF-α	5′-CAGAGGGAAGAGTTCCCCAG-3′ 5′-CCTTGGTCTGGTAGGAGACG-3′	5 s at 95 °C, 6 s at 57 °C, 13 s at 72 °C for 40 cycles	324
IL-6	5′-ATGAACTCCTTCTCCACAAGCGC-3′ 5′-GAAGAGCCCTCAGGCTGGACTG-3′	5 s at 95 °C, 13 s at 56 °C, 25 s at 72 °C for 40 cycles	628
IL-8	5-ATGACTTCCAAGCTGGCCGTG-3′ 5-TGAATTCTCAGCCCTCTTCAAAAACTTCTC-3′	5 s at 94 °C, 6 s at 55 °C, 12 s at 72 °C for 40 cycles	297
TGF-β	5′-CCGACTACTACGCCAAGGAGGTCAC-3′ 5′-AGGCCGGTTCATGCCATGAATGGTG-3′	5 s at 94 °C, 9 s at 60 °C, 18 s at 72 °C for 40 cycles	439

3. Results

In order to have more information about the possibility of producing beauty masks from films obtained by flat die extrusion, which is considered an affordable process for the scaling up of biobased film production, it was necessary to establish if the formulation containing P-PLST/PHA 1/1 with CC could also be extruded. The Minilab was set at 140 °C and 60 rpm, and the torque was recorded during the 1 min extrusion. The torque values were recorded along the extrusion trial (Figure 1a), showing a good melt stability of the prepared blends despite a slightly decreasing trend attributable to some chain scission of the processed polyesters [52]. The ribbon-like extruded strands recovered after extrusion were controlled, and their elasticity was qualitatively evaluated by deforming them using tweezers. The E-BM1 could be extruded, as reported elsewhere [14]. However, the study of its extrusion evidenced that it was not elastic and had a very high torque (Figure 1). Therefore, for this formulation, the scaling up to flat die extrusion is not possible.

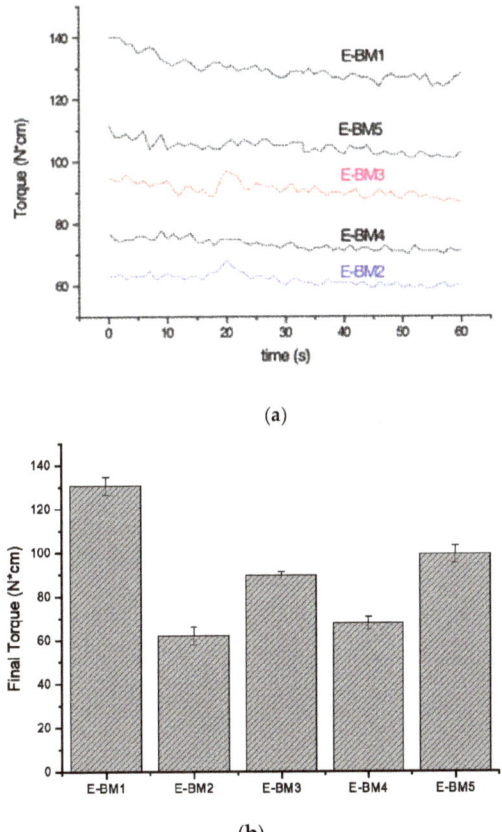

Figure 1. (a) Torque trends as function of extrusion time of the trial reported in Table 1; (b) final torque values of different extruded formulations.

A set of blends containing PBSA or PBAT replacing the PHA fully or at 50% was prepared, as reported in Table 2. All the blends contained 46.5 wt% of preplasticized starch (P-PLST). The replacement of the PHA with PBSA (or PBAT) led to a more elastic melt, with a lower torque, which seemed more suitable for flat die extrusion. The final torque values are reported in Figure 1b, where it is evident that both PBSA and PBAT reduced the torque in the same way.

The extruded blends were thus characterized to evaluate the possibility of scaling up the extrusion by melt flow rate test at 160 °C and weight of 2.16 kg. The obtained results are reported in Table 3. It is possible to observe that the pure PHA did not flow at all at 160 °C. When plasticized starch and calcium carbonate are added (E-BM1), the materials can flow at 160 °C, but the MFR is quite low (close to zero), in good agreement with torque data. The viscosity is thus very high, and this composition is thus not suitable for a possible scaling up in flat die extrusion equipment.

Table 3. Melt flow Rate (MFR) and melt volume rate (MVR) values for the different tested compositions.

Blends	MFR (g/10 min)	MVR (cm^3/10 min)
PHA	0	0
PBSA	2.7 ± 0.1	2.45 ± 0.09
PBAT	5.1 ± 0.1	4.7 ± 0.1
E-BM1	0.53 ± 0.09	0.44 ± 0.08
E-BM2	8.9 ± 0.4	7.9 ± 0.4
E-BM3	2.3 ± 0.2	2.0 ± 0.2
E-BM4	7.2 ± 0.1	5.94 ± 0.09
E-BM5	2.3 ± 0.3	2.2 ± 0,3

When PHA was replaced with PBSA, the viscosity decreased and the MFR increased at the value of 8.9 g/10 min. The values observed for E-BM2 and E-BM4 indicated that the strategy of using these PBSA or PBAT polymers as additives of PHA was valid, as these polymers show a significantly lower melt viscosity than PHA (E-BM1) in the presence of starch. In fact, if only half of PHA was replaced by PBSA (E-BM3 blend) the MFR was 2.3 g/10 min. The use of PBAT to fully or partially replace the PHA led to a similar trend. Therefore, the strategy of partially replacing PHA with PBSA or PBAT seemed to successfully modulate melt viscosity.

The MVR values of the blends were stable as a function of time, as reported in Figure 2. Consequentially, the successive melting and flowing of the blends at 160 °C did not result in a significant chain scission of the biopolyesters, in agreement with torque trends (Figure 1a).

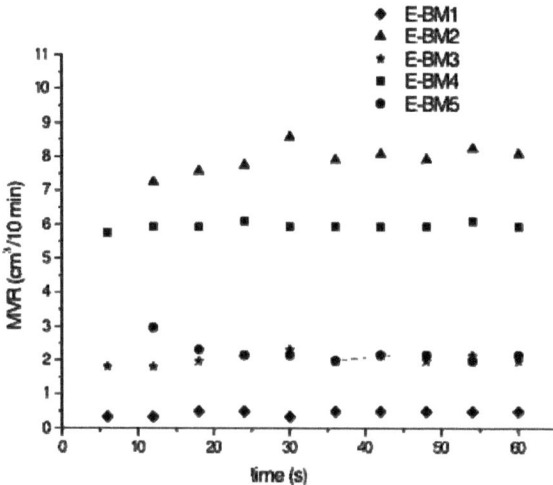

Figure 2. MVR as a function of time, recorded during the MFR tests.

These data can be fundamental for scaling up the flat die extrusion in a semi-industrial plant. A processing temperature range between 140 and 160 °C can be suitable for E-BM3 and E-BM5, obtained by replacing half of the PHA with PBSA and PBAT, respectively, showing an improved processability with respect to the P-PLST/PHA/CC (E-BM1) blend.

Films were obtained by compression molding and were extremely homogeneous and semi-transparent with respect to BM, obtained without extrusion (Figure 3). The composition of BM is the same as E-BM 1. The highest homogeneity was reasonably due to the better mixing achieved in the melt polymer, with the elongation flow generated by screws.

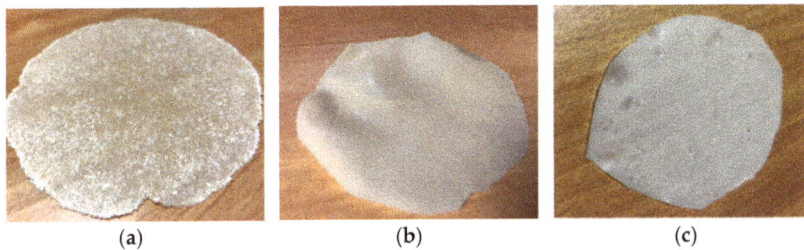

Figure 3. Compression-molded films: (**a**) BM without extrusion; (**b**) E-BM1; (**c**) E-BM5.

At least two replicated square specimens (30 mm side length) of each sample were immersed for 10 s in water and positioned on the forehead of a volunteer in a vertical position, noting the time that the wet pad remained on the subject' forehead. Average resistances for E-BM1, E-BM2, E-BM3, E-BM4 and E-BM5 were 65, 7, 15, 3 and 60 min, respectively. E-BM5 displayed a similar behavior regarding adhesion to skin as BM1 but had an improved processability thanks to the partial replacement of PHA with PBAT. The blends E-BM1 and E-BM5 can be used for making beauty masks that are easy not only to wear but also to remove.

On the whole, taking into account the requirements of the final application, E-BM1 and E-BM5 seemed the best formulations for an easy-to-wear beauty mask, even though E-BM1 cannot be processed easily by flat die extrusion. So, the alternative BM (compression-molded) remains the best alternative.

The release of starch in water was studied for E-BM1 and E-BM5 and compared with that of BM to understand if the extrusion and the different compositions influenced the capability of starch to impart a rapid release to the films (Table 4). Interestingly, it was found that more than 80% of the extractable mass loss was lost in the first 30 min. This result regarding the short release kinetic is positive for the final application as a beauty mask. A slight reduction in the mass release was observed comparing BM and E-BM1, despite their identical composition. The extrusion, which allowed to obtain a more homogeneous sample thanks to a better dispersion of starch domains in the biopolyester matrix, reasonably made it less available to be released in water. However, the released mass amount is quite relevant, suggesting that the water extraction is still efficient in E-BM1. The presence of PBAT only slightly reduced the mass loss at 30 min, attributable to a slight decrease of release velocity that is due to the good affinity observed between PBAT and starch [53].

Table 4. Mass loss test in water for BM, E-BM1 and E-BM5 formulations.

Blends	After 30 min (wt%)	After 16 h (wt%)	After 30 min with Respect to P-PLST (wt%)
BM	24.6	25.0	49.0
E-BM1	19.3	20.8	41.0
E-BM5	17.6	22.2	38.0

In Figure 4, the infrared spectra of PBAT and PHA are reported. The stretching C=O peak of PBAT has a maximum intensity at 1710 cm^{-1}, whereas that of PHA is at 1720 cm^{-1}. Moreover, the peak at 727 cm^{-1}, typical of bending vibration of CH-plane of the benzene ring [54], was strong in intensity in PBAT and absent in PHA.

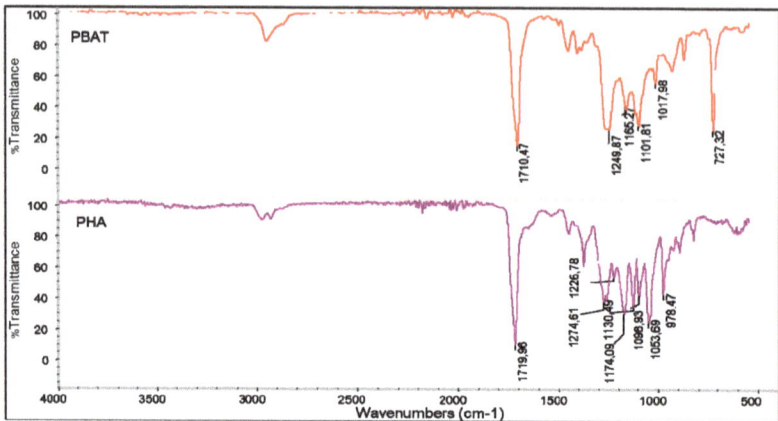

Figure 4. Infrared spectra of PBAT (red) and PHA (purple).

In Figure 5, the spectra of native wheat starch and pregelatinized native wheat starch (P-PLST) are reported. The infrared spectra are quite similar because the macromolecular primary structure is not modified because of the pregelatinization that occurred. The increase of intensity of some bands is due to the addition of glycerol and poly(ethylene glycol) (PEG) to the native starch. In particular, the increase of intensity of the band at 3300 cm^{-1} (O–H stretching) is attributable to the increase of concentration of -OH groups due to the addition of both glycerol and PEG. A similar interpretation for the increase of the band at 2900 cm^{-1}, due to C-H stretching, can be considered. The intensity of the peak at 1000 cm^{-1}, attributable to C-O stretching, was not significantly modified because of the addition of glycerol and PEG, in good agreement with their molecular structure. Interestingly, the peak at 1640 cm^{-1}, attributable to protein content of native starch [55], was decreased thanks to pregelatinization, essentially because of dilution due to the addition of glycerol and PEG.

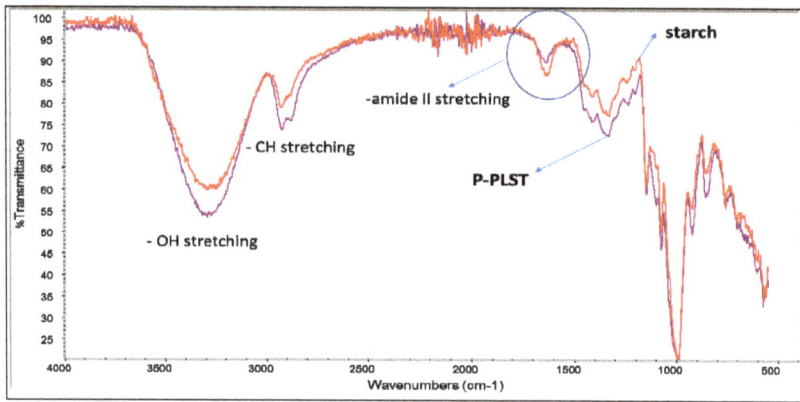

Figure 5. Infrared spectra of starch (red) and plasticized starch (purple).

In Figure 6, the spectra of E-BM1 before and after the immersion tests of 30 min and 16 h are reported. The reduction of the typical peaks of starch at 3350 and 1000 cm^{-1} (Figure 5) in the spectra of the films immersed in water can be easily noticed. The similar profiles of the spectra of E-BM1 after 30 min and 16 h are in agreement with the extraction data (Table 3). Interestingly, the bands typical of PHA (C-H stretching at 2900 cm^{-1}) (Figure 4, PHA spectrum) are more evident in these spectra, as the release of starch from the film surface increased the concentration of biopolyester in the sample.

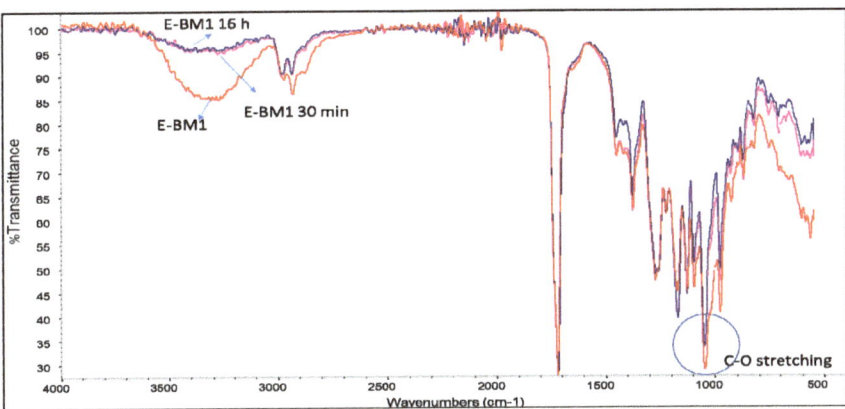

Figure 6. Infrared spectra of E-BM1 (red), E-BM1 immersed in water for 30 min (pink) and E-BM1 immersed in water for 16 h (blue).

A similar conclusion can also be drawn for E-BM5, where only the bands typical of starch decreased due to immersion in water, whereas the bands typical of PHA and PBAT were not modified. Interestingly, in the 1000 cm^{-1} spectrum region, after the release of starch, the two peaks at 1054 and 1102 cm^{-1} of PHA and PBAT, respectively, (Figure 7, circle) appeared quite evident.

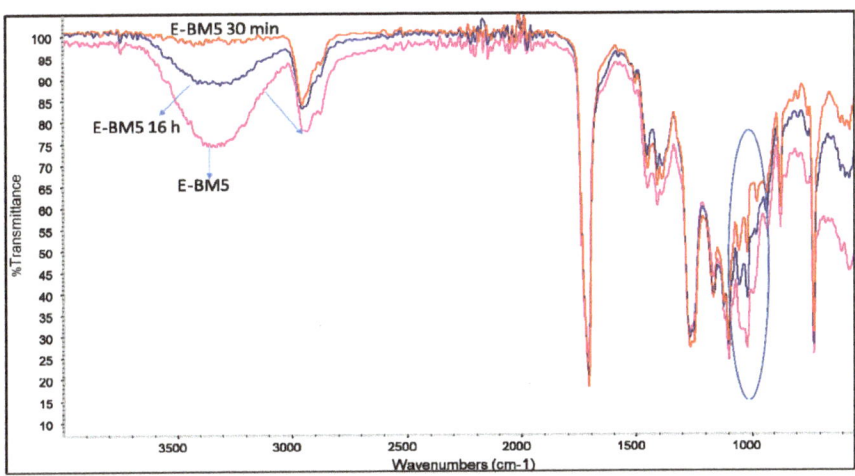

Figure 7. Infrared spectra of E-BM5 (pink), E-BM5 immersed in water for 30 min (red) and E-BM1 immersed in water for 16 h (blue).

Regarding the analysis of the surface of E-BM1 and E-BM5 films by stereomicroscopy, the surface appeared smooth and homogeneous before the treatment in water. After the treatment for 30 min, some small holes and surface roughness changes can be noticed, and after 16 h the film remained similar (Figure 8). Therefore, as noticed by visually observing the improved homogeneity, the extrusion has made the blend of starch with the other polymers (PHB and PBAT) more homogeneous by dispersing it more finely in the polymeric melt.

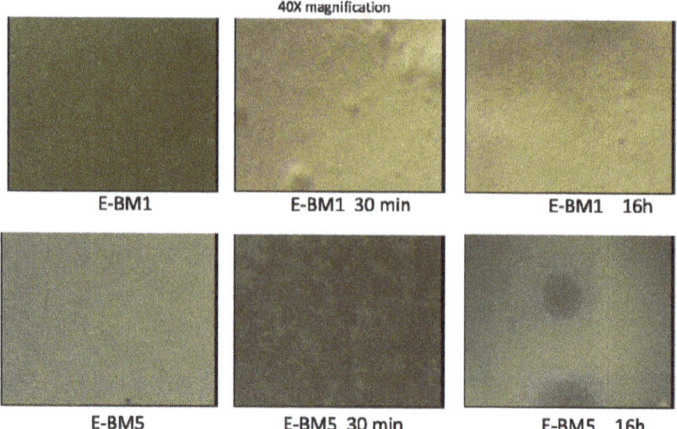

Figure 8. Analysis of E-BM1 and E-BM5 carried out by stereomicroscopy (40× magnification).

The surface properties of BM were different from those of E-BM1 and E-BM5, as the samples obtained without extrusion were less homogenous [16], as evidenced in Figure 3. To better understand the differences in between the films' release behavior, the morphology of the E-BM1 and E-BM5 films was compared with that of BM using scanning electron microscopy (SEM) (Figures 9–11).

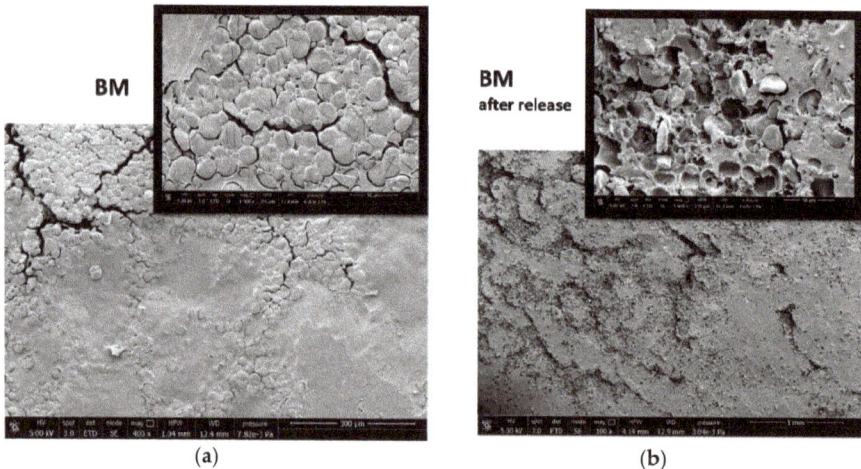

Figure 9. SEM micrographs of the BM film surface: (**a**) before the release in water at magnification 400× and 1500×; (**b**) after 16 h of release in water at magnification 100× and 1600×.

Regarding BM obtained by compression molding the powder consisting of P-PLST, PHA and CC, a very rough morphology can be observed, where the round granules of starch were still present as such. The short processing by compression molding did not allow the full plasticization of the starch. In good agreement in the sample analyzed after the release in water, the presence of round holes evidenced, in agreement with infrared results, the release of almost spherical starch domains. In Figure 9b, at low magnification, it is evident that the starch granules are distributed in clusters. This is reasonably the cause of the inhomogeneity typical of this film (Figure 3a). On the other hand, the fast and efficient release is explained by the low adhesion between the starch granules and the biopolyester matrix.

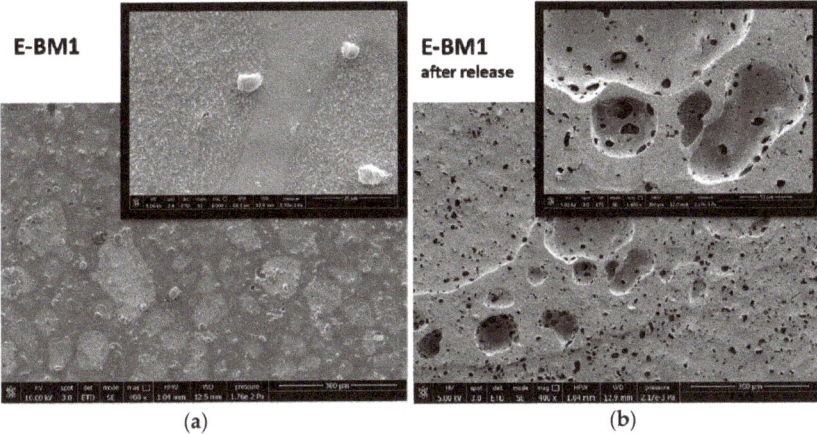

Figure 10. SEM micrographs of the BM film surface: (**a**) before the release in water at magnification 400× and 6000×; (**b**) after 16 h of release in water at magnification 400× and 1600×.

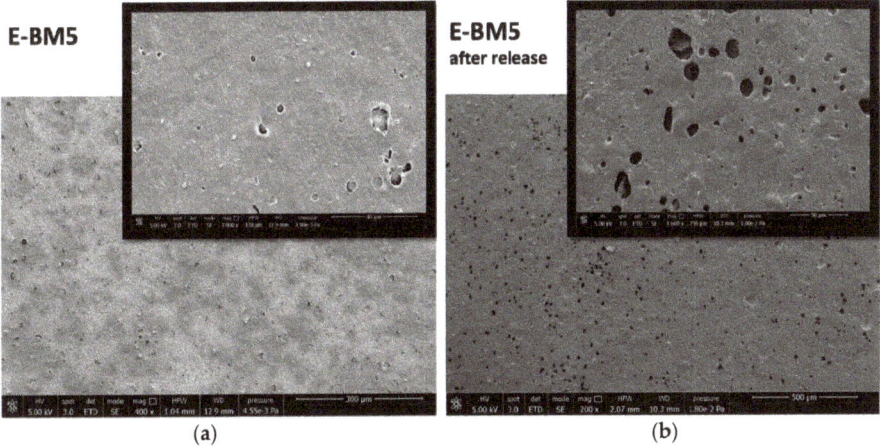

Figure 11. SEM micrographs of the BM film surface: (**a**) before the release in water at magnification 400× and 3000×; (**b**) after 16 h of release in water at magnification 200× and 1600×.

The morphological structure of E-BM1 showed a reduced number of starch granules in good agreement with a very good dispersion and efficient plasticization of starch achieved thanks to the extrusion. The starch was mainly present as big domains of more than 100 µm. A good adhesion between the starch domains and the PHA matrix can be observed (Figure 10a, 6000× magnification). After the extraction, it can be confirmed that the starch domains are very big, as big holes of more than 100 µm can be revealed on the surface. Reasonably, the high viscosity of the blend does not allow a very fine dispersion of the plasticized starch domains into the PHA matrix. On the contrary, a more homogenous morphology in which micrometric starch domains are dispersed in the biopolyester matrix can be observed in the case of E-BM5 film. These domains are certainly more difficult to be extracted from the film and suspended in water because they are finely dispersed in the material bulk, and this can explain the slightly reduced release velocity of E-BM5.

As both PHA and PBAT are insoluble in water, the micrographs (Figure 11b) show the surface of the films consisting of the two polyesters. The two polyesters cannot be easily distinguished, and this can indicate a good compatibility between the two polyesters. However, the conditions to observe the phase morphology in a biphasic blend include the treatment through a cryogenic (brittle) fracture, better evidencing the interfaces. In this investigation, only the morphology of the films surface was considered, as it is specifically related to the beauty mask efficiency (release onto skin).

Regarding the compatibility with skin, the best films (compression-molded BM and E-BM5) sustain a high metabolic activity of keratinocytes (Table 5).

Table 5. Keratinocytes viability tests.

Sample	%AB$_{RED}$
BM	105
EBM-5	98

Moreover, these two films possess a significant immunomodulatory activity; in fact, they are able to upregulate the expression of both pro- and anti-inflammatory cytokines (Figure 12). However, while BM seems to induce a very marked and protracted immunomodulation over time, E-BM5 induces a downregulation of the proinflammatory cytokines after 24 h, suggesting an initial wound healing activity, followed by a resolution of the skin damage with consequent reduction of the inflammatory state.

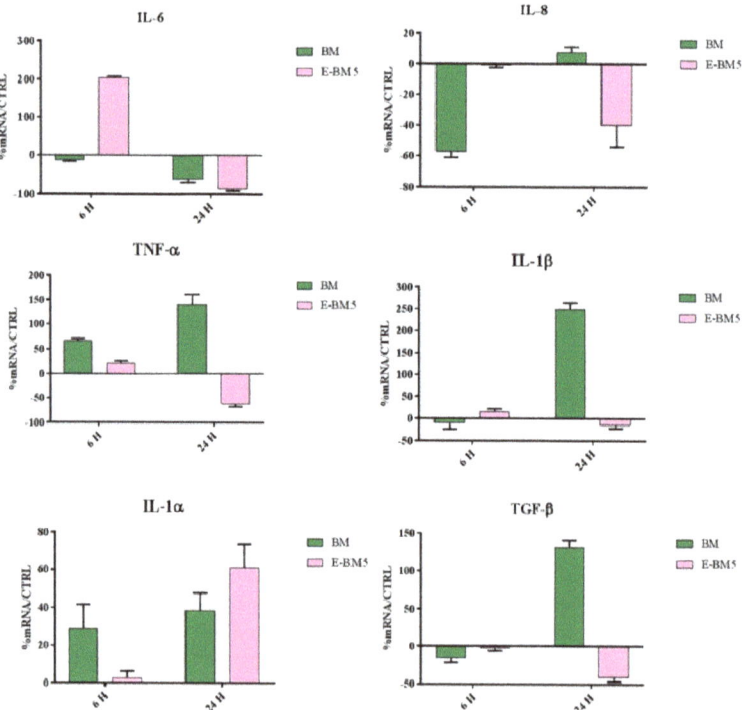

Figure 12. Relative gene expression of TNF-α, IL-8, IL-6, IL-1β, IL-1α and TGF-β in HaCaT cells incubated with BM and E-BM5 for 6 and 24 h. Data are mean ± SD and are expressed as percentage of increment relative to untreated HaCaT cells (control).

4. Discussion

As shown in Section 3, the use of PBSA or PBAT is advantageous to reduce the viscosity of the plasticized starch/PHA blends, thus expanding the processing methodologies applicable to these blends and favoring their use in large scale applications. However, an important difference can be underlined between PBSA and PBAT observing the values related to melt fluidity. In fact, the MVR of pure PBSA was 2.45 cm^3/10 min but the MVR of E-BM2, where PHA was fully replaced by PBSA, was 7.9 cm^3/10 min. This strong increase of MVR can be explained only considering that the PBSA is strongly affected by chain scission [56] with a reduction of the average molecular weight because of the presence of starch and its plasticizers [32] acting as nucleophiles towards the ester groups on the polyester backbone. The MVR of PBAT was 4.7 cm^3/10 min, whereas E-BM4, where PHA was fully replaced by PBAT, showed an MVR value of about 5.94 cm^3/10 min. The increase of MVR due to chain scission was thus reduced with respect to the one observed for PBSA. This can be attributed to the general lower tendency of the aromatic aliphatic copolymers to undergo chain scission during processing, as evidenced by Signori et al. [52]. Interestingly, when PHA was present in the blend, the issue of chain scission seemed completely overcome. On the other hand, PHA also undergoes chain scission in the presence of P-PLST, as evident considering the MVR of E-BM1 (0.54 cm^3/10 min), which is higher than the MVR of pure PHA (0 cm^3/10 min). Thus PHA, having a good compatibility with PBAT and PBSA because of chemical affinity, highly increases the melt viscosity of the binary blends despite of the chain scission due to the presence of P-PLST as third component. The possible modulation of melt viscosity by finely tuning PHA/biopolyesters blend compositions is thus a possibility suggested by the present results that could be exploited for specific applications in the packaging, cosmetic and biomedical fields and connected processing methodologies. The possibility of obtaining films by flat die extrusion could be relevant for producing beauty masks, cosmetic pads or personal care films based on bioplastics [17]. Flat die extrusion can be better in the case that the production of beauty masks should be done considering a bigger scale than the one considered in compression molding. Flat die extrusion, consisting in the production of a film by a continuous methodology, is very fast. In this way, about 20 m of films is easily produced in one minute. Consequentially, the time necessary for producing a mask is more due to cutting and packaging than film production. In the case of compression molding, a maximum amount of 38 thousand pieces/year can be estimated, whereas the production of masks by an automatized flat die extrusion plant can produce more than 600 thousand pieces/year. This makes the process more competitive than the current industrial methodologies to produce films or nonwoven tissues for beauty masks.

The adhesivity to skin of the different films after the immersion in water is completely different. E-BM1 showed the maximum adhesivity to skin, and good performances were observed only in the blends where PHA was present. These results evidence the necessity of using PHA to grant a good skin adhesivity of the film. This is probably linked to the very high flexibility of this elastomer, which has a very irregular primary structure including hydroxybutyrate and hydroxyvalerate repeating units in similar atomic percentage [57]. The other biopolyesters showed a reduced capacity of adhering to the skin. Interestingly, the combination of PBAT and PHA resulted in good performances. PBSA and PBAT resulting as less skin-adhesive with respect to PHA has never been observed and explained before. On the other hand, the process of adhering to skin, slightly drying and successive detaching of the film is complex. The skin biochemical affinity, the flexibility of the films linked to their structure and viscoelastic behavior and their surface area can be considered as some important factors influencing this effect.

Regarding the release, the most efficient film is the one obtained by compression molding (named BM). Its most inhomogeneous and rough nature and the analysis of morphology performed by SEM integrated with other results (infrared spectroscopy, stereomicroscopy) allowed to explain the different release mechanisms (Figure 13). BM released rapidly and efficiently thanks to high interconnections, despite its inhomogeneity, the E-BM1 released rapidly but less efficiently due to a decrease in interconnectivity of P-PLST domains and E-BM5 showed an efficiency similar to E-BM1 but with a slightly slower mechanism of release due to the lower dimension of the P-PLST domains. Nevertheless, in general, all the morphologies seemed quite efficient in releasing, thanks to the high P-PLST content, allowing the interconnection of different starch granules or domains.

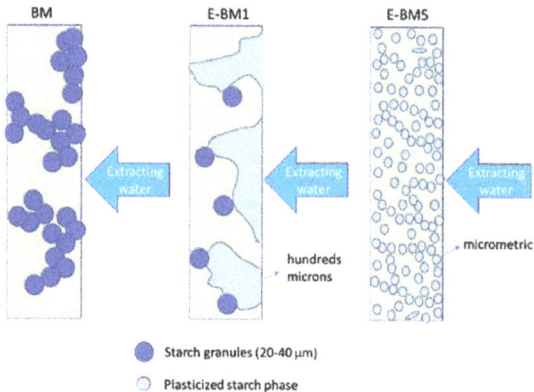

Figure 13. Scheme regarding morphologies of films linked to their release mechanism.

E-BM1 is the blend with the highest melt viscosity. In a paper regarding extrusion of starch and biopolyesters [36], it is reported that the higher the mechanical energy during the extrusion process, the greater the likelihood of interactions between starch and biopolyesters, which may be related to the increased resistance and decreased hygroscopicity found for these films. Thus, by increasing the compatibility between starch and PBAT there is a reduction of starch–water interactions. This change in hygroscopicity can explain the differences observed between BM and EBM1 in the mechanism of release in water.

Immunomodulatory activity tests performed on human keratinocytes (HaCaT cells) have shown that BM and E-BM5 are able to modulate the expression of both pro- and anti-inflammatory cytokines.

This behavior can be explained by giving the films a role in the wound repair.

Wound healing is a complex process characterized by a series of overlapping events which, starting from an inflammatory state, through reepithelization and matrix formation, lead to a remodeling of the tissue.

Wound healing is composed of two main phases. In the first phase, the production of proinflammatory cytokines occurs. TNF-α and IL-1α represent the primary cytokines for proinflammatory responses, immediately released by keratinocytes upon wound healing.

TNF-α can induce the production of fibroblast growth factor FGF-7, suggesting that it can indirectly promote reepithelization [58,59], while IL-1 increases migration and proliferation of keratinocytes [60]. In addition, a direct effect of IL-1 release is the upregulation of IL-6, important in initiating the healing response, and/or IL-8 production, a chemokine which contributes to the regulation of reepithelization, tissue remodeling and angiogenesis [59].

The second phase is associated with growth-oriented cytokines and factors, including TGF-β. This anti-inflammatory cytokine helps initiate granulation tissue formation by increasing the expression of genes associated with ECM formation including fibronectin, the fibronectin receptor and collagenane protease inhibitors [61–63] by preventing hyperproliferation of keratinocytes after wound closure. It is also involved in up-regulating the angiogenic growth factor VEGF [64] and in collagen production (particularly type I and III) and is also a potent inhibitor of metalloproteinases MMP-1, MMP-3 and MMP-9 [61,63,65]. Interestingly, once the wound field is sterilized, TGF-β is able to deactivate superoxide production from macrophages in vitro [66]. This helps to protect the surrounding healthy tissue and prepares the wound for granulation tissue formation [67].

However, our results show that BM seems to have a much stronger immunomodulatory activity compared to E-BM5. This is probably ascribable to the rougher surface, with inhomogeneous concentration of starch, that is more suitable for cell adhesion and proliferation, but more studies should investigate eventual specific effects of PBAT in the E-BM5 formulation.

5. Conclusions

Blends of PHA, plasticized starch and biopolyesters (PBSA and PBAT) were investigated as possible films for preparing beauty masks. The addition of PBAT or PBSA resulted as a good strategy for an improved processability by extrusion of the blends at 140 °C. PHA/starch blends or ternary PBAT/PHA/starch blends resulted in a good adhesion to skin after wetting the film. The release of starch from the films was between 38% and 49% of the total amount, indicating that these films can release more than 80% of the released mass in 30 min, hence with a fast kinetic. Some observed differences in the release behavior were explained by considering the different morphologies, investigated by scanning electron microscopy, obtained by changing blends composition.

The biocompatibility and immunomodulatory behaviors of PHA/starch blends or ternary PBAT/PHA/starch blends showed that both films are not detrimental for keratinocyte viability. However, the compression-molded versions, more inhomogeneous in terms of surficial morphology, resulted in having a much stronger immunomodulatory activity compared to E-BM5.

The results were interesting for selecting materials suitable for obtaining beauty masks by adopting different processing methodologies. Further work should investigate the way to add functional additives suitable to be released onto skin to these films.

Author Contributions: Conceptualization, M.-B.C. and G.D.; Methodology, M.-B.C. and G.D.; Validation, M.-B.C., L.P. and A.F.; formal analysis, M.-B.C. and A.F.; Investigation, M.-B.C. and G.D.; Data curation, M.-B.C., A.F. and L.P.; Writing—Original draft preparation, M.-B.C., G.D. and A.F.; Writing—Review and editing, P.M., G.D., L.P., M.-B.C. and S.D.; Supervision, A.L., A.B. and G.D.; Project administration, M.-B.C. and A.L.; Funding acquisition, M.-B.C. and A.L. All authors have read and agreed to the published version of the manuscript.

Funding: This research was funded by the Bio-Based Industries Joint Undertaking under the European Union Horizon 2020 research program (BBI-H2020), PolyBioSkin project, grant number G.A. 745839.

Acknowledgments: Claudio Cantoni of OMYA is thanked for helpful discussion regarding calcium carbonate. Eng. Alessandro Vannozzi is thanked for his support in sample preparation for SEM analysis. The CISUP—Centre for Instrumentation Sharing—University of Pisa is thanked for its support.

Conflicts of Interest: The authors declare no conflict of interest.

References

1. UN. *World Population Aging 2017: Highlights*; United Nations Report; UN: New York, NY, USA, 2017.
2. Research Report. 15 Trends Changing the Face of the Beauty Industry in 2020. Available online: https://www.cbinsights.com/research/report/beauty-trends-2019/ (accessed on 6 January 2020).
3. Global Cosmetics Products Market—Segmented by Product Type, Distribution Channel (Direct Selling, Supermarket, Specialty Stores), and Region—Growth, Trend and Forecasts (2018–2023). 360 Research Report. 20 February 2018. Available online: https://www.360researchreports.com/global-cosmetics-products-market-13100793 (accessed on 2 April 2020).
4. Rembiesa, J.; Ruzgas, T.; Engblom, J.; Holefors, A. The Impact for pollution on skin and proper testing for anti-pollution claims. *Cosmetics* **2018**, *5*, 4. [CrossRef]
5. Morganti, P.; Coltelli, M.B. A New Carrier for Advanced Cosmeceuticals. *Cosmetics* **2019**, *6*, 10. [CrossRef]
6. Pacheco, G.; Vespoli de mehlo, C.; Galdorfini, B.; Borges Isaac, V.-L.; Lima Ribeiro, S.J.; Pecoraro, E.; Trovatti, E. Bacterial cellulose skin masks—Properties and sensory tests. *J. Cosmet. Dermatol.* **2018**, *17*, 840–847. [CrossRef] [PubMed]
7. Shogren, R.L.; Swanson, C.L.; Thompson, A.R. Extrudates of Cornstarch with Urea and Glycols: Structure/Mechanical Property Relations. *Starch Stärke* **1992**, *44*, 335–338. [CrossRef]
8. Garrido-Miranda, K.A.; Rivas, B.L.; Pérez-Rivera, M.A.; Sanfuentes, E.A.; Peña-Farfal, C. Antioxidant and antifungal effects of eugenol incorporated in bionanocomposites of poly(3-hydroxybutyrate)-thermoplastic starch. *LWT Food Sci. Technol.* **2018**, *98*, 260–267. [CrossRef]
9. Bugnicourt, E.; Cinelli, P.; Lazzeri, A.; Alvarez, V. Polyhydroxyalkanoate (PHA): Review of synthesis, characteristics, processing and potential applications in packaging. *eXPRESS Polym. Lett.* **2014**, *8*, 791–808. [CrossRef]

10. Williams, S.F.; Martin, D.P. Applications of PHAs in Medicine and Pharmacy. In *Biopolymers Online*; Wiley: Hoboken, NJ, USA, 2005; Chapter 20. [CrossRef]
11. Chen, Y.; Tsai, Y.-H.; Chou, I.-N.; Tseng, S.-H.; Wu, H.-S. Application of Biodegradable Polyhydroxyalkanoates as Surgical Films for Ventral Hernia Repair in Mice. *Int. J. Polym. Sci.* **2014**, *2014*, 789681. [CrossRef]
12. Yu, J.; Chen, L.X.L. The Greenhouse Gas Emissions and Fossil Energy Requirement of Bioplastics from Cradle to Gate of a Biomass Refinery. *Environ. Sci. Technol.* **2008**, *42*, 6961–6966. [CrossRef]
13. Seggiani, M.; Cinelli, P.; Balestri, E.; Mallegni, N.; Stefanelli, E.; Rossi, A.; Lardicci, C.; Lazzeri, A. Novel Sustainable Composites Based on Poly(hydroxybutyrate-co-hydroxyvalerate and Seagrass Beach-CAST Fibers: Performance and Degradability in Marine Environments. *Materials* **2018**, *11*, 772. [CrossRef]
14. Sashiwa, H.; Fukuda, R.; Okura, T.; Sato, S.; Nakayama, A. Microbial Degradation Behavior in Seawater of Polyester Blends Containing Poly(3-hydroxybutyrateco-3-hydroxyhexanoate)(PHBHHx). *Mar. Drugs* **2018**, *16*, 34. [CrossRef]
15. Carvalho, A.J.F. Starch: Major Sources, Properties and Applications as Thermoplastic Materials. In *Monomers, Polymers and Composites from Renewable Resources*; Belgacem, M.N., Gandini, A., Eds.; Elsevier: Amsterdam, The Netherlands, 2008; Chapter 15; pp. 321–342.
16. Coltelli, M.-B.; Danti, S.; Trombi, L.; Morganti, P.; Donnarumma, G.; Baroni, A.; Fusco, A.; Lazzeri, A. Preparation of Innovative Skin Compatible Films to Release Polysaccharides for Biobased Beauty Masks. *Cosmetics* **2018**, *5*, 70. [CrossRef]
17. Gigante, V.; Coltelli, M.-B.; Vannozzi, A.; Panariello, L.; Fusco, A.; Trombi, L.; Donnarumma, G.; Danti, S.; Lazzeri, A. Flat Die Extruded Biocompatible Poly(Lactic Acid) (PLA)/Poly(ButyleneSuccinate) (PBS) BasedFilms. *Polymers* **2019**, *11*, 1857. [CrossRef]
18. Coltelli, M.B.; Gigante, V.; Cinelli, P.; Lazzeri, A. Flexible Food Packaging Using Polymers from Biomass. In *Bionanotechnology to Save the Environment. Plant and Fishery's Biomass as Alternative to Petrol*; Morganti, P., Ed.; MDPI: Basel, Switzerland, 2018; pp. 272–296.
19. Peelman, N.; Ragaert, P.; De Meulenaer, B.; Adons, D.; Peeters, R.; Cardon, L.; Van Impe, F.; Devlieghere, F. Application of bioplastics for food packaging. *Trends Food Sci. Technol.* **2013**, *32*, 128–141. [CrossRef]
20. Pachekoski, W.M.; Dalmolin, C.; Marcondes Agnelli, J.A. The Influence of the Industrial Processing on the Degradation of Poly(hydroxybutyrate)–PHB. *Mater. Res.* **2013**, *16*, 327–332. [CrossRef]
21. Mohanty, A.K.; Misra, M.; Drzal, L.T. Sustainable biocomposites from renewable resources: Opportunities and challenges in the green materials world. *J. Polym. Environ.* **2002**, *10*, 19–26. [CrossRef]
22. Hoffmann, A.; Kreuzberger, S.; Hinrichsen, G. Influence of thermal degradation on tensile strength and Young's modulus of poly(hydroxybutyrate). *Polym. Bull.* **1994**, *33*, 355–359. [CrossRef]
23. Li, Z.; Yang, J.; Loh, X.J. Polyhydroxyalkanoates: Opening doors for a sustainable future. *NPG Asia Mater.* **2016**, *8*, e265. [CrossRef]
24. Parulekar, Y.; Mohanty, A.K. Extruded Biodegradable Cast Films from Polyhydroxyalkanoate and Thermoplastic Starch Blends: Fabrication and Characterization. *Macromol. Mater. Eng.* **2007**, *292*, 1218–1228. [CrossRef]
25. Godbole, S.; Gote, S.; Latkar, M.; Chakrabarti, T. Preparation and characterization of biodegradable poly-3-hydroxybutyrate–starch blend films. *Bioresour. Technol.* **2003**, *86*, 33–37. [CrossRef]
26. Willett, J.L.; Kotnis, M.A.; O'Brien, G.S.; Fanta, G.F.; Gordon, S.H. Properties of starch-graft-poly(glycidyl methacrylate)–PHBV composites. *J. Appl. Polym. Sci.* **1998**, *70*, 1121–1127. [CrossRef]
27. Lai, S.-M.; Sun, W.-W.; Don, T.-M. Preparation and characterization of biodegradable polymer blends from poly(3-hydroxybutyrate)/poly(vinyl acetate)-modified corn starch. *Polym. Eng. Sci.* **2015**, *55*, 1321–1329. [CrossRef]
28. De Paula, F.C.; De Paula, C.B.C.; Contiero, J. Prospective Biodegradable Plastics from Biomass Conversion Processes. In *BioFuels-State of Development*; Biernat, K., Ed.; IntechOpen: London, UK, 2018; Chapter 12. [CrossRef]
29. Huang, Z.; Qian, L.; Yin, Q.; Yu, N.; Liu, T.; Tian, D. Biodegradability studies of poly(butylene succinate) composites filled with sugarcane rind fiber. *Polym. Test.* **2018**, *66*, 319–326. [CrossRef]
30. Ma, P.; Hristova-Bogaerds, D.G.; Lemstra, P.J.; Zhang, Y.; Wang, S. Toughening of PHBV/PBS and PHB/PBS blends via In Situ compatibilization using dicumyl peroxide as a free-radical grafting initiator. *Macromol. Mater. Eng.* **2011**, *297*, 402–410. [CrossRef]

31. Rajendra, K.; Krishnaswamy, R.K.; Sun, X. PHA Compositions Comprising PBS and PBSA and Methods for Their Production. U.S. Patent US9056947B2, 16 June 2015.
32. Zhang, G.; Xie, W.; Wu, D. Selective localization of starch nanocrystals in the biodegradable nanocomposites probed by crystallization temperatures. *Carbohydr. Polym.* **2020**, *227*, 115341. [CrossRef]
33. Ayu, R.S.; Khalina, A.; Harmaen, A.S.; Zaman, K.; Jawaid, M.; Lee, C.H. Effect of Modified Tapioca Starch on Mechanical, Thermal, and Morphological Properties of PBS Blends for Food Packaging. *Polymers* **2018**, *10*, 1187. [CrossRef]
34. Li, J.; Luo, X.; Lin, X.; Zhou, Y. Comparative study on the blends of PBS/thermoplastic starch prepared from waxy and normal corn starches. *Starch Stärke* **2013**, *65*, 831–839. [CrossRef]
35. Zhong, Y.; Godwin, P.; Jin, Y.; Xiao, H. Biodegradable polymers and green-based antimicrobial packaging materials: A mini-review. *Adv. Ind. Eng. Polym. Res.* **2020**, *3*, 27–35. [CrossRef]
36. Herrera Brandelero, R.P.; Eiras Grossmann, M.V.; Yamashita, F. Effect of the method of production of the blends on mechanical and structural properties of biodegradable starch films produced by blown extrusion. *Carbohydr. Polym.* **2011**, *86*, 1344–1350. [CrossRef]
37. Olivato, J.B.; Müller, C.M.O.; Carvalho, G.M.; Yamashita, F.; Grossmann, M.V.E. Physical and structural characterisation of starch/polyester blends with tartaric acid. *Mater. Sci. Eng. C* **2014**, *39*, 35–39. [CrossRef]
38. Larsson, M.; Markbo, O.; Jannasch, P. Melt processability and thermomechanical properties of blends based on polyhydroxyalkanoates and poly(butylene adipateco-terephthalate). *RSC Adv.* **2016**, *6*, 44354–44363. [CrossRef]
39. De Matos Costa, A.R.; Marques Santos, R.; Noriyuki Ito, E.; Hecker de Carvalho, L.; Luís Canedo, E. Melt and cold crystallization in a poly(3-hydroxybutyrate) poly(butylene adipate-co-terephthalate). *J. Therm. Anal. Calorim.* **2019**, *137*, 1341–1346. [CrossRef]
40. Lin, X.; Fan, X.; Li, R.; Li, Z.; Ren, T.; Ren, X.; Huang, T.-S. Preparation and characterization of PHB/PBAT–based biodegradable antibacterial hydrophobic nanofibrous membranes. *Polym. Adv. Technol.* **2018**, *29*, 481–489. [CrossRef]
41. Zarrinbakhsh, N.; Mohanty, A.K.; Misra, M. Improving the interfacial adhesion in a new renewable resourcebasedbiocomposites from biofuel coproduct and biodegradable plastic. *J. Mater. Sci.* **2013**, *48*, 6025–6038. [CrossRef]
42. Belyamani, I.; Kim, K.; Rahimi, S.K.; Sahukhal, G.S.; Elasri, M.O.; Otaigbe, J.U. Creep, recovery, and stress relaxation behavior of nanostructured bioactive calcium phosphate glass–POSS/polymer composites for bone implants studied under simulated physiological conditions. *J. Biomed. Mater. Res. Part B* **2019**, *107B*, 2419–2432. [CrossRef]
43. Tabasi, R.Y.; Ajji, A. Selective degradation of biodegradable blends in simulated laboratory composting. *Polym. Degrad. Stab.* **2015**, *120*, 435–442. [CrossRef]
44. Sprague, A.H.; Khalil, R.A. Inflammatory Cytokines in Vascular Dysfunction and Vascular Disease. *Biochem. Pharmacol.* **2009**, *78*, 539–552. [CrossRef]
45. Gabay, C.; Lamacchia, C.; Palmer, G. IL-1 pathways in inflammation and human diseases. *Nat. Rev. Rheumatol.* **2010**, *6*, 232–241. [CrossRef]
46. Esposito, E.; Cuzzocrea, S. TNF-alpha as a therapeutic target in inflammatory diseases, ischemia-reperfusion injury and trauma. *Curr. Med. Chem.* **2009**, *16*, 3152–3167. [CrossRef]
47. Scheller, J.; Chalaris, A.; Schmidt-Arras, D.; Rose-John, S. The Pro- and Anti-Inflammatory Properties of the Cytokine Interleukin-6. *Biochim. Biophys. Acta (BBA) Mol. Cell Res.* **2011**, *1813*, 878–888. [CrossRef]
48. Koch, A.E.; Polverini, P.J.; Kunkel, S.L.; Harlow, L.A.; DiPietro, L.A.; Elner, V.M.; Elner, S.G.; Strieter, M.R. Interleukin-8 as a macrophage-derived mediator of angiogenesis. *Science* **1992**, *258*, 1798–1801. [CrossRef]
49. Sanjabi, S.; Zenewicz, L.A.; Kamanaka, M.; Flavell, R.A. Anti-inflammatory and pro-inflammatory roles of TGF-beta, IL-10, and IL-22 in immunity and autoimmunity. *Curr. Opin. Pharmacol.* **2009**, *9*, 447–453. [CrossRef]
50. Breulmann, M.; Künkel, A.; Philipp, S.; Reimer, V.; Siegenthaler, K.O.; Skupin, G.; Yamamoto, M. Polymers, Biodegradable. In *Ullmann's Encyclopedia of Industrial Chemistry*; Wiley-VCH: Weinheim/Berlin, Germany, 2009. [CrossRef]
51. Suhartini, M.; Mitomo, H.; Yoshii, F.; Nagasawa, N.; Kume, T. Radiation Crosslinking of Poly(Butylene Succinate) in the Presence of Inorganic Material and Its Biodegradability. *J. Polym. Environ.* **2001**, *9*, 163–171. [CrossRef]

52. Signori, F.; Coltelli, M.-B.; Bronco, S. Thermal degradation of poly(lactic acid) (PLA) and poly(butylene adipate-co-terephthalate) (PBAT) and their blends upon melt processing. *Polym. Degrad. Stab.* **2009**, *94*, 74–82. [CrossRef]
53. Ren, H.; Fu, T.; Ren, W.Y. Preparation, characterization and properties of binary and ternary blends with thermoplastic starch poly(lactic acid) and poly(butylene adipate-co-terephthalate). *Carbohydr. Polym.* **2009**, *77*, 576–582. [CrossRef]
54. Weng, Y.-X.; Jin, Y.-J.; Meng, Q.-Y.; Wang, L.; Zhang, M.; Wang, Y.-Z. Biodegradation behavior of poly(butylene adipate-co-terephthalate) (PBAT), poly(lactic acid) (PLA), and their blend under soil conditions. *Polym. Test.* **2013**, *32*, 918–926. [CrossRef]
55. Warren, F.J.; Gidley, M.J.; Flanagan, B. Infrared spectroscopy as a tool to characterise starch ordered structure-a joint FTIR-ATR, NMR, XRD and DSC study. *Carbohydr. Polym.* **2015**, *139*, 35–42. [CrossRef]
56. Puchalski, M.; Szparaga, G.; Biela, T.; Gutowska, A.; Sztajnowski, S.; Krucińska, I. Molecular and Supramolecular Changes in Polybutylene Succinate (PBS) and Polybutylene Succinate Adipate (PBSA) Copolymer during Degradation in Various Environmental Conditions. *Polymers* **2018**, *10*, 251. [CrossRef]
57. Wu, C.S.; Liao, H.T. Fabrication, characterization, and application of polyester/wood flour composites. *J. Polym. Eng.* **2017**. [CrossRef]
58. Brauchle, M.; Angermeyer, K.; Hubner, G.; Werner, S. Large induction of keratinocyte growth factor expression by serum growth factors and pro-inflammatory cytokines in cultured fibroblasts. *Oncogene* **1994**, *9*, 3199–3204.
59. Kristensen, M.; Chu, C.Q.; Eedy, D.J.; Feldmann, M.; Brennan, F.M.; Breathnach, S.M. Localization of tumour necrosis factor-alpha (TNF-alpha) and its receptors in normal and psoriatic skin: Epidermal cells express the 55-kD but not the 75-kD TNF receptor. *Clin. Exp. Immunol.* **1993**, *94*, 354–362. [CrossRef]
60. Raja Sivamani, K.; Garcia, M.S.; Isseroff, R.R. Wound reepithelialization: Modulating keratinocyte migration in wound healing. *Front. Biosci.* **2007**, *12*, 2849–2868. [CrossRef]
61. White, L.A.; Mitchell, T.I.; Brinckerhoff, C.E. Transforming growth factor beta inhibitory element in the rabbit matrix metalloproteinase-1 (collagenase-1) gene functions as a repressor of constitutive transcription. *Biochim. Biophys. Acta* **2000**, *1490*, 259–268. [CrossRef]
62. Greenwel, P.; Inagaki, Y.; Hu, W.; Walsh, M.; Ramirez, F. Sp1 is required for the early response of alpha2(I) collagen to transforming growth factor-beta1. *J. Biol. Chem.* **1997**, *272*, 19738–19745. [CrossRef]
63. Mauviel, A.; Chung, K.Y.; Agarwal, A.; Tamai, K.; Uitto, J. Cellspecific induction of distinct oncogenes of the Jun family is responsible for differential regulation of collagenase gene expression by transforming growth factor-beta in fibroblasts and keratinocytes. *J. Biol. Chem.* **1996**, *271*, 10917–10923. [CrossRef]
64. Riedel, K.; Riedel, F.; Goessler, U.R.; Germann, G.; Sauerbier, M. Tgf-beta antisense therapy increases angiogenic potential in human keratinocytes In Vitro. *Arch. Med. Res.* **2007**, *38*, 45–51. [CrossRef]
65. Zeng, G.; McCue, H.M.; Mastrangelo, L.; Millis, A.J. Endogenous TGF-beta activity is modified during cellular aging: Effects on metalloproteinase and TIMP-1 expression. *Exp. Cell Res.* **1996**, *228*, 271–276. [CrossRef]
66. Mitra, R.; Khar, A. Suppression of macrophage function in AK-5 tumor transplanted animals: Role of TGF-beta1. *Immunol. Lett.* **2004**, *91*, 189–195. [CrossRef]
67. Tsunawaki, S.; Sporn, M.; Ding, A.; Nathan, C. Deactivation of macrophages by transforming growth factor-beta. *Nature* **1988**, *334*, 260–262. [CrossRef]

© 2020 by the authors. Licensee MDPI, Basel, Switzerland. This article is an open access article distributed under the terms and conditions of the Creative Commons Attribution (CC BY) license (http://creativecommons.org/licenses/by/4.0/).

Article

Antimicrobial Materials with Lime Oil and a Poly(3-hydroxyalkanoate) Produced via Valorisation of Sugar Cane Molasses

Pooja Basnett [1], Elena Marcello [1], Barbara Lukasiewicz [1], Rinat Nigmatullin [1,2], Alexandra Paxinou [1], Muhammad Haseeb Ahmad [1], Bhavana Gurumayum [1] and Ipsita Roy [3,*]

1. Faculty of Science and Technology, University of Westminster, London W1W 6UW, UK; P.Basnett@westminster.ac.uk (P.B.); w1614733@my.westminster.ac.uk (E.M.); barbara.lukasiewicz@gmail.com (B.L.); rn17541@bristol.ac.uk (R.N.); w1621777@my.westminster.ac.uk (A.P.); Mhaseebahmadd@gmail.com (M.H.A.); gurumayumbhavana@gmail.com (B.G.)
2. Bristol Composites Institute (ACCIS), University of Bristol, Bristol BS8 1TR, UK
3. Department of Material Science and Engineering, Faculty of Engineering, University of Sheffield, Sheffield S1 3JD, UK
* Correspondence: I.Roy@sheffield.ac.uk; Tel.: +44-114-2225962

Received: 9 March 2020; Accepted: 6 April 2020; Published: 10 April 2020

Abstract: A medium chain-length polyhydroxyalkanoate (PHA) was produced by *Pseudomonas mendocina* CH50 using a cheap carbon substrate, sugarcane molasses. A PHA yield of 14.2% dry cell weight was achieved. Chemical analysis confirmed that the polymer produced was a medium chain-length PHA, a copolymer of 3-hydroxyoctanoate and 3-hydroxydecanoate, P(3HO-*co*-3HD). Lime oil, an essential oil with known antimicrobial activity, was used as an additive to P(3HO-*co*-3HD) to confer antibacterial properties to this biodegradable polymer. The incorporation of lime oil induced a slight decrease in crystallinity of P(3HO-co-3HD) films. The antibacterial properties of lime oil were investigated using ISO 20776 against *Staphylococcus aureus* 6538P and *Escherichia coli* 8739, showing a higher activity against the Gram-positive bacteria. The higher activity of the oil against *S. aureus* 6538P defined the higher efficiency of loaded polymer films against this strain. The effect of storage on the antimicrobial properties of the loaded films was investigated. After one-year storage, the content of lime oil in the films decreased, causing a reduction of the antimicrobial activity of the materials produced. However, the films still possessed antibacterial activity against *S. aureus* 6538P.

Keywords: polyhydroxyalkanotes; sugarcane molasses; antibacterial materials; essential oils

1. Introduction

Polyhydroxyalkanoates (PHAs) are a family of natural polymers produced by bacterial fermentation using renewable carbon sources under nutrient-limiting conditions. PHAs can be broadly classified into short chain length (SCL: C_3-C_5) and medium-chain length (MCL: C_6-C_{14}) based on the number of carbon atoms within their monomer units [1]. SCL PHAs have a high melting temperature and are brittle, whereas MCL PHAs are elastomeric and have a low-melting and glass-transition temperature. PHAs can match the functional properties of common petroleum-based plastics. For example, the mechanical properties of poly(3-hydroxybutyrate) (P(3HB)) are similar to those of polystyrene, while its copolymers with 3-hydroxyvalerate (P(3HB-*co*-3HV)) are characterised by an improved ductility resulting in a combination of mechanical properties close to polypropylene. One the other hand, MCL-PHAs can emulate the properties of soft thermoplastic elastomers such as thermoplastic polyurethanes, olefinic thermoplastic elastomers. The functional properties of PHA-based materials can be further modified and tuned via blending with other polymers [2]. Thanks to the presence of hydrolysable ester linkage connecting the monomer units of PHAs, PHAs are biodegradable in nature unlike most of conventional plastics. Their degradation rate is affected by the chemical composition of the polymer, and by environmental factors [3]. Based on life-cycle assessment (LCA) analysis, PHAs have been recognised as one of the safest bioplastics for the environment [4]. PHAs are also known to be biocompatible and have been widely investigated for various biomedical applications [5].

Although PHAs represent a family of sustainable polymers with attractive properties, they are yet to replace petroleum based conventional plastics used in most industrial applications. One of the major factors limiting the commercialisation of PHAs is its high production cost. Several cheap carbon sources, waste materials, genetically modified bacterial strains, fermentation conditions and recovery methods have been employed either on their own or in combination, to allow economical production of PHAs [3]. Since the cost of carbon accounts for 31% of the overall PHA production cost, the use of inexpensive substrates is one of the effective strategies to produce PHAs in an economical manner [6,7]. Simple sugars, triglycerides (vegetable oils) are the most widely used carbon sources for the microbial production of PHAs [8]. In addition to their high cost, the utilization of such raw materials can be in competition with the use of these resources as food. Therefore, the valorisation of cheap agri-food by-products can address both issues; the cost of carbon source and the competition with food. Although potentially biomass rich in polysaccharides can be used for generating simple sugars, this requires additional chemical and enzymatic treatment. Molasses and whey, on the other hand, contain up to 60 and 4.5 wt.% of simple sugars, respectively. Currently, a significant part of molasses is used in bioethanol production. The production of PHAs with the use of molasses as the carbon feedstock is a feasible alternative for the valorisation of molasses. Many studies demonstrated that molasses is a good substrate for microbial production of poly(3-hydroxybutyrate) [9].

PHAs have diverse physical properties which makes them suitable for various applications such as packaging, medical devices, bio-implants and tissue engineering [10]. For all these applications, microbial contamination is a major issue. Bacterial adhesion and colonisation on the polymer surface form a reservoir for pathogens which leads to various nosocomial infections, skin related issues, degradation of food and cosmetic products, and food contamination [11]. For several decades, antibiotics have been used to treat and prevent such infections. However, with the rise in the number of multidrug-resistant pathogens, antibiotic resistance has now become a global health risk. Annual costs of treating antibiotic resistant infections is estimated to be between $21,000 and $34,000 million in the United States alone and around €1500 million in Europe [12]. In such a scenario, there is a high demand for natural antibacterial agents to prevent or mitigate bacterial contamination.

Traditionally, essential oils, plant extracts and other natural products were used to treat infectious diseases. According to the World Health Organization (WHO), the majority of the world population is dependent on treatment using natural products as a primary care. Various essential oils have shown to possess antibacterial, antifungal and antioxidant properties [13–15]. These oils are now being screened to find alternative remedies to fight bacterial infections in a wide range of applications. Generally, a higher antimicrobial activity has been observed for oxygenated constituents of essential oils [16]. However, as antimicrobial additives to polymer materials, non-polar terpene hydrocarbons can bring other benefits such as improved compatibility with hydrophobic polymer matrices, enhanced water vapour barrier properties and plasticising effect [17]. Lime oil is an example of essential oils rich of terpene hydrocarbons such as limonene (ca. 40%), β-pinene (ca. 25%), and γ-terpinene (ca. 10%) [18]. The antimicrobial activity of lime oil is well documented [18–20].

The aim of this study was the development of natural antimicrobial polymer films via the incorporation of an essential oil into biodegradable PHA produced from agri-food by-products. We report a feasible way of valorisation of sugarcane molasses for the production of PHA by *Pseudomonas mendocina* CH50, a bacterial strain which has been previously confirmed as an efficient producer of MCL PHAs [5,20,21]. It was demonstrated that in the presence of sugarcane molasses *P. mendocina* CH50 was able to accumulate an MCL PHA, which was identified as a copolymer of 3-hydroxyoctanoate and 3-hydroxydecanoate, or P(3HO-co-3HD). Lime oil was incorporated into P(3HO-co-3HD) films to produce a natural antimicrobial material. The oil content in the films was monitored during long storage, and loss of oil was observed during the storage. Both fresh and aged films exhibited antimicrobial activity against Gram-positive S. aureus. The films were less efficient against Gram-negative E. coli reflecting lower activity of lime oil against this bacterial strain.

2. Material and Methods

2.1. Bacterial Strains and Culture Conditions

Pseudomonas mendocina CH50 was obtained from the University of Westminster culture collection. The strain was cultured at 30 °C for 16 h in a shaking incubator (200 rpm). After incubation, the strain was stored as a glycerol stock at −80 °C. For the antibacterial characterization, *Staphylococcus aureus* 6538P and *Escherichia coli* 8739 were bought from the American Type Culture Collection (ATCC). They were cultured in sterile nutrient broth at 37 °C and 200 rpm for 16 h in a shaking incubator and stored as a glycerol stock at −80 °C.

2.2. Chemicals

All the chemicals for the production and characterization of PHAs were purchased from Sigma Aldrich Ltd., England and VWR international, England. Lime oil used as an antibacterial agent was purchased from Holland and Barrett, London, UK.

2.3. Production of Polyhydroxyalkanoates by Pseudomonas mendocina CH50 Using Sugarcane Molasses as the Sole Carbon Source

Production of PHAs by *Pseudomonas mendocina* CH50 using 20 g/L of sugarcane molasses was carried out in 15 L bioreactors, with 10 L working volume (Applikon Biotechnology, Tewkesbury, UK). Batch fermentation was carried out in two stages. The seed culture was prepared using a single colony of *P. mendocina* CH50 to inoculate sterile nutrient broth. This was incubated for 16 h at 30 °C and at 200 rpm. Inoculum of 10 vol.% of the final volume was used to inoculate the second stage seed culture (mineral salt medium-MSM) which was incubated at 30 °C, 200 rpm for 24 h. Second stage seed culture was used to inoculate the final PHA production media [21,22]. The culture was grown for 48 h, at 30 °C, 200 rpm and air flow rate of 1 vvm. Temporal profile of the production of PHAs by *P. mendocina* CH50 using 20 g/L sugarcane molasses was obtained by monitoring parameters such as optical density, biomass, nitrogen and glucose concentrations, pH and dissolved oxygen tension (DOT) at regular intervals during the fermentation.

2.4. Analytical Studies

Bacterial growth was measured by monitoring the optical density (OD) at 450 nm. Biomass yield was determined by weighing the dried pellet of the cells obtained after centrifugation of 1 mL at 12,000 rpm for 10 min (Heraeus Pico 17 Centrifuge, Thermofisher Scientific, MA, USA). pH of the supernatant was measured using Seven Compact pH meter (Mettler Toledo Ltd., Leicester, UK). Nitrogen in the form of ammonium ions was estimated by the phenol-hypochlorite method [21,22].

Polymer was extracted from the dried biomass using the soxhlet extraction method. To remove organic soluble impurities, dried biomass was refluxed in methanol for 24 h followed by refluxing in chloroform for 48 h to extract the polymer. Polymer solution in chloroform was concentrated using a rotary evaporator. Polymer was precipitated using ice-cold methanol under continuous stirring and dried at room temperature. Polymer yield was calculated as a percentage of dry cell weight (DCW), using the formula:

$$\%DCW = \frac{\text{Polymer mass}}{\text{Mass of biomass}} \times 100 \quad (1)$$

2.5. Preparation of Film Samples

Solvent cast films were prepared by pouring 5 wt.% polymer solution in chloroform in 60 mm diameter glass Petri dishes. The polymer was dried in the fume hood at the room temperature until the solvent evaporated completely. For oil-incorporated films, the oils were added to 5 wt.% polymer solution in chloroform. The loaded films were stored at room temperature in sealed Petri dishes for one year at room temperature and identified as aged samples. Quantification of the oil content in the films after complete solvent evaporation was carried out by thermogravimetric analysis (TGA) using the STA 449 F3 Jupiter® instrument (Netzsch, Germany); 10–15 mg film samples were heated at 10 degrees per minute from 30 to 150 °C under nitrogen flow and kept at 150 °C for 2 h. The cumulative mass loss in dynamic and isothermal steps was defined as oil amount left in the films after solvent evaporation.

2.6. Polymer Characterization

The monomeric composition of polymer was determined using gas chromatography–mass spectrometry (GC–MS) analysis and ^{13}C, ^{1}H nuclear magnetic resonance (NMR) spectroscopy. For GC–MS analysis (Chrompack CP-3800 gas chromatograph and Saturn 200 MS/MS block), 20 mg of the polymer sample was subjected to methanolysis as described in Basnett et al. [5]. For NMR analysis, the polymer was dissolved in deuterated chloroform (20 mg/mL).

Thermal analysis of polymers was conducted using a differential scanning calorimeter (DSC) 214 Polyma (Netzsch, Germany) equipped with Intracooler IC70 cooling system. 5 mg of polymer was heated from −50 to 100 °C at a heating rate of 20 °C per min. DSC thermograms were analysed using Proteus 7.0 software.

2.7. Antibacterial Characterization

2.7.1. MIC (Minimum Inhibition Concentration) and MBC (Minimum Bactericidal Concentration)

The minimum inhibition concentration (MIC) was estimated according to the ISO 20776 against *Staphylococcus aureus* 6538P and *Escherichia coli* 8739. Stock solution of the lime oil was prepared in Mueller Hinton Broth and 1% dimethyl sulfoxide (added as a co-solvent which enables mixing the oils with aqueous medium). A range of concentrations from 100 to 5 μL/mL essential oils was tested in a 96-well plate. Each well was inoculated to a final volume of 100 μL using a microbial suspension adjusted to achieve a final concentration of 5×10^5 CFU/mL and incubated at 37 °C for 24 h at 100 rpm. Studies were performed in triplicate. The MIC was determined by measuring the absorbance at 600 nm. Following the MIC test, the entire volume of the MIC well and the concentrations above were removed and 10-fold dilution was performed in PBS and plated onto agar plates. The plates were incubated at 37 °C for 24 h. The minimum bactericidal concentration (MBC) was determined as the lowest concentration of the compound that induced 99.9% killing of the bacteria.

2.7.2. Halo Test

The films loaded with essential oils were tested for antibacterial activity using the halo test against *Staphylococcus aureus* 6538P and *Escherichia coli* 8739. The test was conducted according to the European Committee on Antimicrobial Susceptibility Testing (EUCAST) Disc Diffusion Test Methodology. A single colony was used to prepare a suspension of 0.5 McFarland turbidity standard (approx. 10^8 colony-forming units (CFU)/mL. A sterile cotton swab was immersed into the suspension and was spread evenly on the agar plates. The agar plates as well as the antimicrobial discs were kept at room temperature prior to inoculation. Polymer films of 0.7 cm diameter containing active agents, sterilized under ultraviolet (UV) light for 30 min (15 min each side) were placed on the inoculated agar surface. The plates were incubated at 37 °C for 18 h. The inhibition zones formed were measured and noted. For the antibacterial test, both freshly prepared samples and aged samples were investigated. Each material was tested in triplicates. Films without active agents were used as negative controls, antibiotic discs containing oxacillin (1 µg/disc) and streptomycin (300 µg/disc) were used as positive controls against *Staphylococcus aureus* 6538P and *Escherichia coli* 8739, respectively.

3. Results and Discussion

3.1. Temporal Profile of Polyhydroxyalkanoate (PHA) Production by Pseudomonas mendocina CH50 Using Sugarcane Molasses as the Sole Carbon Source

Figure 1 represents the fermentation profile of the PHA production by *P. mendocina* CH50 using sugarcane molasses as the carbon source in a 15 L bioreactor. Fermentation was performed as described in Section 2.3., and OD, pH, % DOT, biomass, nitrogen concentration and polymer yield were monitored.

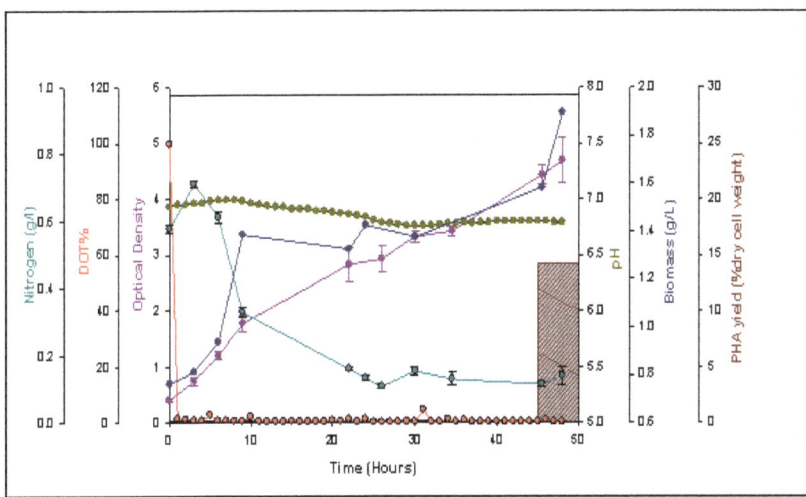

Figure 1. Temporal profile of P(3HO-*co*-3HD) production by *P. mendocina* CH50 using sugarcane molasses as the sole carbon source. The profiling illustrates the polyhydroxyalkanoate (PHA) yield (▨), optical density (—◆—), pH (—▲—), DOT% (—■—), biomass (—●—) and the nitrogen concentration in the form of the ammonium ions throughout 48 h of fermentation.

The optical density increased gradually until the first 3 h (lag phase) of the fermentation and continued to increase steadily until the end of process, reaching a value of 4.7. Biomass concentration increased until 48 h reaching a value of 1.9 g/L. Prior to the fermentation, pH of the production media was set to 7.0. The pH of the culture did not change significantly during the fermentation process whereas the DOT% reduced from 100% to 0% within few hours of the fermentation and remained very low until the end of the process, maintaining an oxygen-limiting environment. Nutrient limitation plays an important role in the production of PHAs. Furthermore, it is known that nutrient limitation coupled with an excess of carbon provides an ideal environment to produce PHAs [23]. In this study, nitrogen was the limiting factor. As illustrated in Figure 1, the nitrogen concentration decreased from 0.58 g/L to 0.14 g/L at the end of 48-h fermentation. The PHA yield (% dcw) at 48 h was found to be was 14.22%. As mentioned above, optical density and biomass concentration increased steadily during the fermentation indicating cell growth in the presence of nitrogen in the media. The absence of nitrogen limitation could have influenced the overall accumulation of PHAs. Sugarcane molasses have been previously used as a carbon substrate to produce SCL-PHA by *Bacillus cereus* SPV with a PHA yield of 51.37% dcw [24]. In another study, Naheed and Jamil obtained a maximum PHA yield of 80.5% dcw by *Enterobacter* sp. SEL2, a bacterium isolated from soil of contaminated sites, with the use 3% of sugarcane molasses after fermentation for 24 h at pH 5.0 [25]. Chaijamrus and Udpuay reported a 43% yield of P(3HB) by *Bacillus megaterium* ATCC 6748 after 45 h of fermentation when 4% sugarcane molasses was used as the carbon source [26]. With regards to MCL-PHA production using sugarcane molasses, a limited investigation has been conducted. This is due to the fact that the majority of the *Pseudomonas* species (i.e., the main producer of MCL-PHAs) cannot utilize sucrose, which is one of the main components of sugarcane molasses along with glucose and fructose [27]. Therefore, the substrate is usually modified in order to obtain equimolar concentrations of glucose and fructose by either ionic exchange or acid hydrolysis [27]. To date, only two species of the *Pseudomonas* family have been shown to metabolize sucrose directly, *P. fluorescens* A2a5 and *P. corrugata* 388.

P. fluorescens A2a5 produced P(3HB) with a yield of 70% dcw when cultured using sugarcane liquor [28]. In contrast, Solaiman et al. showed that *P. corrugata* 388 was able to consume sucrose (present in soy molasses used as carbon source) producing an MCL-PHA with a yield of 17% dcw [27]. *P. mendocina* CH50, the strain used in this study is not able to utilize the sucrose present in the media. The PHA produced by *P. mendocina* CH50 using sugarcane molasses as the carbon source was further characterised.

3.2. Polymer Characterisation

3.2.1. Polymer Composition Analysis

As can be seen from Figure 2, GC–MS chromatograms of the methanolysed polymer demonstrated two peaks identified using the National Institute of Standards and Technology (NIST) library. The retention peak at 7.7 and 9.2 min were identified as the methyl ester of 3-hydroxyoctanoic acid and methyl ester of 3-hydroxydecanoic acid respectively. Peak at 6.64 min corresponded to the methyl benzoate, used as an internal standard. Thus, the polymer produced by *P. mendocina* CH50 using sugarcane molasses as the sole carbon source was identified as the copolymer of 3-hydroxyoctanoate and 3-hydroxydecanoate, Poly(3-hydroxyoctanoate-*co*-3-hydroxydecanoate) or P(3HO-*co*-3HD). The mole percentage of 3-HO and 3-HD was found to be 25.6% and 74.4% respectively. P(3HO-co-3HD) copolymers with 3-HD-dominant composition are usually produced by *Pseudomonas* species when grown on structurally unrelated simple sugars [29].

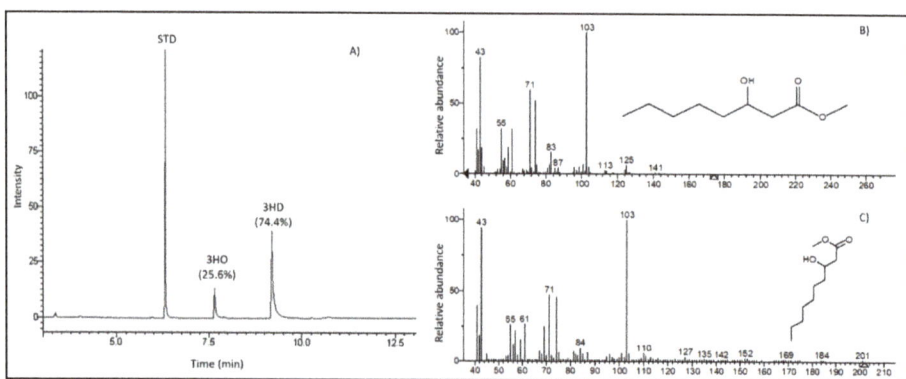

Figure 2. Gas chromatography–mass spectrometry (GC–MS) data of the PHA copolymer produced with sugarcane molasses as the carbon source: (**A**) Gas chromatogram, (**B**) Mass spectrum of a peak with R_t = 7.7 min identified using the National Institute of Standards and Technology (NIST) library as methyl ester of 3-hydroxyoctanoic acid, and (**C**) mass spectrum of a peak with R_t = 9.2 min identified using NIST library as methyl ester of 3-hydroxydecanoic acid. Methyl benzoate was used as an internal standard (STD).

In order to confirm the structure of synthesised PHA, NMR analysis has been performed. ^{1}H and ^{13}C NMR spectra are presented in Figure 3. All peaks expected for P(3HO-*co*-3HD) where observed in

^{1}H and ^{13}C NMR spectroscopy. The chemical shift at 169, 70.9, 39.1 ppm corresponds to C_1 (C=O), C_3 (–CH), C_2 (–CH$_2$) respectively, forming the backbone of the copolymer. The resonance in the range between 23–35 ppm related to olefinic methylene groups: C_4,*C_4 (33.86, 33.92 ppm), C_5, *C_5 (24.84, 25.20 ppm), C_6,*C_6 (31.64, 29.48 ppm), C_7,*C_7 (22.63, 29.32 ppm), *C_8 (31.91 ppm), *C_9 (22.76 ppm) (carbon atoms of 3-hydroxydecanoate market with asterisk). The terminal methyl groups (–CH$_3$) of C_8, *C_{10} were observed at 14.12 and 14.22 ppm, respectively. All observed peaks in the ^{13}C spectrum (Figure 3A) are in agreement with spectral characteristic reported for 3-hydroxyoctanoate and 3-hydroxydecanoate [30].

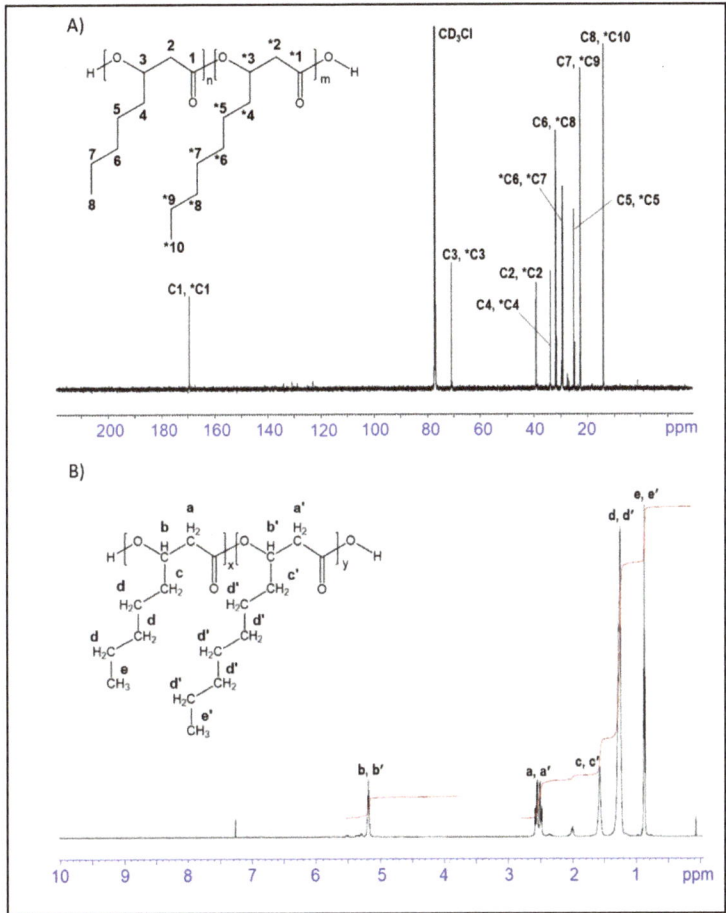

Figure 3. ^{13}C nuclear magnetic resonance (NMR) (**A**), and ^{1}H NMR (**B**) spectra of PHA copolymer produced with sugarcane molasses as carbon source. The structure of P(3HO-co-3HD) is shown as an insert within the spectra.

The ^{1}H NMR spectrum (Figure 3B) shows 5 groups of overlapping peaks corresponding to 5 different proton environments characteristic for medium chain 3-hydroxyalakanoates. Three different environments of the olefinic proton manifested at chemical shifts 0.8, 1.2, and 1.5 ppm. The chemical shift at 0.8 corresponds to the methyl groups of both monomers. The protons of the methylene group at carbon atoms 5, 6, 7 of 3-hydroxyoctanoate and carbon atoms 5, 6, 7, 8, 9 of 3-hydroxydecanoate were observed at 1.2 ppm, while proton resonance of the methylene group at carbon 4 for both monomers shifted further downfield to 1.5 ppm. Two types of backbone protons for both monomers generated signals at chemical shifts of 2.52 (methylene group of carbon 2) and 5.2 (methine group of carbon 3) ppm.

The monomeric composition of PHAs is strictly influenced by the carbon source utilized. In presence of structurally related carbon sources (e.g., fatty acids), the final composition of the polymer reflects that of the substrate. When unrelated substrates (e.g., carbohydrates) are utilized, the monomeric unit of the PHAs are independent of the structure of the carbon source [1–3]. As mentioned before, sugarcane molasses is composed of up to 60% of sugars which are classified as unrelated substrates. *Pseudomonas* sp. can convert sugars into PHAs through the *de novo* fatty acid pathway. One of the enzymes involved in this pathway has a high affinity to the substrate containing 10 carbon atoms [31]. For this reason, *Pseudomonas* sp. have been shown to produce PHAs with a high content of the 3-HD monomer when cultured using glucose or glycerol [5,31]. This study is in agreement with previous studies, as the MCL-PHA produced showed a high content of 3-HD.

3.2.2. Properties of Copolymer and Films Containing Lime Oil

The DSC of P(3HO-*co*-3HD) produced with sugarcane molasses as the carbon source revealed both low glass transition and melting temperature (Figure 4A, Table 1) which are characteristic properties of MCL-PHAs [5]. With glass-transition temperature ca. −45 °C, the amorphous phase of this semi-crystalline polymer is in a rubbery state at room temperature and polymer is expected to be a flexible and stretchable material. Also, such low glass transition would prevent transition of the polymer into a brittle material at sub-zero temperatures which can be required for a packaging material. The melting temperature is around 55 °C. This could be beneficial for melt processing at relatively low temperatures. Melting is not observed in a DSC thermogram of second heating demonstrating slow kinetics of the crystallisation process for this copolymer.

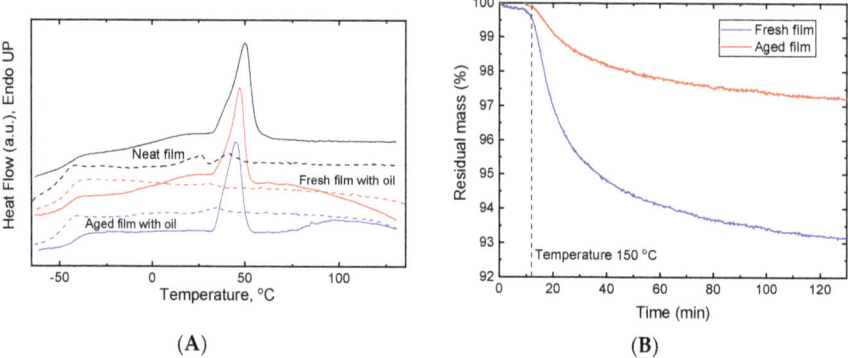

Figure 4. Thermal analysis of P(3HO-*co*-3HD) neat film and films with incorporated lime oil. (**A**) Differential scanning calorimetry (DSC) thermograms: neat film (black lines), fresh film with lime oil (red lines), and aged films with lime oil (blue lines). First heating—solid line, second heating—dashed line. Thermograms were shifted vertically for better visibility. (**B**) Thermogravimetric analysis (TGA) curves for determination of oil content after isothermal evaporation at 150 °C for 2 h.

Table 1. Thermal properties of P(3HO-co-3HD) neat films and films with incorporated lime oil.

Sample	Thermal Properties		
	T_g, °C	T_m, °C [a]	ΔH, J/g
Neat film	−45.6	55.6	18.2
Fresh film with lime oil	−44.3	51.4	14.7
Aged film with lime oil	−40.8	50.5	15.8

[a] Determined as the end of melting peak.

Incorporating lime oil into P(3HO-co-3HD) films decreased the melting temperature by approximately 4 °C. For an evaluation of the effect of lime oil on the crystallinity of the copolymer, an observed enthalpy of melting of P(3HO-co-3HD) in films containing lime oil needs to be normalised to the actual mass fraction of the copolymer. The real content of oil in the films does not correspond to oil/polymer composition in the casting solution due to oil loss as a result of evaporation along with solvent during the film formation as well as during the storage of the prepared film. Residual amounts of lime oil in P(3HO-co-3HD) films were determined from TGA experiments (Figure 4B) and found to be 6.9 wt.% for the new film and 2.8 wt.% for the aged one. The enthalpy of melting of the P(3HO-co-3HD) crystalline phase decreased from 18.2 J/g to 14.7 and 15.8 J/g for fresh and aged films, respectively. Thus, the addition of lime oil induced a decrease in crystallinity of the copolymer. This implies that lime oil components had a plasticising effect on the copolymer. Slight increase in enthalpy of polymer melting (and thereby crystallinity degree) for aged films compared with the fresh one might be explained by continuation of crystallisation during 1-year storage and by promotion of crystallisation with the loss of plasticising lime oil. Also, with storage, enthalpy of melting shifted to higher temperatures indicating rigidification of the amorphous phase.

3.3. Antibacterial Characterization

3.3.1. MIC (Minimum Inhibition Concentration) and MBC (Minimum Bactericidal Concentration)

Lime oil was tested against *S. aureus* 6538P and *E. coli* 8739, chosen as an example of Gram-positive and Gram-negative bacteria, respectively. The values obtained are shown in Table 2. The MIC and MBC values against *S. aureus* 6538P were 60 and 100 µL/mL respectively. While against *E. coli* 8739, MIC value of 80 µL/mL was obtained, no MBC could be determined at the concentration investigated. The MBC is defined as the lowest concentration of compound inducing 99.9% killing of the bacteria, corresponding to 3 log reduction in CFU. Therefore, the results obtained for lime oil showed that the compound had a bacteriostatic rather than bactericidal effect against *E. coli* 8739 at the concentrations investigated. In literature, MIC and MBC values for lime oil and generally for all essential oils comprise a very large range. For lime oil MIC values were reported from 0.1 mg/mL [18] to few hundreds mg/mL [32]. Such wide variations can be caused by plant sources, time of harvesting, methods of oil extraction, etc. Lime oil used in this study was of relatively low antimicrobial activity.

Table 2. Minimum inhibition concentration (MIC) and minimum bactericidal concentration (MBC) values of lime against *S. aureus* 6538P and *E. coli* 8739.

S. aureus 6538P		*E. coli* 8739	
MIC (µL/mL)	MBC (µL/mL)	MIC (µL/mL)	MBC (µL/mL)
60	100	80	-

The antibacterial mechanism of such essential oils is usually related to the activity of their main components, terpenes and terpenoids. Lime oil (*Citrus aurantifolia*) is composed mainly of limonene [18,20]. The mechanism of action of such organic compounds is linked to their ability to modify the integrity of the bacterial cell wall, increasing its permeability and causing membrane disruption [15,33]. Moreover, a higher effect against Gram-positive bacteria compared to negative ones is usually reported, due to protection of Gram-negative bacteria by the outer membrane composed of lipopolysaccharides [34]. Such a response was confirmed in this study as overall lime oil showed a higher effect against *S. aureus* 6538P than *E. coli* 8739.

3.3.2. Halo Test

The antimicrobial materials were tested against both *S. aureus* 6538P and *E. coli* 8739. Freshly prepared and aged films were analysed to investigate whether the material was able to preserve their antimicrobial efficacy. As it was discussed earlier, the content of lime oil decreased from 74 mg/g of copolymer for fresh film to 30 mg/g for the aged sample. The freshly prepared samples showed activity against *S. aureus* 6538P, inducing the formation of inhibition zones as shown in Figure 5 and Table 3. Moreover, the samples obtained after a one-year storage period still possessed activity against the microorganism, even though a 15% reduction of the zone of inhibition could be detected. By contrast, both the freshly prepared and aged films did not show activity against *E. coli* 8739, as reported in Table 3.

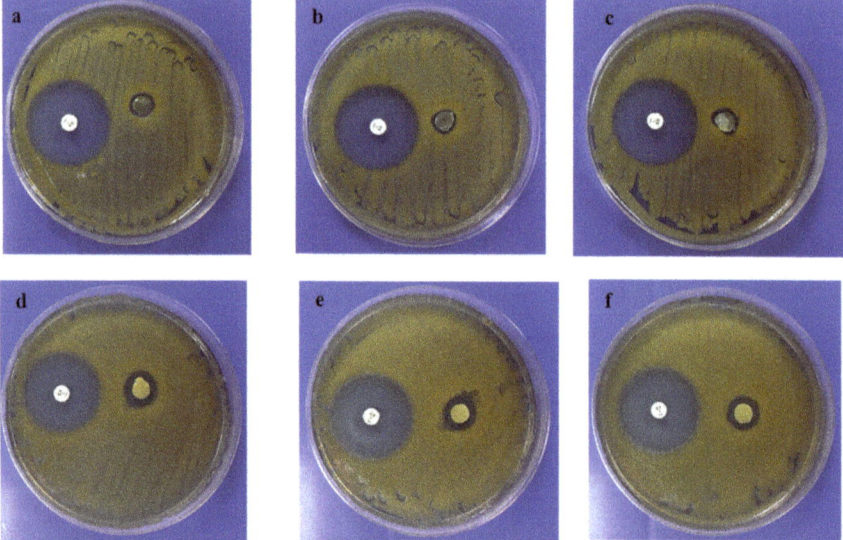

Figure 5. Halo test showing the antibacterial activity of aged lime oil containing P(3HO-*co*-3HD) films (a–c) and fresh lime oil containing films (d–f) against *S. aureus* 6538P (control: antibiotic disc containing 1 µg of oxacillin, left hand side).

Table 3. Halo diameters showing the effect of P(3HO-*co*-3HD) films containing lime oil against *S. aureus* 6538P and *E. coli* 8739.

Sample	Zone of Inhibition (ZOI) (cm)	
	S. aureus 6538P	*E. coli* 8739
Aged P(3HO-*co*-3HD)/lime films	1.06 ± 0.02	-
Fresh P(3HO-*co*-3HD)/lime films	1.23 ± 0.05	0.13 ± 0.04
Control	3.13 ± 0.10 (oxacillin)	1.99 ± 0.09 (streptomycin)

Limited research has been conducted on the incorporation of essential oils into PHAs and the studies conducted have focused only on SCL-PHAs. Films for antibacterial food packaging were produced using P(3HB-*co*-3HV) and two different essential oils such as oregano and clove, showing antibacterial activity against common food pathogens *Listeria innocua* 910 and *E. coli* 101 [35]. In another study, P(3HB-*co*-3HV) films were loaded with coconut fibres impregnated with oregano essential oil to produce composite antibacterial packaging materials, showing activity against *S. aureus* 6538P [36]. Alternatively, electrospinning has been investigated to produce fibrous mats based on P(3HB) or P(3HB-*co*-3HV) fibres containing essential oil. The natural compounds were either adsorbed onto the fibres after the spinning process (cinnamon, clove, oregano and oak bark) [37] or incorporated into the solution prior to electrospinning (rosemary and green tea extracts) [38]. The material produced showed activity against *Micrococcus luteus* 1569, *Serratia marcescens* 8587, *S. aureus* 6538P and *E. coli* 25922. Finally, Mendelez-Rodriguez et al. produced P(3HB-*co*-3HV) antibacterial fibres by loading eugenol essential oil onto mesoporous silica nanoparticles, followed by their incorporation into the fibres by electrospinning. The encapsulation was employed to maintain and protect the antibacterial effect of the agent, as essential oils are volatile compounds. The films obtained showed antibacterial activity against the *S. aureus* 6538P and *E. coli* 25922 up to 15 days after the day of fabrication and storage in a tightly closed system [39].

4. Conclusions

This study demonstrated the valorisation of sugarcane molasses, an agri-food by-product, through the production of a high-value MCL-PHA. The use of sugarcane molasses as the single carbon source and *Pseudomonas mendocina* CH50, a known MCL PHA producer, resulted in the synthesis of P(3HO-*co*-3HD) copolymer. P(3HO-*co*-3HD) is a polyester with low glass-transition and melting temperature, high elasticity and flexibility. In order to expand its potential applications, in this study, we explored the possibility of making P(3HO-*co*-3HD) antimicrobial through the incorporation of lime oil, an essential oil known for its antimicrobial activity. The antibacterial properties of the lime oil were investigated using the ISO 20776 against two bacterial strains, Gram-positive *S. aureus* 6538P and Gram-negative *E. coli* 8739. In line with general observations on the activity of essential oils, the lime oil used in this study was more active against *S. aureus* 6538P than *E. coli* 8739. Incorporation of the lime oil into the P(3HO-*co*-3HD) film had a minor effect on the material properties, slightly decreasing its crystallinity. As expected, the content of lime oil decreased with the storage due to migration out and evaporation from the polymer films. After one-year storage, the content of lime oil in the films decreased from *ca* 7 wt.% to 2.9 wt.%. Although the antimicrobial activity decreased after prolonged storage, both fresh and aged films of P(3HO-*co*-3HD) with incorporated lime oil showed activity against *S. aureus* 6538P. On the contrary, no antimicrobial activity was observed against *E. coli* 8739 for both fresh and aged films. The feasibility of imparting microbial activity to PHAs can broaden the application of this biodegradable polymer in several sectors, including coatings, adhesives, biodegradable rubbers, implantable materials. In the last sector, the developed P(3HO-*co*-3HD)/lime oil films could be employed as an antibacterial material for the regeneration of soft tissues, like skin.

Author Contributions: Conceptualization, P.B. and I.R.; methodology, P.B., E.M., R.N. and A.P.; formal analysis, P.B., E.M., R.N., A.P., B.G., M.H.A.; investigation, P.B., E.M., R.N., A.P., B.G., M.H.A.; resources, I.R.; data curation, P.B., E.M., R.N., A.P., B.G., M.H.A.; writing—P.B., E.M., R.N., A.P., B.G., M.H.A., B.L.; writing—review and editing, I.R.; supervision, I.R. and P.B.; project administration, I.R.; funding acquisition, I.R. All authors have read and agreed to the published version of the manuscript.

Funding: This research was funded by POLYBIOSKIN (H2020), Grant Agreement Number 745839, EM was funded by HyMedPoly, Marie Sklodowska-Curie Grant Agreement No 643050. The APC was funded by the University of Sheffield using the POLYBIOSKIN (H2020), Grant Agreement Number 745839.

Acknowledgments: The authors will like to acknowledge technical support by Neville Antonio and Zhi Song at the University of Westminster.

Conflicts of Interest: The authors declare no conflict of interest.

References

1. Khanna, S.; Srivastava, A.K. Recent advances in microbial polyhydroxyalkanoates. *Process Biochem.* **2005**, *40*, 607–619. [CrossRef]
2. Li, Z.; Yang, J.; Loh, X. Polyhydroxyalkanoates: Opening doors for a sustainable future. *NPG Asia Mater.* **2016**, *8*, 265. [CrossRef]
3. Raza, Z.A.; Abid, S.; Banat, I.M. Polyhydroxyalkanoates: Characteristics, production, recent developments and applications. *Int. Biodeterior. Biodegrad.* **2018**, *126*, 45–56. [CrossRef]
4. Álvarez-Chávez, C.R.; Edwards, S.; Moure-Eraso, R.; Geiser, K. Sustainability of bio-based plastics: General comparative analysis and recommendations for improvement. *J. Clean. Prod.* **2012**, *23*, 47–56. [CrossRef]
5. Basnett, P.; Lukasiewicz, B.; Marcello, E.; Gura, H.K.; Knowles, J.C.; Roy, I. Production of a novel medium chain length poly(3-hydroxyalkanoate) using unprocessed biodiesel waste and its evaluation as a tissue engineering scaffold. *Microb. Biotechnol.* **2017**, *10*, 1384–1399. [CrossRef] [PubMed]
6. Singh, A.K.; Mallick, N. Exploitation of inexpensive substrates for production of a novel SCL-LCL-PHA co-polymer by *Pseudomonas aeruginosa* MTCC 7925. *J. Ind. Microbiol. Biotechnol.* **2009**, *36*, 347–354. [CrossRef]
7. Chee, J.; Yoga, S.; Lau, N.; Ling, S.; Abed, R.M.M. Bacterially produced Polyhydroxyalkanoate (PHA): Converting renewable resources into bioplastics. In *Current Research, Technology and Education Topics in Applied Microbiology and Microbial Biotechnology*; Méndez-Vilas, A., Ed.; Formatex Research Center: Badajoz, Spain, 2010; pp. 1395–1404.
8. Jiang, G.; Hill, D.J.; Kowalczuk, M.; Johnston, B.; Adamus, G.; Irorere, V.; Radecka, I. Carbon sources for Polyhydroxyalkanoates and an integrated biorefinery. *Int. J. Mol. Sci.* **2016**, *17*, 1157. [CrossRef]
9. Nikodinovic-Runic, J.; Guzik, M.; Kenny, S.T.; Babu, R.; Werker, A.; Connor, K.E. Carbon-rich wastes as feedstocks for biodegradable polymer (Polyhydroxyalkanoate) production using bacteria. In *Advances in Applied Microbiology*, 1st ed.; Sariaslani, S., Gadd, G.M., Eds.; Academic Press: Cambridge, MA, USA, 2013; Volume 84, pp. 139–200.
10. Basnett, P.; Ravi, S.; Roy, I. Natural bacterial biodegradable medical polymers: Polyhydroxyalkanoates. In *Science and Principles of Biodegradable and Bioresorbable Medical Polymers*; Zhang, X., Ed.; Woodhead Publishing: Cambridge, UK, 2017; pp. 257–277.
11. Paris, J.B.; Seyer, D.; Jouenne, T.; Thébault, P. Elaboration of antibacterial plastic surfaces by a combination of antiadhesive and biocidal coatings of natural products. *Colloids Surf. B* **2017**, *156*, 186–193. [CrossRef]
12. Roca, I.; Akova, M.; Baquero, F.; Carlet, J.; Cavaleri, M.; Coenen, S.; Cohen, J.; Findlay, D.; Gyssens, I.; Heure, O.E.; et al. The global threat of antimicrobial resistance: Science for intervention. *NMNI* **2015**, *6*, 22–29. [CrossRef]
13. Prabuseenivasan, S.; Jayakumar, M.; Ignacimuthu, S. *In vitro* antibacterial activity of some plant essential oils. *BMC Complement. Altern. Med.* **2006**, *6*, 39. [CrossRef]
14. Kalemba, D.; Kunicka, A. Antibacterial and Antifungal Properties of Essential Oils. *Curr. Med. Chem.* **2003**, *10*, 813–829. [CrossRef] [PubMed]
15. Hyldgaard, M.; Mygind, T.; Meyer, R.L. Essential oils in food preservation: Mode of action, synergies, and interactions with food matrix components. *Front. Microbiol.* **2012**, *3*, 12. [CrossRef] [PubMed]
16. Di Pasqua, R.; Hoskins, N.; Betts, G.; Mauriello, G. Changes in membrane fatty acids composition of microbial cells induced by addiction of thymol, carvacrol, limonene, cinnamaldehyde, and eugenol in the growing media. *J. Agric. Food Chem.* **2006**, *54*, 2745–2749. [CrossRef] [PubMed]
17. Arrieta, M.P.; López, J.; Hernández, A.; Rayón, E. Ternary PLA–PHB–Limonene blends intended for biodegradable food packaging applications. *Eur. Polym. J.* **2014**, *50*, 255–270. [CrossRef]
18. Hsouna, A.B.; Halima, N.B.; Smaoui, S.; Hamdi, N. *Citrus lemon* essential oil: Chemical composition, antioxidant and antimicrobial activities with its preservative effect against *Listeria monocytogenes* inoculated in minced beef meat. *Lipids Health Dis.* **2017**, *16*, 146. [CrossRef]
19. Hąc-Wydro, K.; Flasiński, M.; Romańczuk, K. Essential oils as food eco-preservatives: Model system studies on the effect of temperature on limonene antibacterial activity. *Food Chem.* **2017**, *235*, 127–135. [CrossRef]

20. Costa, R.; Bisignano, C.; Filocamo, A.; Grasso, E.; Occhiuto, F.; Spadaro, F. Antimicrobial activity and chemical composition of *Citrus aurantifolia* (Christm.) Swingle essential oil from Italian organic crops. *J. Essent. Oil Res.* **2014**, *26*, 400–408. [CrossRef]
21. Basnett, P.; Marcello, E.; Lukasiewicz, B.; Panchal, B.; Nigmatullin, R.; Roy, I. Biosynthesis and characterization of novel, biocompatible medium chain length polyhydroxyalkanoates by *Pseudomonas mendocina* CH50 using coconut oil as the carbon source. *J. Mater. Sci. Mater. Med.* **2018**, *29*, 179. [CrossRef]
22. Rai, R.; Yunos, D.M.; Boccaccini, A.R.; Knowles, J.C.; Barker, I.A.; Howdle, S.M.; Tredwell, G.D.; Keshavarz, T.; Roy, I. Poly-3-hydroxyoctanoate P(3HO), a Medium Chain Length Polyhydroxyalkanoate Homopolymer from *Pseudomonas mendocina*. *Biomacromolecules* **2001**, *12*, 2126–2136. [CrossRef]
23. Gumel, A.M.; Annuar, M.S.M.; Heidelberg, T. Growth kinetics, effect of carbon substrate in biosynthesis of MCL-PHA by *Pseudomonas putida* Bet001. *Braz. J. Microbiol.* **2014**, *45*, 427–438. [CrossRef]
24. Akaraonye, E.; Moreno, C.; Knowles, J.C.; Keshavarz, T.; Roy, I. Poly(3-hydroxybutyrate) production by *Bacillus cereus* SPV using sugarcane molasses as the main carbon source. *Biotechnol. J.* **2012**, *7*, 293–303. [CrossRef] [PubMed]
25. Naheed, N.; Jamil, N. Optimization of biodegradable plastic production on sugar cane molasses in *Enterobacter* sp. SEL2. *Braz. J. Microbiol.* **2014**, *45*, 417–426. [CrossRef] [PubMed]
26. Chaijamrus, S.; Udpuay, N. Production and characterization of polyhydroxybutyrate from molasses and corn steep liquor produced by *Bacillus megaterium* ATCC 6748. *Agric. Eng. Int. CIGR J.* **2008**, *10*, 1–12.
27. Solaiman, D.K.Y.; Ashby, R.D.; Hotchkiss, A.T.; Foglia, T.A. Biosynthesis of medium-chain-length Poly(hydroxyalkanoates) from soy molasses. *Biotechnol. Lett.* **2006**, *28*, 157–162. [CrossRef]
28. Jiang, Y.; Song, X.; Gong, L.; Li, P.; Dai, C.; Shao, W. High poly (β-hydroxybutyrate) production by *Pseudomonas fluorescens* A2a5 from inexpensive substrates. *Enzym. Microb. Technol.* **2008**, *42*, 167–172. [CrossRef]
29. Chen, Y.-J.; Huang, Y.-C.; Lee, C.-Y. Production and characterization of medium-chain-length polyhydroxyalkanoates by *Pseudomonas mosselii* TO7. *J. Biosci. Bioeng.* **2014**, *118*, 145–152. [CrossRef]
30. De Rijk, T.C.; Van de Meer, P.; Eggink, G.; Weusthuis, R.A. Methods for Analysis of Poly(3-hydroxyalkanoate) Composition. In *Biopolymers: Polyesters. II Properties and Chemical Synthesis*; Doi, Y., Steinbuchel, A., Eds.; Wiley-Blackwell: Hoboken, NJ, USA, 2005; Volume 3b, pp. 1–12.
31. Poblete-Castro, I.; Rodriguez, A.L.; Chi Lam, C.M.; Kessler, W. Improved production of medium-chain-length polyhydroxyalkanoates in glucose-based fed-batch cultivations of metabolically engineered *Pseudomonas putida* strains. *J. Microbiol. Biotechnol.* **2014**, *24*, 59–69. [CrossRef]
32. Aibinu, I.; Adenipekun, T.; Adelowotan, T.; Ogunsanya, T.; Odugbemi, T. Evaluation of the antimicrobial properties of different parts of *Citrus Aurantifolia* (Lime Fruit) as used locally. *Afr. J. Tradit. Complement. Altern. Med.* **2007**, *4*, 185–190.
33. Khameneh, B.; Iranshahy, M.; Soheili, V.; Fazly Bazzaz, B.S. Review on plant antimicrobials: A mechanistic viewpoint. *Antimicrob. Resist. Infect. Control* **2019**, *8*, 118. [CrossRef]
34. Nazzaro, F.; Fratianni, F.; De Martino, L.; Coppola, R.; De Feo, V. Effect of essential oils on pathogenic bacteria. *Pharmaceuticals* **2013**, *6*, 1451–1474. [CrossRef]
35. Requena, R.; Jiménez, A.; Vargas, M.; Chiralt, A. Poly [(3-hydroxybutyrate)-co-(3-hydroxyvalerate)] active bilayer films obtained by compression moulding and applying essential oils at the interface. *Polym. Int.* **2016**, *65*, 883–891. [CrossRef]
36. Torres-Giner, S.; Hilliou, L.; Melendez-Rodriguez, B.; Figueroa-Lopez, K.J.; Madalena, D.; Cabedo, L.; Covas, J.A.; Vicente, A.A.; Lagaron, J.M. Melt processability, characterization, and antibacterial activity of compression-molded green composite sheets made of poly(3-hydroxybutyrate-*co*-3-hydroxyvalerate) reinforced with coconut fibers impregnated with oregano essential oil. *Food Packag. Shelf Life* **2018**, *17*, 39–49. [CrossRef]
37. Kundrat, V.; Matouskova, P.; Marova, I. Facile Preparation of Porous Microfiber from Poly-3-(R)-Hydroxybutyrate and Its Application. *Materials* **2019**, *13*, 86. [CrossRef] [PubMed]

38. Figueroa-Lopez, K.J.; Vicente, A.A.; Reis, M.A.; Torres-Giner, S.; Lagaron, J.M. Antimicrobial and antioxidant performance of various essential oils and natural extracts and their incorporation into biowaste derived poly (3-hydroxybutyrate-co-3-hydroxyvalerate) layers made from electrospun ultrathin fibers. *Nanomaterials* **2019**, *9*, 144. [CrossRef]
39. Melendez-Rodriguez, B.; Figueroa-Lopez, K.J.; Bernardos, A.; Martínez-Máñez, R.; Cabedo, L.; Torres-Giner, S.; M Lagaron, J. Electrospun antimicrobial films of poly(3-hydroxybutyrate-*co*-3-hydroxyvalerate) containing eugenol essential oil encapsulated in mesoporous silica nanoparticles. *Nanomaterials* **2019**, *9*, 227. [CrossRef]

© 2020 by the authors. Licensee MDPI, Basel, Switzerland. This article is an open access article distributed under the terms and conditions of the Creative Commons Attribution (CC BY) license (http://creativecommons.org/licenses/by/4.0/).

Journal of
Functional Biomaterials

Article

Modification of PLA-Based Films by Grafting or Coating

Aleksandra Miletić [1], Ivan Ristić [1], Maria-Beatrice Coltelli [2,3] and Branka Pilić [1,*]

[1] Faculty of Technology Novi Sad, University of Novi Sad, 21000 Novi Sad, Serbia; alexm@uns.ac.rs (A.M.); ivan.ristic@uns.ac.rs (I.R.)
[2] Department of Civil and Industrial engineering, University of Pisa, 56122 Pisa, Italy; maria.beatrice.coltelli@unipi.it
[3] National Inter University Consortium of Materials Science and Technology (INSTM), 50121 Florence, Italy
* Correspondence: brapi@uns.ac.rs; Tel.: +381-63537232

Received: 1 March 2020; Accepted: 24 April 2020; Published: 7 May 2020

Abstract: Recently, the demand for the use of natural polymers in the cosmetic, biomedical, and sanitary sectors has been increasing. In order to meet specific functional properties of the products, usually, the incorporation of the active component is required. One of the main problems is enabling compatibility between hydrophobic and hydrophilic surfaces. Therefore, surface modification is necessary. Poly(lactide) (PLA) is a natural polymer that has attracted a lot of attention in recent years. It is bio-based, can be produced from carbohydrate sources like corn, and it is biodegradable. The main goal of this work was the functionalization of PLA, inserting antiseptic and anti-inflammatory nanostructured systems based on chitin nanofibrils–nanolignin complexes ready to be used in the biomedical, cosmetics, and sanitary sectors. The specific challenge of this investigation was to increase the interaction between the hydrophobic PLA matrix with hydrophilic chitin–lignin nanoparticle complexes. First, chemical modification via the "grafting from" method using lactide oligomers was performed. Then, active coatings with modified and unmodified chitin–lignin nanoparticle complexes were prepared and applied on extruded PLA-based sheets. The chemical, thermal, and mechanical characterization of prepared samples was carried out and the obtained results were discussed.

Keywords: coating; poly(lactide); chitin–lignin nanocomplex; grafting from; lactide oligomers

1. Introduction

Following the latest EU policies and strategies on plastics use, the implementation of bio-based materials is more than welcome in all industrial fields, especially for single-use products. The healthcare sector, including biomedical, cosmetics, and sanitary products, is one of the big consumers of single-use polymer-based products, and significant effort is put into changing fossil-based materials with bio-based materials [1]. Besides that, healthcare products need to fulfill many other requirements like safety, durability, efficacy, and biocompatibility. High-value products in this area have some functionalities, like anti-inflammatory, antimicrobial, and antioxidant functions, and represent an improved version of the previous ones [2]. Due to increased antimicrobial resistance problems, nowadays, natural active compounds are under investigation for use in the healthcare sector [3–5]. Thus, in the development of a new product for this purpose, researchers are facing two main issues: the right choice of bio-based material and suitable functionalization methods.

Biopolyesters, with poly(lactide) (PLA) as the most prominent representative, are an excellent alternative to fossil-based materials [6–8]. Sugar-based raw materials like corn and starch can be used for the production of PLA, which makes it sustainable. PLA is biocompatible with the human organism and can be used in the production of implants as well as in other biomedical applications [9,10]. Also, it is biodegradable, with adjustable biodegradability according to the end-use requirements, and after degradation, only water and carbon-dioxide remains. It is certified as safe-to-use in healthcare products. PLA is a thermoplastic material, inert, with low interaction with cells, no functionality, and is resistant to acids, alkali, and fats. PLA is often used in drug-delivery systems, like carriers, for the controlled delivery of different medicines [11].

As evidenced in works regarding PLA bionanocomposites, one of the challenges in the development of functional films or medical devices is to achieve a good dispersion of active components within the polymer matrix [12,13] and, at the same time, maintaining their activity. Additionally, it is essential to ensure enough surfaceavailability of the component in order to achieve the desired activity [14,15]. Two main directions can be followed in the functionalization of materials: direct incorporation of active molecules into the material and physical or chemical bonding of actives onto the surface of the material. Since natural compounds are thermosensitive and deteriorate when exposed to high temperatures, direct incorporation is very challenging. Moreover, the difference in surface properties and hydrophilic–hydrophobic interactions makes successful incorporation even more difficult. One of the possible routes can be a surface modification of active compounds using some components with better compatibility with PLA, such as the "grafting from" method of modification [16]. The grafting of the polymer chain on a solid surface is a very adaptable method for surface modification and functionalization. Polymer chains can be grafted to the solid substrate (grafting to), or the grafting reaction can be proceeded by polymerization from the surface (grafting from). Both methods are suitable for forming a thin layer on the solid surface with the desired physical and chemical surface properties [17,18].

Bonding active molecules onto the surface often requires pretreatment of the surface for activation, as PLA surface is known as inert, and it is hard to attach something on it chemically. It is usually undertakenby treatment with aggressive chemicals [19,20], resulting in the introduction of new chemical groups like OH-, which are more reactive. This approach is suitable in low-volume products, where safety concerns are not so important. However, in medical products, theycan be a problem due to possible residues of chemicals that can induce irritations or allergies. Another possibility is plasma surfaceactivation, but this method is expensive for high-volume products, and the effect of the modification is generally modest [21–23].

A more straightforward method of surface activation, which is suitable for high-volume products, is cost-effective and still gives results, is coating with bio-based polymer active coating. It is favorable, but not limited to waterborne coatings, like polyurethane dispersions [24].

To obtain a very high adhesion between PLA-based films and the coating on it, the use of a PLA-based coating can be advantageous. PLA-based coatings were widely investigated because of their biocompatibility and biodegradability on several material surfaces.

For example, slow drug delivery systems based on biodegradable poly-lactic acid and antibiotic loaded hydroxyapatite microspheres were developed to be applied as coating on metal implants to prevent post-operative infections [25]. Chang et al. [26] prepared biodegradable corn starch films and used a PLA coating to improve theirwaterproof performance. Compared to the starch film without PLA coating, the films coated with PLA significantly reduced water solubility and increased the mechanical stability. Sputtering deposition and plasma assisted atomic layer deposition were also used for surface modification and the functionalization of PLA films, but those processes are not easily scalable to the industrial level [27,28].

This paper represents an attempt to select simple methodologies for modifying PLA-based film surfaces by including anti-microbial and anti-oxidant complexes based on chitin nanofibrils and lignin containing glycyrrhetic acid. In slightly acidic water suspension, chitin nanofibrils are positive, whereas the lignin is negative, hence they form a chitin–lignin (CN–LG) complex eventually entrapping selected functional molecules. This complex became cytocompatible, showing anti-inflammatory activity, and may serve for the delivery of biomolecules for skin care and regeneration [29]. In the present paper, the complex containing niacinamide is also considered. Niacinamide anti-inflammatory properties make it an attractive treatment for skin conditions marked by inflammations [30].

This work can be divided into three parts. First, surface modification of chosen active compounds was carried out by a "grafting from" method. Second, unmodified and modified active compounds were incorporated into solution cast PLA-based films, and properties were compared. Furthermore, in the third part, PLA-based coatings with unmodified and modified active compounds were prepared and applied to PLA-extruded film. The morphological, mechanical, thermal, and chemical properties were investigated.

2. Materials and Methods

2.1. Materials

High molecular poly(lactide) (PLA, Nature Works 2003 D, Minnetonka, MN, USA) and low molecular poly(lactide) (PLLA, Condensia Quimica, Barcelona, Spain, Mw 2000) were used as the polymer matrix. L-lactide (Sigma Aldrich, St. Louis, MI, USA) was used as a monomer for poly(lactide) synthesis, dichloromethane (Fisher Scientific, Hampton, NH, USA) as the solvent, and reaction medium for synthesis, H_2SO_4 acid (Sigma Aldrich, Taufkirchen, Germany) was used as the initiator of cationic polymerisation of lactide. Chitin–lignin nanoparticles loaded with niacinamide and glycyrrhetic acid were received from MAVI, produced in Texol, Alanno Scalo, Italy and were used as the initiator. They were produced according to patented technology [31]. All chemicals were used as received. The PLA-based extruded film was prepared according to a previous methodology by flat die extrusion [32]. The specific film was obtained by plasticizing with acetyl tri-n-butyl citrate (ATBC) a PLA/poly(butylene adipate co-terephathalate) (PBAT) blend [33].

2.2. Chitin–Lignin Nanoparticles Surface Modification

The chitin–lignin nanoparticle surface was modified by the "grafting from" method forming the PLA oligomers (OLA) at the particles surfaces by cationic polymerizationusing L-lactide as the monomer.

First, 1 g of chitin–lignin nanoparticles were dispersed in 60 mL of dichloromethane. Then, 10 g of the lactide monomer and 0.1 wt% (calculated on PLA weight) of a strong acid, sulfuric acid (H_2SO_4), as the initiator of the polymerization, were added. The reaction mixture was placed in a 250 mL balloon and connected to a reflux condenser. The synthesis was carried 5 h at 36 °C until all the particles become soluble in dichloromethane, which confirmed that all the particles were grafted by PLA oligomer. The grafted chitin–lignin powder was obtained after evaporation of solvent and drying at 40 °C for 10 h.

2.3. Preparation of PLA-Based Films by Solution Casting Method

PLA-based films (based on PLA 2003 D) with unmodified and surface-modified Chitin–lignin complex nanoparticles were prepared by the solution casting method. First, 10 wt% solution of PLA in dichloromethane was prepared, and then, in the second step, 1 wt% (calculated on PLA weight) of nanoparticles was added. After mixing at room temperature, the dispersion was poured out in Petri dish, and films were obtained after drying 24 h on the air in the room temperature.

2.4. Preparation of Poly(L-Lactide) (PLLA)-Based Coating

Coatings containing unmodified and surface-modified Chitin–lignin complex nanoparticles were prepared using low molecular weight PLA (PLLA) by dissolving it in dichloromethane in the concentration of 30 wt% and adding 1 wt% (calculated on PLLA weight) of appropriate chitin–lignin complex. The prepared coating was applied to PLA-based extruded film using a brushing technique.

2.5. Characterization Methods

The light microscope was used for onsite examination of sample morphology.

Chemical properties of samples were examined using Attenuated Total Reflectance-Fourier Transform Infrared spectroscopy (ATR-FTIR) Shimadzu IRaffinity equipment, Kyoto, Japan, by scanning samples from 4000 to 400 cm^{-1}, with a resolution of 4 cm^{-1}, and following changes in obtained spectra.

Thermal properties were determined using the differential scanning calorimetry (DSC) method on TA Instruments Q20 equipment (New Castle, DE, USA) by heating samples with a heat flow of 10 °C/min in one cycle. The samples were sampled from the casted films and from the coated film made of plasticized PLA/PBAT including all the film thickness.

Shimadzu EZ-Test, Kyoto, Japan instrument was used for mechanical properties assessment. Samples were cut in a rectangular shape of dimensions 1 cm × 4 cm, 0.3 mm thickness and tested with a clamp speed of 10 mm/min.

3. Results and Discussion

The cationic solution polymerization was selected for the grafting of lactide oligomers on the chitin–lignin–glycyrrhetic acid complex because it is possible to perform it even at a temperature of 40 °C, yielding PLA with narrow molecular weight distribution in a very short period of time (less than 6 h), on the basis of our previously work [34]. It was shown that a different macroinitiator could be used to promote lactide polymerization enabling desired control of the properties of the final product [35].

The grafting was performed onto two different CN–LG systems: one containing glycyrrhetic acid [36] and the second with niacinamide (Figure 1). These two different trials were performed to consider the reliability of the polymerization despite of the different chemical structure of the functional molecule.

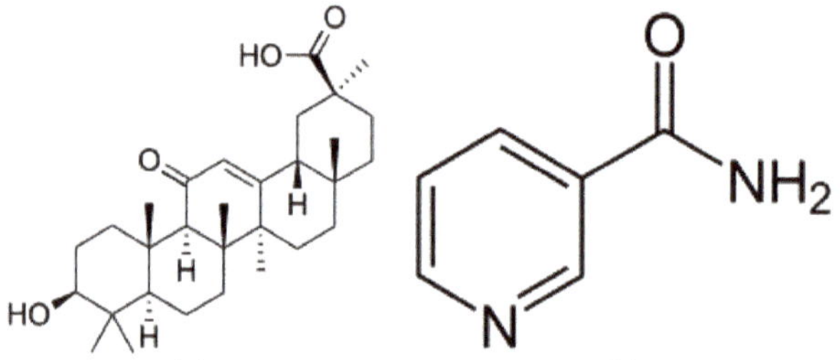

Figure 1. Molecular structure of (**a**) glycyrrhetic acid; (**b**) niacinamide.

In both cases, grafting was successful, and the chemical structure of grafted CN–LG complexes was confirmed by FTIR analysis (Figure 2a,b). In fact, in both PLA grafted CN–LG complexes, characteristic bands for the complex are present with additional bands for PLA (such as C=O carbonyl at 1760 cm^{-1} and 1090 cm^{-1} for C–O–C stretching).

From the DSC results of unmodified and grafted chitin–lignin complexes (Figure 3a,b), it can be concluded that the grafting of PLA onto complex surfaces has a strong influence on their thermal properties. While chitin–lignin complexes show the typical peak of chitin attributed to water loss [37,38] at a temperature at about 50 °C and glass transition of lignin around 150 °C [39], modified complexes grafted with PLA showed thermal properties similar to PLA, with slightly lower Tg at around 26 °C, a crystallization temperature (Tc) value around 80 °C with broad cold-crystallization peak (Tcc), and melting temperature (Tm) around 140 °C. This can indicate that the PLA layer formed around the chitin–lignin nanoparticle protects it from the direct influence of high temperatures and can maintain its activity.

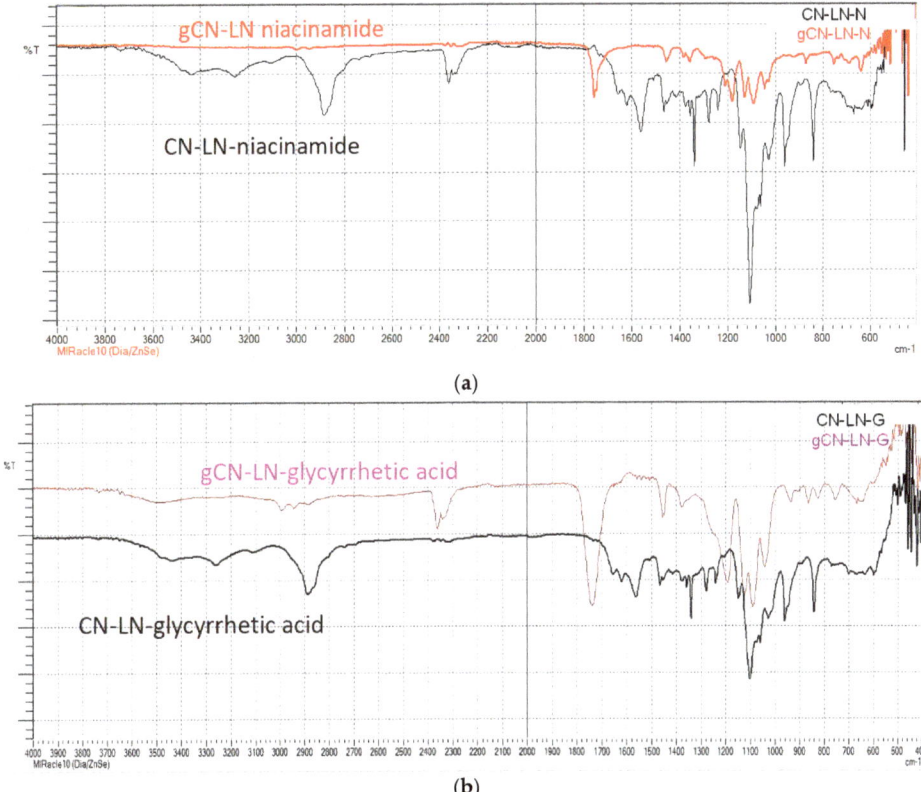

Figure 2. FTIR spectrum of: (**a**) Poly(lactide) (PLA) grafted chitin–lignin complex with niacinamide; (**b**) PLA grafted chitin–lignin complex with glycyrrhetic acid.

Unmodified and modified active compounds were incorporated into solution-cast PLA-based films, and the properties were compared. The PLA used, in this case, is the general grade PLA used for extrusion and various applications. It can be seen that, according to the color of prepared films (Figure 4), the applied grafting method improved the compatibility of PLA and the selected complex. Films with grafted complexes are transparent, similar to the pure PLA film, which confirmed that grafted complexes were well dispersed in the PLA matrix.

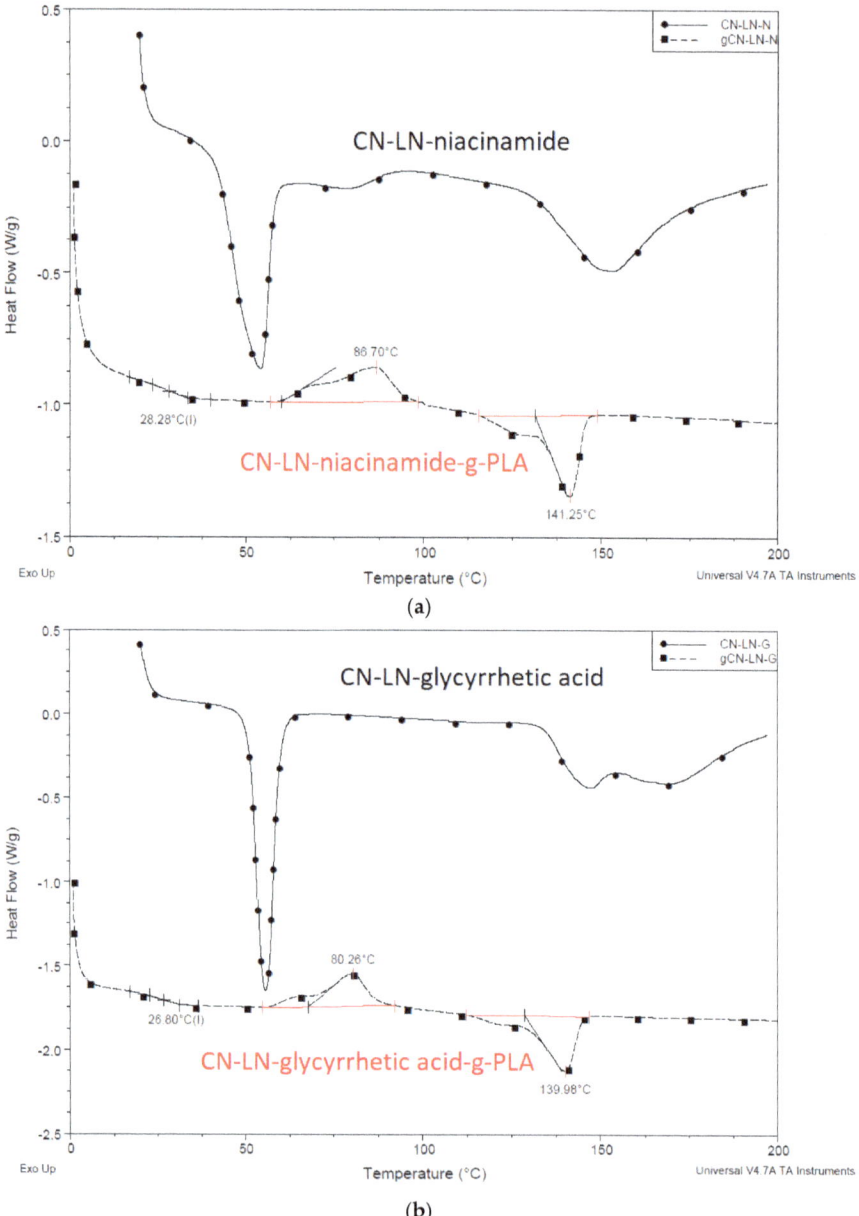

Figure 3. (a) Differential scanning calorimetry (DSC) thermogram of unmodified and modified chitin–lignin complexes by PLA grafting niacinamide complexes (b) glycyrrhetic acid.

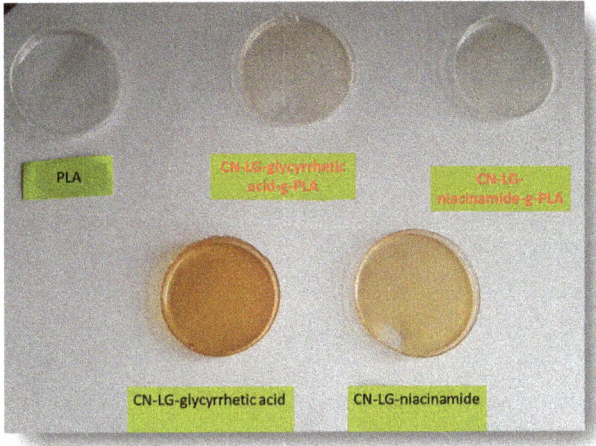

Figure 4. PLA-based films obtained via a solution casting method; first raw left to right PLA, PLA with chitin–lignin complex with glycyrrhetic acid modified by grafting, PLA with chitin–lignin complex with niacinamide modified by grafting, second raw PLA with unmodified chitin–lignin complex with glycyrrheticacid and PLA with unmodified chitin–lignin complex with niacinamide.

Unmodified complex loaded films have intense colors of the complex, which indicated phase separation due to the low compatibility of the PLA and complex. This is confirmed by optical microscopy (Figure 5a,b), where complex aggregates are visible as dark regions. PLA films with grafted chitin–lignin complexes did not show any aggregate, and figures are transparent as pure PLA films.

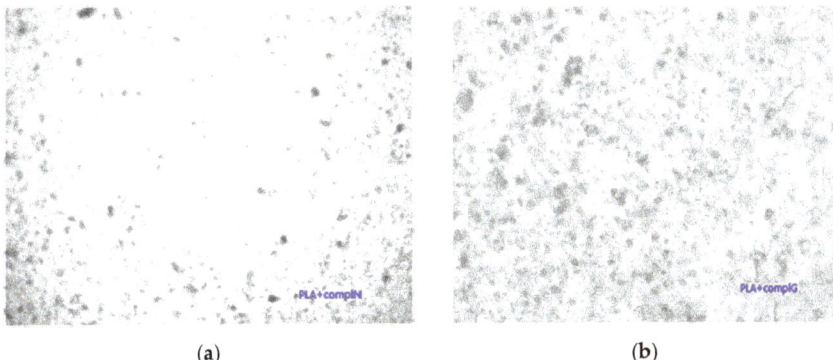

Figure 5. Micrographs (magnification 100×) of solution cast films of PLA with added (**a**) chitin–lignin complex with niacinamide and (**b**) chitin–lignin complex with glycyrrhetic acid.

Changes in band position and intensity in FTIR spectra of PLA films with unmodified and PLA grafted chitin–lignin complexes (Figures 6 and 7) were not detected. These results confirmed that in the grafting method, complexes were not entirely covered by PLA, but in such a way that compatibility was improved without the loss of functionality.

DSC thermograms of PLA films with unmodified chitin–lignin complexes, (Figures 7a and 8b), showed the characteristic transition temperatures of PLA; a glass transition temperature at 57 °C and 40 °C, Tc at 116 °C and 94 °C and Tm at 150 °C for film with niacinamide and glycyrrhetic acid, respectively.

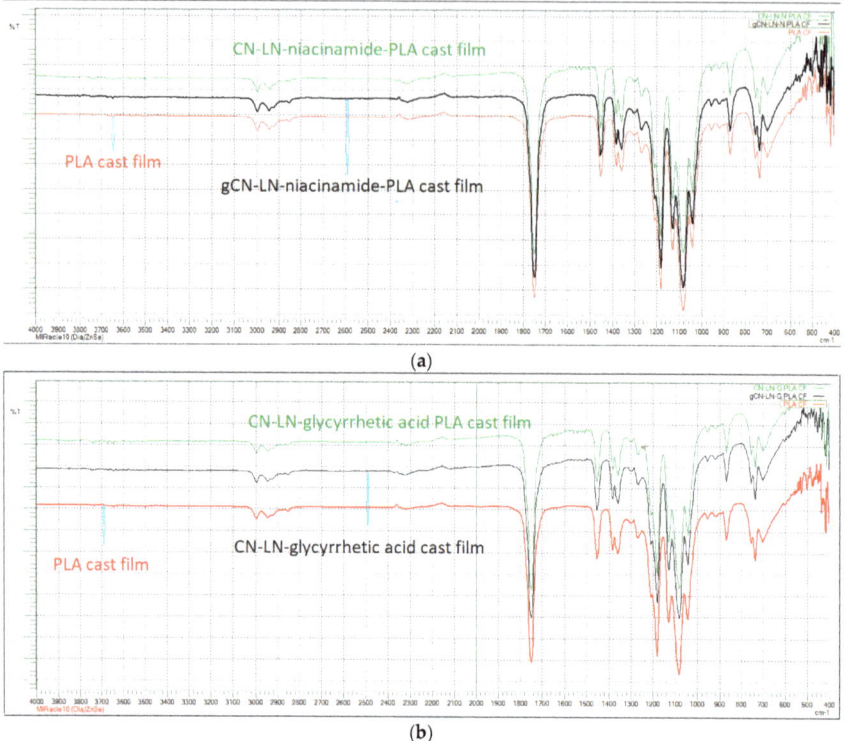

Figure 6. FTIR spectra of PLA cast films with (**a**) unmodified chitin–lignin complexes and (**b**) modified chitin–lignin complexes.

The significant differences in the recorded values can be attributed to the different behavior of glycyrrhetic acid and niacinamide. It is possible that niacinamide is responsible for the formation of hydrogen bonds with PLA and CN–LG, because of its amide groups and electron rich nitrogen in the pyridine ring. This can explain the higher value of Tg, due to reduced chain mobility, and the higher value of cold crystallization temperature, due to the increased disorder and the energy barrier in the material. Moreover, the sample films with unmodified chitin–lignin and glycyrrhetic acid showed a higher crystallinity than the one with niacinamide. Glycyrrhetic acid, having a higher molecular weight and thus a high molecular dimension [33], and a carboxylic function [39], is probably better in inducing heterogeneous nucleation in PLA [40]. In fact, its 3D structure consists of a planar system with a methyl group in opposition, almost perpendicular to the plan. As PLA has CH_3 groups on its chain, the capacity of the glycyrrhetic acid to interact with PLA, inducing its local organization in crystals, is thus significant, and seems similar to the nucleating action due to the formation of PLA stereocomplexes [41].

Chitin–lignin complex grafted with PLA had a strong influence on PLA properties: in the first cycle of the heating, the peak of cold crystallization disappeared, the Tg value decreased by 12 °C, and melting temperature increased up to 156 °C (Figure 7c), typical of the crystal form [42]. These results showed that the presence of the CN–LG-niacinamide-g-PLA induced the crystallization of PLA during the cooling occurred before the studied heating. Hence the CN–LG-niacinamide-g-PLA behaved as a nucleant. The crystallinity of the film is thus very high, and for this reason the cold crystallization did not occur. Interestingly, the high value of melting point can be ascribed to the presence of more regular crystal than in the case of unmodified chitin–lignin complexes. It is reasonable that the PLA chains grafted onto the CN–LG complex co-crystallize with PLA, thus better promoting the development of crystals. PLA block copolymers have shown a similar behavior, inducing an easier crystallization of PLA blocks [43]. In good agreement, the addition of silica-grafted-PLA to PLLA also induced a strong increase in crystallinity due to this nucleating effect [44]. All the improvements can be ascribed to stronger entanglements between PLLA stretched by nano-SiO$_2$ and PLA matrix.

The incorporation of both modified and unmodified chitin–lignin complexes nanoparticles influenced the mechanical properties of PLA. In both cases, mechanical properties were improved. When unmodified chitin–lignin nanoparticles loaded with active compounds were added to PLA, the tensile strength was increased almost ten times, together with the improvement of elasticity for three to four times. Better elasticity was achieved by the incorporation of chitin–lignin complex nanoparticles modified with PLA due to the better dispersion of particles within the PLA matrix, which is a direct consequence of better compatibility of these particles with PLA. Compared to the mechanical properties of PLA, films with modified particles had around four times better tensile strength and six times better elasticity. However, if the mechanical properties of the same samples were compared to ones with unmodified particles, it can be observed that tensile strength values were slightly lower. Nevertheless, the elongation at break was improved—it almost doubled. These results are in accordance with the optical and thermal properties of the samples. The results of mechanical testing are summarized in Table 1.

Table 1. Mechanical properties solution casting film.

Sample	Tensile Strength (MPa)	Elongation at Break (%)
PLA cast film	1.96	44.33
CN-LN-niacinamide PLA cast film	11.75	155.13
CN-LN-glycyrrhetic acid PLA cast film	11.61	112.82
gCN-LN-niacinamide PLA cast film	9.24	264.66
gCN-LN-glycyrrhetic acid PLA cast film	7.92	236.78

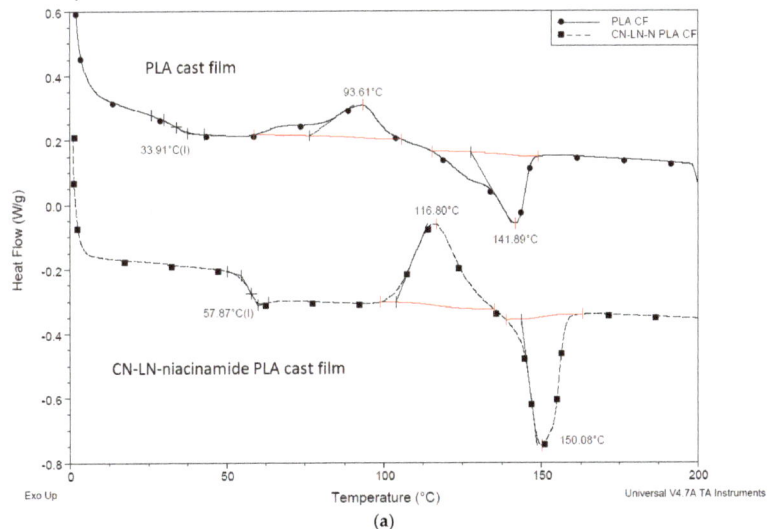

Figure 7. Cont.

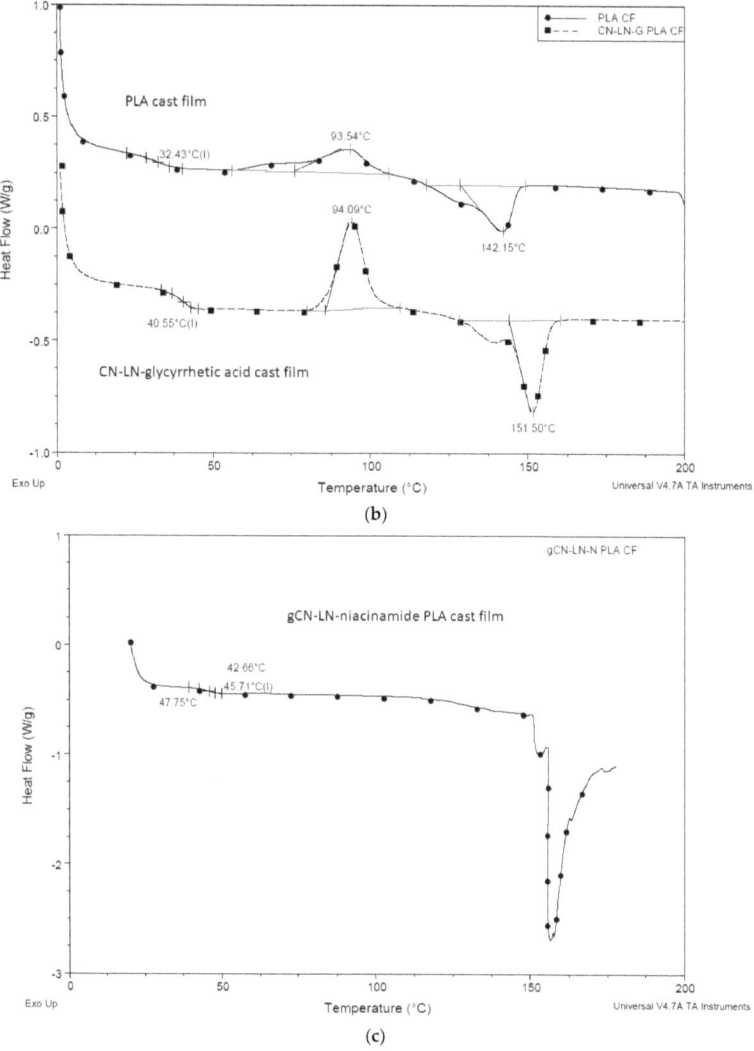

Figure 7. DSC thermogram of PLA film with chitin–lignin complex (a) unmodified with niacinamide, (b) unmodified with glycyrrhetic acid, (c) grafted chitin–lignin complex with niacinamide.

The obtained results showed that CN–LG complexes can be incorporated, grafted or ungrafted, into PLA films by solution casting, and they can allow to strongly affect and modulate mechanical properties of these films.

Nevertheless, the possible application of poly(lactide)-based films as coating onto an extruded film is very interesting. For this activity, a specific oligomeric PLA with a molecular weight of 2000 was used (PLLA). This polymer was selected because it can be potentially used in the future to prepare coatings in common solvents, thanks to its lower molecular weight and thus enhanced solubility.

After this, extruded PLA films were coated with PLLA-based coating containing unmodified and surface modified chitin–lignin complex nanoparticles. As in the previous case, the difference in the color of the two films can be seen. While PLLA loaded with unmodified particles has a brownish color

(Figure 8a), originating from the color of the complex, coating with modified particles has a lighter color (Figure 8b).

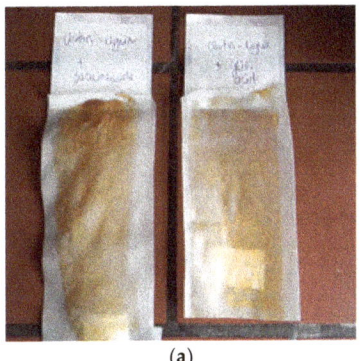 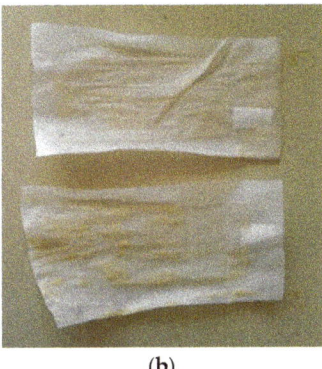

(a) (b)

Figure 8. PLA extruded sheet coated with low molecular poly(lactide) (PLLA)-based coating with unmodified chitin–lignin niacinamidecomplexes (left) and modified chitin–lignin glycyrrhetic complexes (right), (a) and (b) modified chitin–ligninglycyrrhetic complexes (up) and modified chitin–lignin niacinamide complexes (down).

Figure 9a,b illustrate the appearance of the coating with unmodified particles by an optical microscope. It can be seen that particles are mostly in the form of aggregates and that the distribution of particles is extensive, from very small to massive aggregates. By grafting PLA on the surface of chitin–lignin complexes, this aggregation of particles is avoided, and better dispersion was achieved. This is important for the activity of the coating, because in the case where aggregates are present, the number of domains dispending active functionalities is lower than in the case of single particles present.

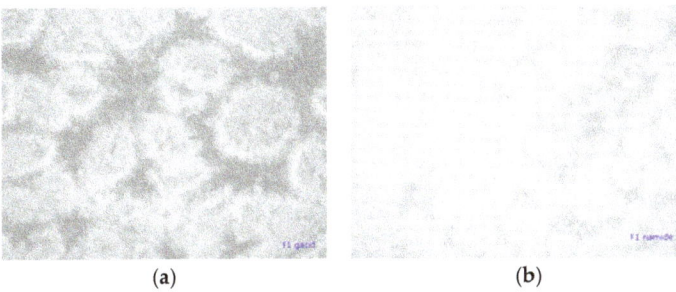

(a) (b)

Figure 9. Micrographs (magnification 100×) of extruded film coated with PLLA-based coating with unmodified chitin–lignin complex loaded with, (a) glycyrrhetic acid (b) niacinamide.

Figure 10 illustrates the FTIR spectra of uncoated and coated PLA extruded films, with unmodified complexes loaded coating. The PLLA-based coating did not induce a change in the infrared spectrum of the plasticized PLA/PBAT extruded film. The same result was obtained when modified complexes were added to the coating. On the other hand, the PLLA spectrum is similar to the one of the film and it is the main component of the coating.

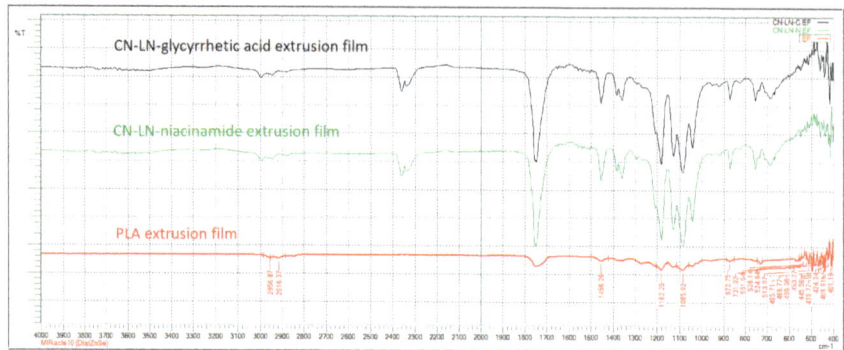

Figure 10. FTIR spectra of PLA extruded film coated unmodified chitin–lignin complex nanoparticles.

Comparing DSC thermograms, related to the first heating of uncoated and coated PLA films (Figure 11a,b), with PLLA-based coating containing unmodified chitin–lignin complexes, significant differences in Tc and Tm can be detected. While the crystallization peak of neat plasticized PLA/PBAT film is regularly shaped, the coated film has a broad crystallization peak, with crystallization occurring at a lower temperature, which indicates the changes in the crystallization behavior of PLA where two peaks, crystallization of the oligomer and crystallization of the film, are merged. On the other hand, the melting point is also lower for the coated films. Neat PLA film has two melting points nicely separated, which is typical for PLA derived from α and α' crystal forms. Within coated PLA, the intensity of the ordered α form decreased, which might indicate that the added coating, rapidly forming a film by solvent evaporation, is contributing with a more disordered α' crystal form. However, all the observed effects are the result of cumulative phase transitions within different PLA-based layers. In fact, we cannot exclude some surficial dissolution or swelling due to the solvent on the surficial layer of the plasticized PLA/PBAT film.

Similar results were obtained when coatings with modified chitin–lignin complexes were applied, as it can be seen from the results summarized in Table 2. The Tg is slightly lower compared to other PLA films, which might be the effect of additional PLA layer grafted on chitin–lignin complexes, adding flexibility and disorder to the coating layer. Moreover, the deposition of the thin coating film containing solvent can slightly affect the film crystallinity.

Table 2. Thermal property results of the extruded film (EF) coated by PLLA-based coating.

Sample	Tg (°C)	Tc (°C)	Tm (°C)
PLA extruded film	46.7	92.14	142.8
CN-LN-niacinamide PLA extruded film	45.2	83.58	137.5
CN-LN-glycyrrhetic acid PLA extruded film	47.6	89.88	138.72
gCN-LN-niacinamide PLA extruded film	44.3	84.7	136.2
gCN-LN-glycyrrhetic acid PLA extruded film	43.1	85.2	136.8

The mechanical properties of uncoated and coated PLA films are summarized in Table 3. Values of tensile strength and elongation at break for uncoated and coated PLA films are almost the same, which indicates that the coating did not affect the mechanical properties of neat PLA. The application of the coating on a preformed extruded film is a good methodology to keep the nano-structured functional molecules on the film surface without modifying its properties.

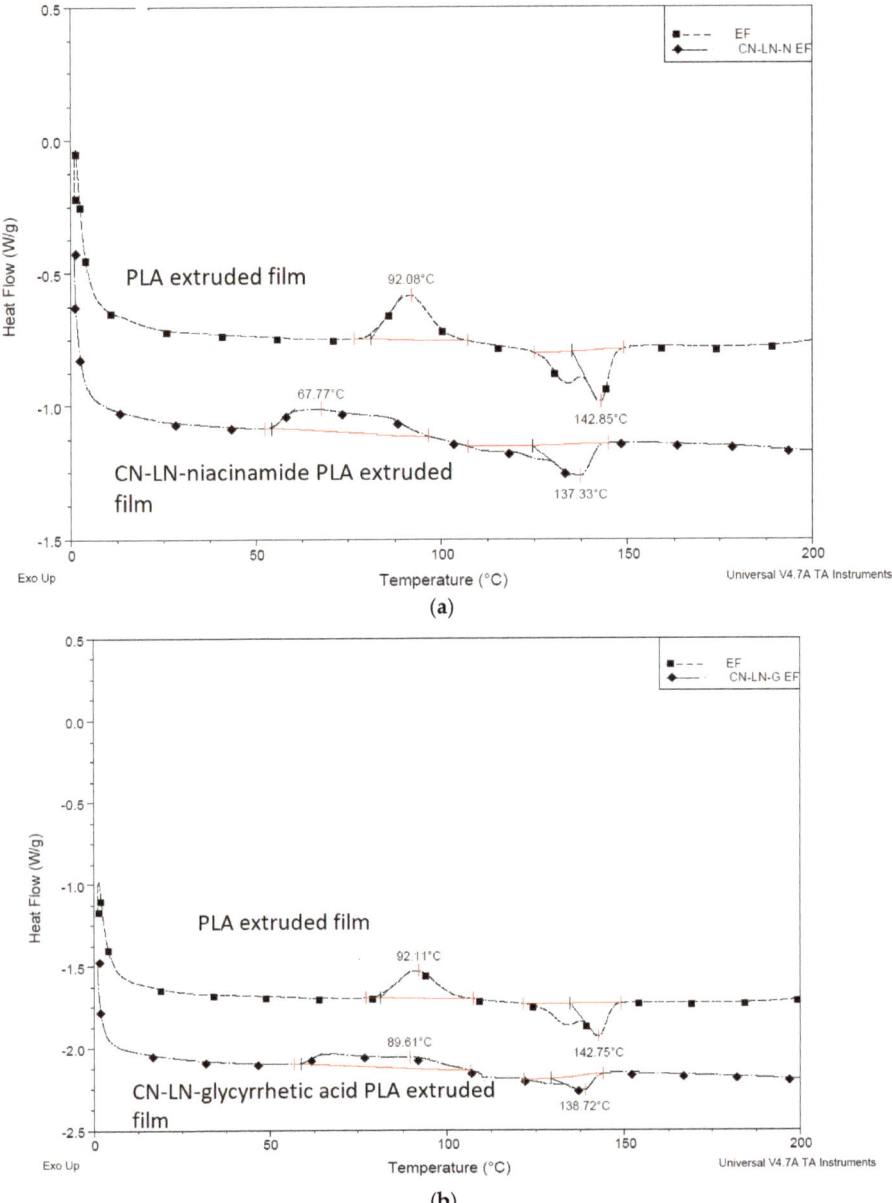

Figure 11. DSC thermogram extruded film coated with PLLA-based coating with (**a**) unmodified chitin–lignin niacinamide complexes and (**b**) unmodified chitin–lignin glycyrrhetic acid complexes.

Table 3. Mechanical results properties of the EF coated by PLLA-based coating.

Sample	Tensile strength (MPa)	Elongation at break (%)
PLA extruded film	0.3	112
CN-LN-niacinamide PLA extruded film	0.26	115.2
gCN-LN-niacinamide PLA extruded film	0.25	116.54
CN-LN-glycyrrhetic acid PLA extruded film	0.24	114.76
gCN-LN-glycyrrhetic acid PLA extruded film	0.24	112.48

4. Conclusions

The modification of chitin–lignin complexes with PLA was successfully carried out by employing the "grafting from" method, grafting PLA chains onto chitin–lignin complexes containing glycyrrhetic acid or niacinamide. These grafted systems were dispersed in an extrusion grade PLA or in an oligomeric PLA (PLLA). The polylactide grafted chains enable a better dispersion and compatibility with PLA. The physical modification of PLA, which was confirmed by FTIR, was undertaken by the direct incorporation of unmodified and modified chitin–lignin complexes into solution casting PLA-based films, and by coating based on low-molecular weight PLLA containing mentioned active compounds. When particles were directly incorporated into the films, the effect on mechanical and thermal properties was evident. PLA-based films with modified particles incorporated had lower Tg and higher Tm values, while they had no cold crystallization peak. In fact, these films showed an increased crystallinity, thanks to the nucleation action of the nanostructured additives. Due to the better dispersion, films with modified particles had better elasticity compared to neat and film containing unmodified particles. When coating was applied on the surface of the extruded film based on PLA, the effect on mechanical and thermal properties was lower, and the uncoated and coated film had similar properties, no matter which active compound they contained. Depending on the desired application, value-to-cost ratio, price of the product, and volume of production, it is possible to exploit the thermo-mechanical or functional (anti-microbial and anti-inflammatory) properties of chitin–lignin complexes, incorporating or depositing them onbiodegradable films. These preliminary studies yielded these findingsand opened several routes for PLA-based films intheir application on contact with skin.

Author Contributions: Conceptualization, B.P. and A.M.; methodology, A.M. and I.R.; investigation, I.R and A.M.; resources, B.P.; data curation, A.M. and I.R.; writing—original draft preparation, B.P. and A.M.; writing—review and editing, M.-B.C.; visualization, M.-B.C.; supervision, M.-B.C. and B.P; project administration, B.P.; funding acquisition, B.P. All authors have read and agreed to the published version of the manuscript.

Funding: This research was funded by European Union's Horizon 2020 research and innovation program under grant agreement No. 745839, project POLYBIOSKIN.

Acknowledgments: Stefano Fiori of Condensia Quimica, Spain, is thanked for providing oligomeric lactic acid samples. Andrea Lazzeri is thanked for helpful discussion. This project has received funding from the Bio Based Industries Joint Undertaking under the European Union's Horizon 2020 research and innovation program under grant agreement No. 745839, project POLYBIOSKIN.

Conflicts of Interest: The authors declare no conflict of interest.

References

1. Klein, F.; Klein, A.E.; Menrad, K.; Möhring, W.; Blesin, J.M. Influencing factors for the purchase intention of consumers choosing bioplastic products in Germany. *Sustain. Prod. Consum.* **2019**, *19*, 33–43. [CrossRef]
2. Park, S.B.; Park, K.S.; Joung, E.K.; Han, D.K. Biopolymer-based functional composites for medical applications. *Prog. Polym. Sci.* **2017**, *68*, 77–105. [CrossRef]
3. Adlhart, C.; Verran, J.; Azevedo, N.F.; Olmez, H.; Keinänen-Toivola, M.M.; Gouveia, I.; Melo, L.F.; Crijns, F. Surface modifications for antimicrobial effects in the healthcare setting: A critical overview. *J. Hosp. Infect.* **2018**, *99*, 239–249. [CrossRef]

4. Poverenov, E.; Klein, M. Formation of contact active antimicrobial surfaces by covalent grafting of quaternary ammonium compounds. *Colloids. Surf. B Biointerfaces* **2018**, *169*, 195–205.
5. Bazaka, K.; Jacob, M.V.; Chrzanowski, W.; Ostrikov, K. Anti-bacterial surfaces: Natural agents, mechanisms of action, and plasma surface modification. *RSC Adv.* **2015**, *5*, 48739–48759. [CrossRef]
6. Zia, K.M.; Noreen, A.; Zuber, M.; Tabasum, S.; Mujahid, M. Recent developments and future prospects on bio-based polyesters derived from renewable resources: A review. *Inter. J. Biol. Macromol.* **2016**, *82*, 1028–1040. [CrossRef] [PubMed]
7. Koller, M. Chemical and Biochemical Engineering Approaches in Manufacturing Polyhydroxyalkanoate (PHA) Biopolyesters of Tailored Structure with Focus on the Diversity of Building Blocks. *Chem. Biochem. Eng. Q.* **2018**, *32*, 413–438. [CrossRef]
8. Ruiz-Ruiz, F.; Mancera-Andrade, E.I.; Parra-Saldivar, R.; Keshavarz, T.; Iqbal, H.M.N. Drug Delivery and Cosmeceutical Applications of Poly-Lactic Acid Based Novel Constructs—A Review. *Curr. Drug Metab.* **2017**, *18*, 914. [CrossRef]
9. Jain, A.; Reddy Kunduru, K.; Basu, A.; Mizrahi, B.; Domb, A.J.; Khan, W. Injectable formulations of poly(lactic acid) and its copolymers in clinical use. *Adv. Drug Deliv. Rev.* **2016**, *107*, 213–227. [CrossRef]
10. Mills, C.A.; Navarro, M.; Engel, E.; Martinez, E.; Ginebra, M.P.; Planell, J.; Errachid, A.; Samitier, J. Transparent micro- and nanopatterned poly(lactic acid) for biomedical applications. *J. Biomed. Mater. Res.* **2006**, *76*, 781–787. [CrossRef]
11. Murariu, M.; Dubois, P. PLA composites: From production to properties. *Adv. Drug Deliv. Rev.* **2016**, *107*, 17–46. [CrossRef] [PubMed]
12. Coltelli, M.B.; Gigante, V.; Panariello, L.; Morganti, P.; Cinelli, P.; Danti, S.; Lazzeri, A. Chitin nanofibrils in renewable materials for packaging and personal care applications. *Adv. Mater. Lett.* **2018**, *10*, 425–430.
13. Coltelli, M.; Cinelli, P.; Gigante, V.; Aliotta, L.; Morganti, P.; Panariello, L.; Lazzeri, A. Chitin Nanofibrils in Poly(Lactic Acid) (PLA) Nanocomposites: Dispersion and Thermo-Mechanical Properties. *Int. J. Mol. Sci.* **2019**, *20*, 504. [CrossRef] [PubMed]
14. Gigante, V.; Coltelli, M.; Vannozzi, A.; Panariello, L.; Fusco, A.; Trombi, L.; Donnarumma, G.; Danti, S.; Lazzeri, A. Flat Die Extruded Biocompatible Poly (Lactic Acid)/Poly(Butylene Succinate) (PBS) Based Films. *Polymers* **2019**, *11*, 1857. [CrossRef]
15. Coltelli, M.B.; Aliotta, L.; Vannozzi, A.; Morganti, P.; Panariello, L.; Danti, S.; Neri, S.; Fernandez-Avila, C.; Fusco, A.; Donnarumma, G.; et al. Properties and skin compatibility of films based on poly(lactic acid) (PLA) bionanocomposites incorporating chitin nanofibrils (CN). *J. Funct. Biomat.* **2020**, *11*, 21. [CrossRef]
16. Braun, B.; Dorgan, J.R.; Hollingsworth, L.O. Supra-Molecular EcoBioNanocomposites Based on Polylactide and Cellulosic Nanowhiskers: Synthesis and Properties. *Biomacromolecules* **2012**, *13*, 2013–2019. [CrossRef]
17. Rasal, R. Surface and Bulk Modification of Poly(Lactic Acid). Ph.D. Thesis, Graduate School of Clemson University, Clemson, SC, USA, 2009; p. 335.
18. Sergiy, M. 8 Clarkson Ave Chapter 11 Grafting on Solid Surfaces: "Grafting to" and "Grafting from" Methods. Master's Thesis, Clarkson University, Potsdam, NY, USA, 2008.
19. Encinas, N.; Pantoja, M.; Abenojar, J.; Martínez, M.A. Control of Wettability of Polymers by Surface Roughness Modification. *J. Adhes Sci. Technol.* **2010**, *24*, 1869–1883. [CrossRef]
20. Tsubokawa, N. Surface Grafting of Polymers onto Nanoparticles in a Solvent-FreeDry-System and Applications of Polymer-grafted Nanoparticles as Novel Functional Hybrid Materials. *Polym. J.* **2007**, *39*, 983–1000. [CrossRef]
21. Wang, S.; Cui, W.; Bei, J. Bulk and surface modifications of polylactide. *Anal. Bioanal. Chem.* **2005**, *381*, 547–556. [CrossRef]
22. Edlund, U.; Källrot, M.; Albertsson, A.C. Single-step covalent functionalization of polylactide surfaces. *J. Am. Chem Soc.* **2005**, *127*, 8865–8871. [CrossRef]
23. Moraczewski, K.; Stepczynska, M.; Malinowski, R.; Rytlewski, P.; Jagodzinski, B.; Zenkiewicz, M. Stability studies of plasma modification effects of polylactide andpolycaprolactone surface layers. *Sustain. Prod. Consum.* **2019**, *19*, 33–43.
24. Cakic, S.M.; Spirkova, M.; Ristic, I.S.; Simendic, J.K.B.; Marinovic-Cincovic, M.; Poreba, R. The waterborne polyurethane dispersions based on polycarbonate diol: Effect of ionic content. *Mater. Chem. Phys.* **2013**, *138*, 277–285. [CrossRef]

25. Karacan, I.; Ben-Nissan, B.; Wang, H.A.; Juritza, A.; Swain, M.V.; Müller, W.H.; Chou, J.; Stamboulis, A.; Macha, I.J.; Taraschi, V. Mechanical testing of antimicrobial biocomposite coating on metallic medical implants as drug delivery system. *Mater. Sci. Eng. C Mater. Biol. Appl.* **2019**, *104*, 109757. [CrossRef] [PubMed]
26. Chang, Q.; Hao, Y.; Cheng, L.; Liu, Y.; Qu, A. Preparation and performance evaluation of biodegradable corn starch film using poly (lactic acid) as waterproof coating. *Surf. Eng.* **2019**, 1743–2944. [CrossRef]
27. Valerini, D.; Tammaro, L.; Villani, F.; Rizzo, A.; Caputo, I.; Paolella, G.; Vigliotta, G. Antibacterial Al-doped ZnO coatings on PLA films. *J. Mater. Sci.* **2020**, *55*, 4830–4847. [CrossRef]
28. Wei, Y.; Guo, H.; Zhou, M.; Yue, L.; Chen, Q. DBD plasma assisted atomic layer deposition alumina barrier layer on selfdegradation polylactic acid film surface. *Plasma Sci. Technol.* **2019**, *21*, 015503. [CrossRef]
29. Danti, S.; Trombi, L.; Fusco, A.; Azimi, B.; Lazzeri, A.; Morganti, P.; Coltelli, M.-B.; Donnarumma, G. Chitin Nanofibrils and Nanolignin as Functional Agents in Skin Regeneration. *Int. J. Mol. Sci.* **2019**, *20*, 2669. [CrossRef]
30. Gehring, W. Nicotinic acid/niacinamide and the skin. *J. Cosmet. Derm.* **2014**, *3*, 88–93. [CrossRef]
31. Morganti, P. Composition and Material Comprising Chitin Nanofibrils, Lignin and A Co-Polymer and Their Uses. International Patent Application WO2016042474A1, 15 September 2015.
32. Coltelli, M.B.; Gigante, V.; Vannozzi, A.; Aliotta, L.; Danti, S.; Neri, S.; Gagliardini, A.; Morganti, P.; Panariello, L.; Lazzeri, A. Poly(Lactic Acid) (Pla) Based Nano-Structured Functional Films For Personal Care Applications. In Proceedings of the AUTEX2019–19th World Textile Conference on Textiles at the Crossroads, Ghent, Belgium, 11–15 June 2019; 1A2_0372. Available online: https://ojs.ugent.be/autex/article/view/11509 (accessed on 22 February 2020).
33. Yang, R.; Ding, L.; Chen, W.; Zhang, X.; Li, J. Molecular-Weight Dependence of Nucleation Effect of a Liquid Crystalline Polyester β Nucleating Agent for Isotactic Polypropylene. *Ind. Eng. Chem. Res.* **2018**, *57*, 6734–6740. [CrossRef]
34. Ristić, I.S.; Tanasić, L.; Nikolić, L.B.; Cakic, S.M.; Ilic, O.; Radicevic, R.; Budinski-Simendic, J. The Properties of Poly(L-Lactide) Prepared by Different Synthesis Procedure. *J. Polym. Environ.* **2011**, *19*, 419. [CrossRef]
35. Ristić, I.S.; Marinović-Cincović, M.; Cakić, S.M.; Tanasic, L.; Budniski-Simendic, J. Synthesis and properties of novel star-shaped polyesters based on L-lactide and castor oil. *Polym. Bull.* **2013**, *70*, 1723–1738. [CrossRef]
36. Shi, D.; Lai, X.; Jiang, Y. Synthesis of Inorganic Silica Grafted Three-arm PLLA and Their Behaviors for PLA Matrix. *Chin. J. Polym. Sci.* **2019**, *37*, 216–226. [CrossRef]
37. Stuani Pereira, F.; da Silva Agostini, D.L.; Job, A.E.; Perez Gonzalez, E.R. Thermal studies of chitin–chitosan derivatives. *Therm. Anal. Calorim.* **2013**, *114*, 321–327. [CrossRef]
38. Lisperguer, J.; Perez, P.; Urizar, S. Structure And Thermal Properties Of Lignins: Characterization By Infrared Spectroscopy And Differential Scanning Calorimetry. *J. Chil. Chem. Soc.* **2009**, *54*, 460–463. [CrossRef]
39. Legras, R.; Mercier, J.P.; Nield, E. Polymer crystallization by chemical nucleation. *Nature* **1983**, *34*, 432–434. [CrossRef]
40. Kawai, T.; Rahman, N.; Matsuba, G.; Nishida, K.; Kanaya, T.; Nakano, M.; Okamoto, H.; Kawada, J.; Usuki, A.; Honma, N.; et al. Crystallization and melting behavior of poly(L-lactic acid). *Macromolecules* **2007**, *40*, 9463–9469. [CrossRef]
41. Anderson, K.S.; Hillmyer, M.A. Melt preparation and nucleation efficiency of polylactide stereocomplex crystallites. *Polymer* **2006**, *47*, 2030–2035. [CrossRef]
42. Aliotta, L.; Cinelli, P.; Coltelli, M.B.; Righetti, M.C.; Gazzano, M.; Lazzeri, A. Effect of nucleating agents on crystallinity and properties of poly (lactic acid) (PLA). *Eur. Polym. J.* **2017**, *93*, 822–832. [CrossRef]
43. Zhou, J.; Jiang, Z.; Wang, Z.; Zhang, J.; Li, J.; Li, Y.; Zhang, J.; Chen, P.; Gu, Q. Synthesis and characterization of triblock copolymer PLA-b-PBT-b-PLA and its effect on the crystallization of PLA. *RSC Adv.* **2013**, *3*, 18464–18473. [CrossRef]
44. Praveen, P.; Rao, V. Synthesis and Thermal Studies of Chitin/AgCl Nanocomposite. *Procedia Mater. Sci.* **2014**, *5*, 1155–1159. [CrossRef]

 © 2020 by the authors. Licensee MDPI, Basel, Switzerland. This article is an open access article distributed under the terms and conditions of the Creative Commons Attribution (CC BY) license (http://creativecommons.org/licenses/by/4.0/).

Article

Determination and Quantification of the Distribution of CN-NL Nanoparticles Encapsulating Glycyrrhetic Acid on Novel Textile Surfaces with Hyperspectral Imaging

Kudirat A. Obisesan, Simona Neri, Elodie Bugnicourt, Inmaculada Campos and Laura Rodriguez-Turienzo *

IRIS Technology Solutions S.L., Parc Mediterrani de la Technologia, Avda.Carl Friedrich Gauss No. 11, Castelldefels, 08860 Barcelona, Spain; kabidemi@iris.cat (K.A.O.); sneri@iris.cat (S.N.); ebugnicourt@iris.cat (E.B.); icampos@iris.cat (I.C.)
* Correspondence: lrodriguez@iris.cat

Received: 31 January 2020; Accepted: 22 April 2020; Published: 20 May 2020

Abstract: Chitin Lignin nanoparticles (CN-NL), standalone and encapsulating glycyrrhetic acid (GA), were applied on novel substrates for textiles to obtain antibacterial, antioxidant properties. Their homogeneous application is an important parameter that can strongly influence the final performance of the investigated textiles for its cosmetic and medical use. In this paper, hyperspectral imaging techniques combined with chemometric tools were investigated to study the distribution and quantification of CN-NL/GA on chitosan and CN-NL on pullulan substrates. To do so, samples of chitosan and pullulan impregnated with CN-NL/GA and CN-NL were analysed through Short Wave Infrared (SWIR) and Visible-Near Infrared (VisNIR) hyperspectral cameras. Two different chemometric tools for qualitative and quantitative analysis have been applied, principal component analysis (PCA) and partial least square regression (PLSR) models. Promising results were obtained in the VisNIR range, which made it possible for us to visualize the CN-NL/GA compound on chitosan and CN-NL on pullulan substrates. Additionally, the PLSR model results had determination coefficient (R_C^2) for calibration and cross-validation (R_{CV}^2) values of 0.983 and 0.857, respectively. Minimum values of root-mean-square error for calibration (RMSEC) and cross-validation (RMSECV) of CN-NL/GA were 0.333 and 0.993 g, respectively. The results demonstrate that hyperspectral imaging combined with chemometrics offers a powerful tool for studying the distribution on chitosan and pullulan substrates and to quantify the content of CN-NL/GA compounds on chitosan substrates.

Keywords: hyperspectral imaging; principal component analysis; spectroscopy; chitosan; pullulan; partial least squares regression; nir; actives substances; cn-nl/ga

1. Introduction

Chitin is a long-chain polymer of N-Acetylglucosamine [1]. It is a primary component of cell walls in fungi and the exoskeletons of arthropods, and it is the most common biomass waste compound after cellulose. Chitin itself presents interesting characteristics, such as biodegradability, predictable degradation rate, structural integrity and biocompatibility, which makes it an ideal candidate to produce polymeric tissues.

Chitin can assemble in the form of nanofibrils (CNs) due to a hierarchical structure, and it represents an interesting new opportunity for the pharmaceutical, cosmetic and textile sectors, thanks to their ability to complex several active ingredients by ionic charges and facilitate their delivery across the skin.

Lignin is a class of complex organic polymers rich in phenolic functional groups obtained from lignocellulosic biomass [2]. It has been reported in the literature that it can be produced as nanostructure and further complexed with CN nanofibrils. Electropositive chitin nanofibrils (CN) can be combined with electronegative nanolignin (NL), leading to microcapsule-like systems suitable for entrapping both hydrophilic and lipophilic molecules. The potential of CN-NL is that their structure can be controlled at the nanoscale, resulting in several interesting properties, such as antibacterial, anti-inflammatory, cicatrizing, and anti-aging effectiveness. Furthermore, CN-NL has also been studied as a carrier for glycyrrhetic acid (GA), a biomolecule with anti-inflammatory activity obtained from liquorice plants (Glycyrrhiza), as a model of bioactive molecule [3].

On the other hand, chitosan has been widely studied as an antimicrobial agent for preventing and treating bacterial and fungal infections. It is obtained commercially from shrimp and crab shell chitin by alkaline deacetylation 2-4 (NaOH, 40–50%). In addition, chitosan has shown cell proliferative activity essential for efficient healing. The blending of chitosan with other biopolymers along with surface modification has shown promising results in wound healing [4]. Likewise, pullulan biomaterials also enable structures for tissue regeneration, ultimately improving quality of life when used in medical and cosmetic applications.

The growing demand for anti-aging, moisturizing, skin-whitening and protective products from sun damage coupled with population ageing and the demand for more functional ingredients, drives the cosmetic market and impels us to study chitosan and pullulan-based substrates impregnated with active nanostructures as potential candidates for skin contact applications.

Over the last few decades, near-infrared spectroscopy (NIR) has arisen as an alternative method of analysis for rapidly determining the composition of a wide variety of materials belonging from many different sectors such as agriculture, food, bioactive, pharmaceuticals, petrochemicals, textiles, cosmetics, medical applications and polymers [5]. This technology can be used for the monitoring of surface composition for many different applications, such as, the analysis of fibres and their properties and the on-line monitoring of different textile coatings [6].

In the cosmetic and biomedical sectors, where the complexity of the various physical forms (e.g., creamy, milky, oily, alcoholic, aqueous, or solid) of the excipients forming emulsions is of variable consistency, the direct analysis of the sample remains challenging. Furthermore, the increasing trend of using natural bio-based components, often expensive, push the companies to look for analytical methods able to minimize the time of analysis and to control the exact amount of components added as well as to allow characterizing the natural variability of these compounds. These types of analytical methods will enable avoid extra costs as well as being able to determine the ingredients' effective concentration. In particular, in one study reported by Blanco et al., the use of NIR spectroscopy combined with multivariate calibration demonstrated the capability of the technique to determine the components of a cosmetic mixture and its hydroxyl value [7].

In the context of the monitoring of textured surfaces composition, an emerging technique that is attracting attention is NIR hyperspectral imaging. It is a widespread, rapid, and non-destructive analytical technology, which combines conventional imaging and infrared spectroscopy to collect spatial and spectra information from a particular object, resulting in a three-way data matrix. In more detail, a traditional optical spectrum instrument commonly presents a simplex spectrum I(l), where the HSI technique offers the two-dimensional distribution of the intensity at each pixel of the image I(x, y), which is described as I(x, y, l) or I(l, x, y), which can be considered as either a detached spatial image I(x, y) at every single wavelength (l), or a spectrum I(l) at each pixel (x, y). Therefore, the main difference between HSI imaging and other imaging techniques is that for each pixel, a whole spectrum is obtained. The obtained spectrum has the function of unveiling the hidden compositional information of that particular pixel. This characteristic is very beneficial for achieving the visualization of the distribution and the composition of different chemical components in a sample. Relying on this capability, hyperspectral imaging has witnessed tremendous growth and has been widely applied in the pharmaceutical [8], food [9], and agriculture [10] sectors; however, only a few studies have been reported for the textile sector.

One recent study applied to cosmetics reports the application of multispectral imaging in the quantification of foundation applied to the skin, evaluated by using the standard deviation of the reflectance, demonstrating its potential use for both industrial and in vivo tests [11]. Due to the large quantity of information generated by these techniques, it is necessary to combine the data structure with a powerful statistical and mathematical treatment to extract the relevant information. Chemometric tools enable all spectra in a data cube to be simultaneously analysed so that the desired information can be extracted. This approach has been applied in the literature to study the heterogeneity and quantify different components of biomass derivatives [12,13].

Moreover, thanks to the ability to provide real-time material composition, NIR and HSI are increasingly being used as an emerging solution for the online separation of waste textiles depending on the nature of the used natural or synthetic fibres, among other parameters [14].

Therefore, in this work, CN-NL and CN-NL embedding GA have been applied on the surface of pullulan and chitosan using two innovative technologies, dry powder impregnation and electrospinning, that have been tested in the framework of the PolyBioSkin project.

To the best of our knowledge, this is the first study applying hsi (in both SWIR and VisNIR range) as a monitoring system for the analysis of the distribution and quantification of CN-NL as a standalone agent or encapsulating GA on pullulan and chitosan textile substrates respectively. This study confirms the potential benefit of using a non-destructive technology that can be easily adapted as inline quality control into different production lines avoiding offline assessment, as well as time and product waste.

2. Materials and Methods

2.1. Samples

Textiles of chitosan and pullulan with a porous structure were used as substrates in this study. The final set of samples were provided by different partners involved in the PolyBioSkin Project and were obtained as follows. Nanochitin, nanolignin, and glycyrrhetic acid were supplied by Mavi Sud, Aprilia (LT), (Rome, Italy). Complexes of CN-NL entrapped with GA were prepared in powder using a gelation method and a BuchiMini B-190 spray dryer technology (Flawil, Switzerland).

In the case of a pullulan-based substrate produced by electrospinning, 8% of CN-NL complexes were co-electrospun. In the case of chitosan used as a substrate, CN-NL/GA was applied through dry powder impregnation and thanks to the use of binders. The binders applied were PEG 8000 from Sigma-Aldrich, while waxy and dewaxed bleached Shellac was purchased from A.F. Suter & Co., Ltd. Complexes of CN-NL/GA were thus prepared following previous works [15,16]. The proportion of the complexes were at least 38 g nanochitin + 6.5 g lignin + 4.4 g glycyrrhetic acid with a nanoparticle size of approximately 59 nm. The complex of CN-NL impregnation on pullulan was carried out using electrospinning technology from water by BIOINICIA (Carrer de l'Algepser, 65, Nave 3, 46,980 Paterna, Valencia). Dry powder impregnation technologies of CN-NL/GA were applied on chitosan textile by FIBROLINE SA (Parc Swen Bât 4C, 1 rue des vergers 69760 LIMONEST–FRANCE) and TEXOL (Via Corradino D'Ascanio 3 65020 Alanno Scalo Pescara Italy).

In this paper, six sets of chitosan and pullulan substrates, with an approximate size of 10×10 cm^2 and 5×5 cm^2 for chitosan and pullulan, respectively, were evaluated. Table 1 lists each of these samples in detail, including the grams of the active substances applied. In the case of the chitosan substrate, five sets of samples were evaluated, which were composed mainly of non-woven chitosan substrates with different quantities of CN-NL/GA as active compounds and attached with different binders together with silica. Binders were applied for better cross-linkers on the chitosan, while silica was added for better fluidity of the complex. All the set samples (containing a total of 11 samples) of the chitosan substrate had the same proportion of CN-NL/GA complexes, but different quantities of the complexes were used to impregnate the chitosan textile. Finally, set 6 consists of two samples of standalone pullulan substrate and pullulan impregnated with CN-NL.

Table 1. List of chitosan and pullulan substrates selected for hyperspectral imaging evaluation. Sets 1 to 5 are chitosan substrates with different quantities of CN-NL/GA complex, and set 6 corresponds to pullulan substrate and pullulan with chitin nanofibrils.

Substrates	Set of Samples	Number of Samples	Substrate Base + Active Compounds + Binders
Chitosan	Set 1	Sample 1	Chitosan + 2.68 g CN-NL/GA + 2% Silica
	Set 2	Sample 2	Chitosan + 6 g CN-NL/GA + 2% Silica
	Set 3	Sample 3	Chitosan + 5 g (50% CN-NL/GA + 50% waxy bleached Shellac) + 2% Silica
		Samples 4	Chitosan + 5.4 g (50% CN-NL/GA + 50% waxy bleached Shellac) + 2% Silica
		Sample 5	Chitosan + 7 g (50% CN-NL/GA + 50% waxy bleached Shellac) + 2% Silica
	Set 4	Sample 6	Chitosan + 2.016 g (50% CN-NL/GA + 50% dewaxed bleached Shellac) + 2% silica
		Sample 7	Chitosan + 5.2 g (50% CN-NL/GA +50% dewaxed bleached Shellac) + 2% silica
		Sample 8	Chitosan + 5.6 g (50% CN-NL/GA + 50% dewaxed bleached Shellac) + 2% silica
	Set 5	Sample 9	Chitosan + 6.5 g (50% CN-NL/GA + 50% PEG) + 2% silica
		Sample 10	Chitosan + 7.6 g (50% CN-NL/GA + 50% PEG) + 2% silica
		Sample 11	Chitosan + 8.2 g (50% CN-NL/GA + 50% PEG) + 2% silica
Pullulan	set 6	Sample 12	Pullulan + 8% CN-NL
		Sample 13	Pullulan only

2.2. NIR Hyperspectral Imaging System

The hyperspectral imaging system, HYPERA®, developed by IRIS Technology Solutions S.L. was used for image collection, equipped with two imaging systems one in the Short Wave Infrared (SWIR) and the other in the Visible-Near Infrared (VisNIR) range, a halogen illumination system and a computer with an image acquisition platform developed by IRIS. To acquire NIR imaging, a line-scan push broom system equipped with a moving belt was used. The wavelength of the SWIR system ranged from 850 to 1600 nm, with a resolution of 4.7 nm. For the VisNIR system, the wavelength ranged from 400 to 1000 nm, with a resolution of 2.2 nm.

Both the chitosan and pullulan substrates were placed directly on the conveyor belt for image acquisition. Prior to the image acquisition, white and dark reference images were captured for image calibration. Reflectance standards are essential for image calibration, to correct pixel-to-pixel variations arising due to inconsistencies in capture and illumination of samples [17].

Each image had approximately 102,400 and 409,600 pixels for SWIR (Bolton, MA, USA) and VisNIR cameras (Bolton, MA, USA), respectively, of which about 63% were chitosan or pullulan (substrate) pixels and 37% represented background pixels. The background information, which corresponded to the pixels representing the moving belt, was eliminated to provide the best possible difference between the sample and the background.

2.3. Principal Component Analysis

Multivariate data analysis is a useful technique to extract relevant and meaningful information from the hyperspectral imaging data [18–20]. Chemometrics was an appealing tool for this study because it allowed us to reduce the dimensionality of the data while retaining essential spectral information and classify areas of interest. In this research paper, an exploratory analysis was applied to study the distribution of active substances on chitosan and pullulan surfaces.

The principal component analysis method is a qualitative method used to reduce the original set of variables into new significant variables called principal components (PC). The principal component analysis is based on bilinear decomposition, which can be mathematically described by Equation (1):

$$X = U V^T + E \quad (1)$$

where X is the unfolded matrix (matrix containing the spectra) decomposed according to Equation (1), the loadings matrix V^T and the scores matrix U. The loadings matrix V^T identifies the primary sources of data variance using their chemical composition (composition loadings), which may eventually be related to the CN-NL and CN-NL/GA complexes. The scores matrix, U, provides sample scores for these data variance patterns. The scores in each PC were re-folded to have the original spatial organization, and one image was built for each significant PC. The first principal component carries the maximum variability of the data, while the second principal component explains all the variability not included in the first component and successively. These components are defined in such a way that they are orthogonal among them. The importance of the original variables in the definition of a principal component is represented by its loadings and the projections of the objects onto the principal components are called the scores of the objects. The performance of the PCA model was evaluated mainly by variance explained, where an optimal model should provide a high variance explained, low numbers of scores, and loading to define the distribution of the active compound.

2.4. Partial Least Square Regression

A calibration model for quantifying the CN-NL/GA complex was performed, applying the partial least squares regression method (PLSR) [21]. PLSR is a bilinear calibration method using data reduction by compressing a large number of measured collinear variables into a few orthogonal principal components, which are known as latent variables (LVs).

The purpose of the PLS regression is to build a linear model enabling prediction of the desired characteristic (Y) from a measured spectrum (X). The model is represented by:

$$Y = Xb + E \quad (2)$$

where Y is the response matrix (nine samples of the calibration set) of chitosan score, X is the set of matrix spectral data (calibration set × wavelength), b is the matrix of regression coefficients (wavelength × 1), E is the residual information matrix that is not explained by the LV. The best number of LVs is chosen according to the lowest root mean square error of cross-validation (RMSECV) and the lowest number of latent variables.

The quality of the calibration model was evaluated using the following statistical parameters: coefficient of determination for calibration and cross-validation (R_C^2 and R_{CV}^2), root mean square error of calibration and cross-validation (RMSEC and RMSECV). The value of R^2 indicates the percentage of the variance in the Y variable (grams of CN-NL/GA on chitosan substrate) that is accounted for by the X variable (spectral data). The statistical parameters were calculated as the following:

$$RMSE = \sqrt{\frac{1}{n}\sum_{i=1}^{n}(y_i - y_{m,i})^2} \quad (3)$$

$$R^2 = 1 - \frac{\sum(y_{m,i} - y_i)^2}{\sum(y_{m,i} - y_{mm})^2} \quad (4)$$

where y_i and $y_{m,i}$ are the model estimated value and the reference value for sample i, respectively, and n is the number of samples. As mentioned by Saeys et al., a calibration model with R^2 value greater than 0.91 is considered to be an excellent calibration, while an R^2 value between 0.82 and 0.90 results in good prediction. A small difference between RMSEC and RMSECV value was also important to avoid "overfitting" in the calibration model.

2.5. Software Tools

For statistical analysis, data pre-processing of the images was performed using routines written in Matlab R2018a. Multivariate data analysis was performed with PLS Toolbox 8.6.2 running on Matlab (version 9.4, R2018a) (The MathWorks, Inc., Natick, MA, USA).

3. Results

3.1. Spectroscopic Analysis

- Chitosan substrates impregnated with CN-NL/GA

A first qualitative approach to understanding the spectral homogeneity of chitosan textiles with and without impregnated CN-NL/GA nanoparticles was evaluated on pure chitosan and chitosan impregnated with 6 g of CN-NL/GA compound (sample 2 from set 2, as described in Table 1). Figure 1 shows the normalised spectra of the CN-NL/GA impregnated chitosan samples extracted from the hyperspectral imaging, which were acquired through SWIR (Figure 1a) and VisNIR (Figure 1b) cameras, respectively.

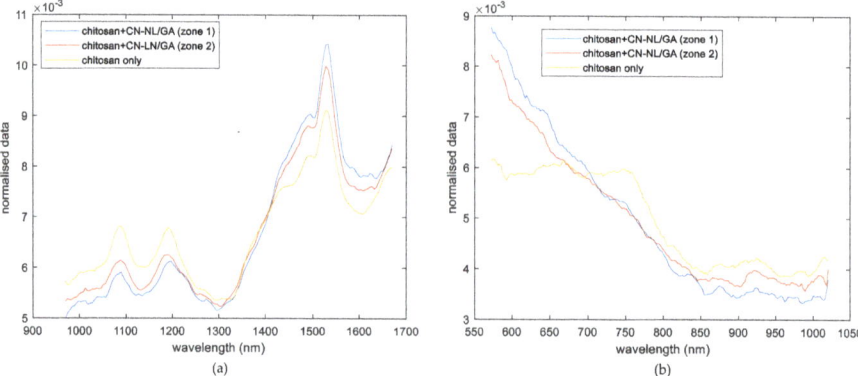

Figure 1. Comparison of hyperspectral imaging signal obtained from (**a**) SWIR range, a spectrum corresponding to pure chitosan substrate and two spectra of chitosan impregnated with CN-NL/GA in two different zones from the same sample (sample 2) and (**b**) VisNIR range, a spectrum corresponding to pure chitosan substrate and two spectra of chitosan impregnated with CN-NL/GA in two different zones from the same sample (sample 2).

With respect to the absorption of molecules in the NIR spectroscopic region resulting from the absorption of overtones and combination bands, types of vibrations are related to the fundamental vibrations seen in this region. Most of the overtone and combination bands involving higher frequency fundamental vibrations are CH, NH, and OH stretching. Taking this fact into account, the signal contribution due to the presence of silica during the impregnation of CN-NL/GA chitosan substrates could be considered minor. However, certain metal complexes have low-lying electronic transitions in the NIR, although this article will focus on vibrational transitions. As can be seen from Figure 1a, the spectra of the reference, and the sample with active compounds show a similar profile with a common peak approximately at 1500 nm. This signal can be related to the similar chemical structures of the substrate (chitosan) and the active compound (chitin-based complexes), which present similar fingerprints in this region. Nevertheless, differences can be detected in two waveband regions, which range from 1000–1300 nm and 1400–1600 nm (SWIR). Comparing these spectroscopy regions, differences from the intensity of the signal could be observed from chitosan reference and chitosan with CN-NL/GA. The chitosan reference spectrum shows an intensity signal higher than chitosan+CN-NL/GA spectra in the range of 1000–1300 nm. Thus, this variation of the intensity might be associated with the presence of CN-NL/GA on the chitosan substrate. On the other hand, two spectra of CN-NL/GA corresponding to different zones of the same samples show closer intensity; this was expected, since they both have the complex. Moreover, a slight deformation of the peak at 1488 nm could also be highlighted, confirming the behaviour reported in the literature, probably related to the first overtone of OH stretching [22,23]. Variations of spectra intensities in some specific areas indicate that this camera could be applied to study the distribution of this compound. Furthermore, evaluating the spectra of the chitosan and chitosan+CN-NL/GA complex provided by the VisNIR ranges (Figure 1b), similar trends could be observed in SWIR range. The chitosan reference spectrum shows an intensity signal higher than the chitosan+CN-NL/GA spectra in the range of 400–800 nm, which corresponds to the visible region that is widely used in spectroscopic techniques for pigment analysis [24]. On the other hand, two spectra of CN-NL/GA corresponding to different zones of the same samples show closer intensity; this was also expected, since they both have the complex.

To visualize and evaluate the homogeneity of CN-NL/GA on chitosan surfaces, a PCA exploratory study was carried out (Section 3.2.1).

- Pullulan substrates impregnated with CN-NL

The qualitative analysis of the presence of CN-NL was also performed on the pullulan and pullulan-impregnated samples (set 6, pullulan with and without CN-NL, samples 12 and 13 respectively, as described in Table 1). The spectra channels obtained from hyperspectral imaging of SWIR and VisNIR ranges are displayed in Figure 2a,b, respectively. These spectra signals corresponded to pullulan alone and two spectra of pullulan impregnated with CN-NL complex. The two impregnated spectra were identified as zone 1 and zone 2. The SWIR spectrum indicates a clear difference between pullulan substrate used as a reference and CN-NL impregnated one. Furthermore, new bands can be observed in the spectra from pullulan+CN-NL that are directly related to the impregnated pullulan, one in the range of 1000–1300 nm and the second one at 1500 nm. Since the C-H band is among the most prominent features influencing the near-infrared spectroscopy in CN-NL, the wavelengths observed at 1078 and 1197 nm could be related to it. Additionally, variation in the signal intensity could also be observed between the two impregnated zones (zone 1 and 2 from sample 12). This variation was expected, since the quantity of CN-NL might not be able to cover the surface of pullulan (sample 12) completely. Figure 2b shows the spectra resulting from the VisNIR camera. Moreover, in this case, three spectra were compared, corresponding to pullulan reference (sample 13) and two spectra of pullulan impregnated with CN-NL (zone 1 and 2, from sample 12). Spectra channels obtained from this spectroscopy range present noisy due to light scattering characteristics in VisNIR spectroscopy range. Savitzky–Golay filters have been previously applied on this signal together with normalization as pretreatments to reduce noise and extract relevant information beyond the signal. The reference pullulan substrate shows differences with the pullulan+CN-NL in the ranges of 650–750 nm and 950–1050 nm. However, the two zones of the pullulan+CN-NL show a similar trend, although slight differences could be observed between wavebands of 650 and 750 nm.

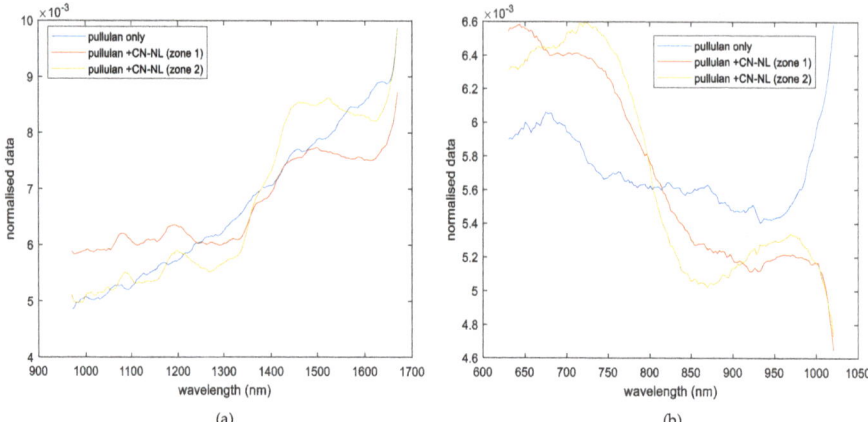

Figure 2. Comparison of hyperspectral imaging signal obtained from (**a**) SWIR range, corresponding to the pullulan substrate (sample 13) and spectra from pullulan impregnated with CN-NL in two different zones from sample 12 (set 6), and (**b**) VisNIR range, corresponding to the pullulan substrate (sample 13) and spectra of pullulan impregnated with CN-NL in two different zones from sample 12 (set 6).

3.2. Qualitative Distribution Analysis (PCA Method)

3.2.1. Evaluation of Chitosan Substrates

- Score Results

Hyperspectral imaging combined with PCA was conducted to study the distribution of CN-NL/GA as an active substance on chitosan substrates. In this study, the background and the insignificant pixels were eliminated, and pixels corresponding to chitosan samples were obtained, which was followed by PCA. Different preprocesses were applied to the spectral data prior to analysis, suppressing scattering effects and undesirable spectral variability. These included the application of SNV (Standard Normal Variate), detrend, Savitzky–Golay, and first and second derivatives. Optimised pretreatment was achieved after applying noise correction with a smoothing filter (Window 7, polynomial 0 and orders 2), followed by normalization of each channel and a mean centre. Images are colour-coded corresponding to infrared intensity scale bars. The homogeneity of CN-NL/GA substances was evaluated based on the contribution of each pixel to the PC1, where the deep blue colour refers to the conveyor, light blue to the chitosan substrate, and the red colour shows the highest intensity of the active substances.

Figure 3a shows the RGB images of samples evaluated with the hyperspectral imaging, corresponding to two different sets of samples. These samples were chitosan impregnated with 2.68 and 6 g of CN-NL/GA compound for set 1 (sample 1) and set 2 (sample 2), respectively. Samples were evaluated with both SWIR and VisNIR ranges, and the score results of the PCA models are illustrated in Figure 3b,c.

In the case of sample 1 (set 1), the first principal component scores were chosen to discriminate between the chitosan substrate and CN-NL/GA compound, since it had the major contribution of the original information. This component described approximately 78.03% of the original information for the SWIR and 72.13% for the VisNIR cameras. In the overall images, both SWIR and VisNIR (Figure 3b,c) cameras showed a clear difference between the substrate and CN-NL/GA compound. The chitosan substrate in the absence of CN-NL/GA can be identified by the blue colour with negative values. The regions with CN-NL/GA compounds are yellow or red, indicating a change in the intensity of the reference peak. A more intense colour can be observed from the VisNIR camera, confirming the results obtained previously. Indeed, CN-NL/GA absorbs in the VisNIR range due to their pigmentation.

Based on these results, hyperspectral cameras at SWIR and VisNIR spectroscopy ranges can be valuable for visualization and to study the presence of CN-NL/GA + silica distribution on chitosan substrates.

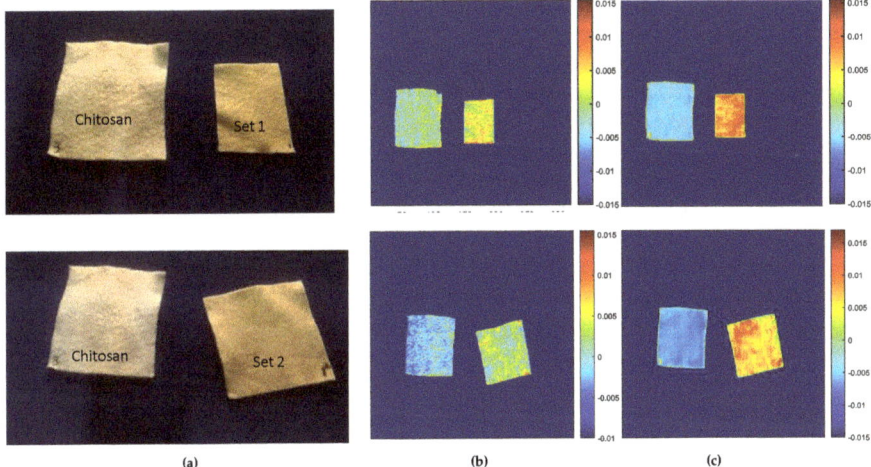

Figure 3. PCA score images for the first principal components: (**a**) Picture of the chitosan and chitosan impregnated with CN-NL/GA (set 1, top images: chitosan with 2.68 g CN-NL/GA and set 2, down images: chitosan with 6 g of CN-NL/GA); (**b**) PC1 scores image for SWIR camera; and (**c**) PC1 scores image for VisNIR camera.

Furthermore, the score results obtained from the second sample (Set 2) from both the SWIR and VisNIR cameras are reported, and they confirmed the previous results. Indeed, in both cases, the first components explained the maximum variation of the original information with scores of PC1 78.47% and 72.34% for the SWIR and VisNIR cameras, respectively. With both cameras, we could detect a clear difference between the chitosan and chitosan impregnated with CN-NL/GA compounds. Again, higher signal intensity can be viewed from the VisNIR camera, which is related to the CN-NL/GA pigment being more intense in the visible range. The elaboration of the data related to the chitosan substrate used as a reference, results in blue colour, reinforcing the accuracy of the previous results. Based on the overall PCA score images obtained for both hyperspectral imaging cameras, this technique could probably be useful for studying the homogeneity distribution of active compounds on chitosan substrates.

Another set (set 3) of samples with chitosan and chitosan impregnated with CN-NL/GA compounds attached to waxy bleached Shellac was also studied. Three quantities of the CN-NL/GA compound—5, 5.4 and 7 g for samples 3, 4, and 5, respectively—were evaluated (see Figure 4a). In this set of samples, SWIR was also studied to see if the presence of the waxy bleached Shellac as a binder enhanced the NIR spectroscopy signal, which may improve the correct visualization of the distribution of CN-LG/GA compounds. The pretreatment applied was a smoothing filter (Window 7, polynomial 0 and orders 2) together with normalised and mean centre, which allows the elimination of the noise on the spectra and the correction of the scattering effect that is typical in NIR radiation. Hypercube data was evaluated with PCA, and the scores for PC1, explaining the maximum and relevant variability from the original information, are displayed in Figure 4b,c. The explained variance for each of the PC1s for SWIR and VisNIR is 74.15% and 61.97%, respectively. However, the results attained from the score images suggested that SWIR might not be suitable for studying the distribution on the chitosan substrate. The SWIR score results (see Figure 4b) show an incoherent output assigning different colours to CN-NL/GA-impregnated sample 5 compared to sample 3 and 4 (both with CN-NL/GA). These variations cannot be associated with the quantities of the CN-NL/GA compound, bringing us to the conclusion that SWIR might not be suitable to study the homogeneity of CN-NL/GA compound due to its lack of sensitivity for these specific compounds. More promising results were achieved from the VisNIR camera (Figure 4c). The blue colour indicates the reference, the two samples containing similar amounts of CN-NL/GA (Samples 3 and 4; 5 and 5.4 g of CN-NL/GA, respectively) can be identified by the red colour spotted with green. This colour variation can be ascribed to the impregnation technique used, since these quantities of active compounds might not be able to cover the surface completely. Additionally, when the quantities of CN-NL/GA were higher (Sample 5; 7 g of CN-NL/GA), the red colour, which is directly related to the presence of CN-NL/GA, is clearly more homogeneous, indicating that the impregnation technique is more accurate due to the higher amount of actives. These results demonstrate that HSI in VisNIR range can be applied not only for the detection of the presence of the active compounds, but could also be considered as a powerful tool for the inline quality control of the impregnation step. Chitosan with a high quantity of CN-NL/GA could be observed with more intensity (see sample 5 in Figure 4c). Furthermore, the analysis showed that samples 3 and 4, with 5 and 5.4 g of the active substance, respectively, exhibit similar distributions, indicating the heterogeneity of the distribution of CN-NL/GA compounds. This result demonstrates that hyperspectral imaging in the range of VisNIR could be a suitable technique for the detection of the composition and distribution of actives (CN-NL/GA) on chitosan substrate.

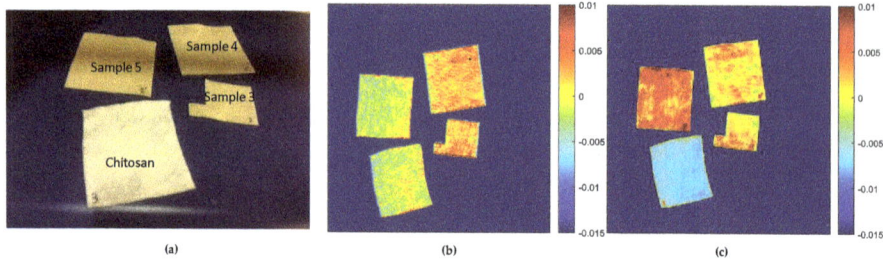

Figure 4. PCA scores images for the first principal components: (**a**) Picture of chitosan and chitosan impregnated with different quantities of CN-NL/GA compound (chitosan: 0 g, samples 3: 5 g; sample 4: 5.4 g; sample 5: 7 g; from set 3); (**b**) PC1 scores for SWIR camera; and (**c**) PC1 scores for VisNIR camera.

- Loading Results

An important aspect of NIR hyperspectral image analysis is spectral interpretation. A preliminary PCA analysis, after applying smoothing together with normalised and mean centre, showed that the first component gives interesting details related to the chemical composition of the substrate, as well as the distribution of CN-NL/GA on chitosan. Afterward, a PCA model was developed with samples, where we have the chitosan substrate alone and chitosan impregnated with 6 g of CN-NL/GA compounds. The loading results from the PCA analysis do not give information on the standalone compounds of chitosan and CN-NL/GA, but provide information about mixed signals. Taking into account that better visualization of the distribution of CN-LN/GA compounds was obtained from the VisNIR spectroscopic range, the loading results were based on those regions.

The first loadings that explain the maximum amount of information for this region are approximately 73% of the original information, as depicted in Figure 5.

A peak in the range of 400–700 nm gives a positive contribution for the distribution of chitosan and chitosan encapsulated with CN-NL/GA. This was expected, since these regions are associated with the visible region, a typical range for many pigments [25]. Furthermore, considering the approximate proportion of the CN-NL/GA complex (68 g nanochitin + 3.5 g nanolignin + 4.5 g of glycyrrhetic acid, as mentioned in Section 2.1), chitin has the majority composition and signal contribution in the VisNIR spectroscopy. The waveband at 400–700 nm due to pigment of CN-NL/GA complex might be due to chitin composition. As mentioned in [26], chitin obtained from crustacean shells could contain pigments of fungi; thanks to the presence of these pigments, the visualization of CN-NL/GA complex could be possible when applying chemometric tools. Additionally, peaks at wavebands near 850 and 950 nm could be associated with the vibration of the NH and OH (third overtone), respectively, of the chitin substances.

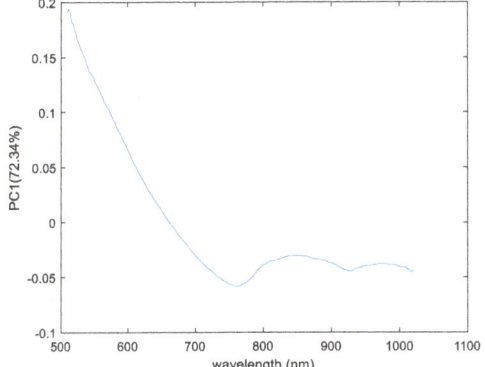

Figure 5. The PC1 Loadings plot result for chitosan and chitosan impregnated with CN-LN/GA for VisNIR camera.

3.2.2. Evaluation of Pullulan Substrates

- Score Results

The distribution of CN-NL compounds was also evaluated on pullulan textiles used for cosmetic applications. NIR spectroscopy has a high level of sensitivity to noise, as observed in Figure 2. However, the PCA model was applied to this data, since chemometric tools are characteristic in reducing and extracting the relevant information in addition to the signals. Taking into account all these factors, the PCA model was applied on the pullulan substrate and the impregnated substrates for both SWIR and VisNIR spectroscopy ranges. Different data pretreatments were tested to eliminate some sources of variation related to light scattering, or at least to reduce the noise. The data pretreatments applied included smoothing, together with normalised and mean centre to centralise all the data. For a better interpretation of the results, images obtained from SWIR and VisNIR ranges were adjusted according to the same intensity. Figure 6 illustrates pullulan samples analysed through hyperspectral imaging, together with the scores for the PC1 images obtained from SWIR and VisNIR cameras. These samples included pullulan alone (used as a reference, sample 13) and pullulan impregnated with CN-NL complex (sample 12) (see Figure 6a). The explained variance of the PCA models built for each of the ranges were 75.82% and 74.4%, respectively, demonstrating that the scores of the first components described the maximum amount of information of the original hypercubes.

The discrimination of CN-NL substances on the pullulan substrate was also evaluated based on the PCA model. The score results from the SWIR range can be observed in Figure 6b, where the deep blue colour is assigned to the conveyor belt, while the reference pullulan substrate sample (see sample 13) can be identified by its light blue colour with negative score values. Nevertheless, the pullulan substrate impregnated with active substances is yellow, derived from changes in the intensity in that region. As illustrated in Figure 6b, SWIR was not able to discriminate between the presence of CN-NL on the pullulan substrate. Furthermore, the PC1 score image results of VisNIR are also displayed in Figure 6c, and differences can be observed in the pullulan with and without the active substances. In this case, a clear change can be observed from the reference and pullulan impregnated with CN-NL substances. The reference substrate can be identified by its red colour, and the pullulan impregnated with CN-NL compounds can be identified by their yellow or light blue colours. Thus, this work proved the possibility of applying hyperspectral imaging techniques in the range of VisNIR for the detection of the composition and distribution of actives (CN-NL) on a pullulan substrate, even though the chemical composition shows a proper difference between the pullulan and the impregnated sample.

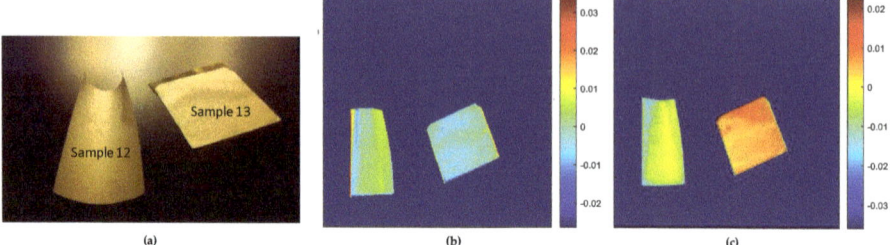

Figure 6. PCA scores images for the first principal components: (**a**) Picture of pullulan reference (sample 13) and pullulan impregnated with CN-NL complex (sample 12); (**b**) PC1 scores for SWIR camera; and (**c**) PC1 scores for VisNIR camera.

- Loadings Result

To study the chemical changes that occur on pullulan substrate with respect to pullulan impregnated with CN-NL compounds, the loadings of the first principal components were evaluated. Figure 7 shows the loading plots of the PC1 for the VisNIR camera that contribute to the changes that occur between the reference substrate (pullulan) and pullulan impregnated with the active compound, respectively. The loading variables displayed in Figure 7 show two broadening bands that contribute to these changes found both in positive and negative regions. Loading results from the display variables ranged between 550–650 nm and 950 nm. The wavebands in the range of 550–650 nm were mainly associated with the visible range. Additionally, the peak at the waveband near 850 nm might be related to the OH stretching of the third overtone from the chitin substances.

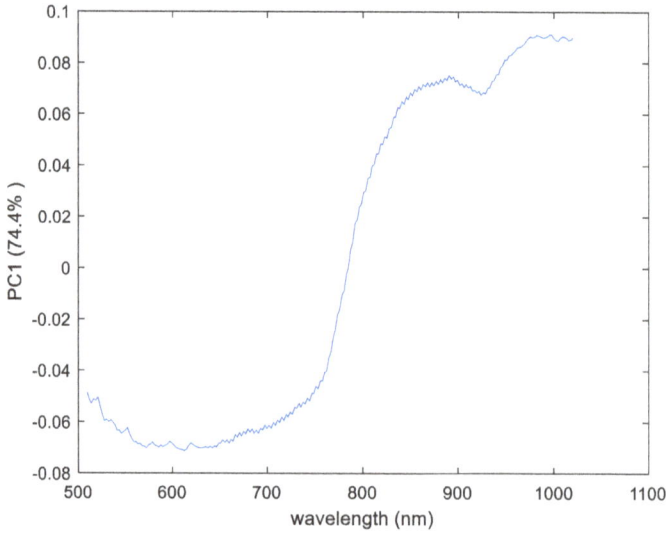

Figure 7. The PC1 for loadings plot result of pullulan and pullulan impregnated with CN-NL for the VisNIR camera.

3.3. Quantitative Distribution Analysis (PCA Method)

Taking into account the promising results obtained from the PCA model to study the distribution of the CN-NL/GA compounds on chitosan substrate with VisNIR, PLS regression models were implemented to quantify the CN-NL/GA compound on chitosan substrate with hypercubes obtained with this camera. To do so, nine samples, including chitosan alone and chitosan+CN-LN/GA, listed previously in Table 1, were used. The Chitosan+CN-NL/GA complex used included samples from set 1 (sample 1), set 2 (sample 2), set 3 (sample 3), set 4 (samples 5, 6, and 7) and set 5 (samples 8, 9 and 10), with a range of 0–8 g for the CN-NL/GA compound. The mean spectra for each of the samples are plotted in Figure 8a, where slight differences can be observed from one of the samples, which can be associated with chitosan substrate without CN-NL/GA compound. Slight variation in the intensity of CN-NL/GA compound signals can also be perceived between the range of 400–800 nm. Although these changes in absorbance can seem irrelevant, the differences among them are suitable enough for a multidimensional analysis such as PLS. With the aim of building robust models capable of predicting totally unknown samples, the original dataset of spectra was used to perform the training set as well as an internal cross-validation of the model.

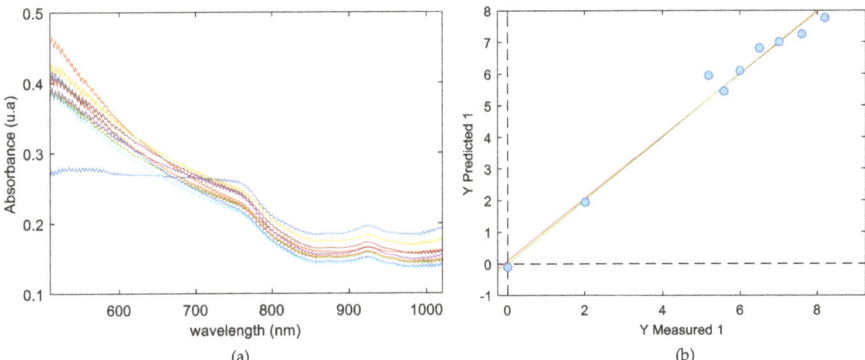

Figure 8. (a) Mean spectral for each of the samples; (b) PLS regression results for CN-NL/GA quantification, showing the actual grams of CN-NL/GA vs. the predicted grams of CN-NL/GA complex.

To build the PLS regression model, the mean NIR spectra from each of the samples were correlated with the grams of CN-NL/GA impregnated on the chitosan substrate. All spectroscopic ranges were selected to build the calibration. Table 2 shows several pretreatments tested together with coefficients of determination calibration, leave-one-out-cross-validation (R^2_{Cal} and R^2_{cv}), and the root mean square error for calibration and cross-validation (RMSEC, RMSECV) respectively for each of the pretreatments tested. In all cases, three latent variables were needed to build the regression models. Based on the results in Table 2, when no pretreatment was performed, the values of RMSEC and RMSECV were 1.335 and 2.435, respectively, indicating that large prediction errors would be caused if data was directly applied to the spectra measured. For smoothing the spectra, Savitzky–Golay (SG) was applied to the calibration model together with normalisation, SNV, baseline, and mean centre to improve the model. Among all the pretreatments, preprocessing applying smoothing together with baseline and mean centre attained the minimum values of RMSEC and RMSECV, which are 0.333 and 0.993 g of CN-NL/GA, respectively. Additionally, high values for the coefficient of determination both for calibration and cross-validation were also assigned for this pretreatment, with the values obtained being 0.983 and 0.857, respectively.

Table 2. Regression statistics results of latent variables, LVs; coefficient of determination for calibration and cross-validation (R^2_{Cal} and R^2_{cv}); the root mean square error for calibration and cross-validation (RMSEC and RMSECV), to quantify CN-NL/GA compound on chitosan substrate.

Pretreatments	LVs	R^2_{Cal}	R^2_{CV}	RMSEC	RMSECV
non pretreatment	3	0.721	0.278	1.335	2.435
smoothing + normalize + mean centre	3	0.973	0.743	0.418	1.485
smoothing + SNV + mean centre	3	0.977	0.624	0.384	4.820
smoothing + baseline + mean centre	3	0.983	0.857	0.333	0.993

4. Conclusions

This study shows the possibility of applying hyperspectral imaging techniques in both the SWIR and VisNIR ranges, coupled with chemometric tools, to study the distribution of CN-NL/GA and CN-NL on chitosan and pullulan substrates, respectively, and the quantification of CN-LN/GA on chitosan substrate. The overall results obtained from the PC score images on the chitosan textiles indicate that the VisNIR spectroscopy range is a useful support for the correct visualization of the distribution between chitosan with CN-NL/GA complex. Thus, the quantification of CN-NL/GA on chitosan using the PLS model developed for this purpose, shows a satisfactory result with an accuracy of 98.3%, indicating that the model is suited to the quantification of CN-NL/GA compounds on chitosan. In addition, the pullulan substrate also showed promising results for visualizing the presence of the CN-NL complex; however, further studies should be performed on this type of textile. The results show the feasibility of developing a simple, rapid, non-destructive, and reliable method for determining the homogeneity distribution of CN-NL/GA nanoparticles on chitosan and CN-NL on pullulan substrates after multivariate analysis. This approach can be easily applied for the in-line monitoring of the quality of the textiles used for cosmetic and biomedical applications, as well as for quality control and optimization of the impregnation process, in order to reduce time and resource waste.

5. Future Perspectives

Future work could involve the application of analytical reference techniques to confirm the differences in the quantity of CN-NL/GA and CN-NL complexes on different surfaces of chitosan and pullulan substrates. This chemical information would make it possible to quantify the differences in the application results observed from the HSI technique. This information could be useful to demonstrate that HSI could be a reliable homogeneity monitoring tool for the textile industry.

Furthermore, concerning the pullulan substrate, more samples of pullulan impregnated with CN-NL complex should be evaluated with the HSI techniques. These could include samples with different quantities of CN-NL impregnated on pullulan surfaces to ensure that the VisNIR range is suitable to study the presence of chitin complex and its distribution on the pullulan surface.

Author Contributions: Conceptualization, K.A.O., L.R.-T.; methodology, K.A.O., L.R.-T.; formal analysis, K.A.O.; investigation, K.A.O.; writing—original draft preparation, K.A.O., S.N.; I.C., writing—review and editing, K.A.O., S.N., I.C., E.B.; L.R.-T.; supervision, L.R.-T.; project administration, S.N. All authors have read and agreed to the published version of the manuscript.

Funding: This project has received funding from the Bio Based Industries Joint Undertaking under the European Union's Horizon 2020 research and innovation programme under grant agreement No 745839.

Acknowledgments: A first acknowledgment goes to Maria Beatrice Coltelli (INSTM) for her valuable scientific insights in particular regarding the chemical characteristics of CN-NL nanoparticles. We acknowledge also Vincent Bonin and Nancy Hummel from FIBROLINE for providing the samples impregnated with CN-NL/GA through their dry impregnation technology. José Maria Lagaron, Maria Pardo Figuerez, Jorge Teno Díaz from BIOINICIA provided the samples of pullulan after the optimization of the process carried out by Karen de Clerck (UGent). Alessandro Gagliardini and Pietro Febo of ATERTEK company (Pescara, Italy) are thanked for the synthesis of the CN-NL/GA nanoparticles and Pierfrancesco Morganti for the scientific support.

Conflicts of Interest: The authors declare no conflict of interest.

References

1. Elieh-Ali-Komi, D.; Hamblin, M.R. Chitin and Chitosan: Production and Application of Versatile Biomedical Nanomaterials. *Am. Math. Mon.* **2004**, *111*, 915.
2. Morganti, P.; Danti, S.; Coltelli, M.B. Chitin and lignin to produce biocompatible tissues. *Res. Clin. Dermatol.* **2018**, *1*, 5–11. [CrossRef]
3. Danti, S.; Trombi, L.; Fusco, A.; Azimi, B.; Lazzeri, A.; Morganti, P.; Coltelli, M.-B. Chitin nanofibrils and nanolignin as functional agents in skin regeneration. *Int. J. Mol. Sci.* **2019**, *20*, 2669. [CrossRef]
4. Sahana, T.G. Biopolymers: Applications in wound healing and skin tissue engineering. *Mol. Biol. Rep.* **2018**, *45*, 2857–2867. [CrossRef] [PubMed]
5. Manley, M. Near-infrared spectroscopy and hyperspectral imaging: Non-destructive analysis of biological materials. *Chem. Soc. Rev.* **2014**, *43*, 8200–8214. [CrossRef]
6. Cleve, E.; Bach, E.; Schollmeyer, E. Using chemometric methods and NIR spectrophotometry in the textile industry. *Anal. Chim. Acta* **2000**, *420*, 163–167. [CrossRef]
7. Blanco, M.; Alcalá, M.; Planells, J.; Mulero, R. Quality control of cosmetic mixtures by NIR spectroscopy. *Anal. Bioanal. Chem.* **2007**, *389*, 1577–1583. [CrossRef]
8. Lyon, R.C.; Lester, D.S.; Lewis, E.N.; Lee, E.; Lawrence, X.Y.; Jefferson, E.H.; Hussain, A.S. Near-Infrared Spectral Imaging for Quality Assurance of Pharma- ceutical Products: Analysis of Tablets to Assess Powder Blend Ho- mogeneity. *AAPS PharmSciTech* **2002**, *3*, 1–15. [CrossRef]
9. Siche, R.; Vejarano, R.; Aredo, V.; Velasquez, L.; Saldaña, E.; Quevedo, R. Evaluation of Food Quality and Safety with Hyperspectral Imaging (HSI) Evaluation of Food Quality and Safety with Hyperspectral Imaging (HSI). *Food Eng. Rev.* **2016**, *8*, 306–322. [CrossRef]
10. Plant, P. Hyperspectral Imaging for Mapping of Total Nitrogen Spatial Distribution in pepper plant. *PLoS ONE* **2014**, *9*, e116205.
11. Nagaoka, T.; Kimura, Y. Quantitative cosmetic evaluation of long-lasting foundation using multispectral imaging. *Ski. Res. Technol.* **2019**, *25*, 318–324. [CrossRef]
12. Ge, Y.; Atefi, A.; Zhang, H.; Miao, C.; Ramamurthy, R.K.; Sigmon, B.; Yang, J.; Schnable, J.C. High-throughput analysis of leaf physiological and chemical traits with VIS-NIR-SWIR spectroscopy: A case study with a maize diversity panel. *Plant. Methods* **2019**, *15*, 1–12. [CrossRef]
13. Mäkelä, M.; Geladi, P. Hyperspectral near infrared imaging quantifies the heterogeneity of carbon materials. *Sci. Rep.* **2018**, *8*, 1–7. [CrossRef] [PubMed]
14. Lambrechts, A.; Blanch-Perez-del-Notario, C. *Hyperspectral Imaging for Textile Sorting and Recycling in Industry*; 2016; Available online: http://www.resyntex.eu/images/downloads/HSI_2016_paper_final.pdf (accessed on 5 March 2020).
15. Bureau, I. A Consumer Goods Product Comprising Chitin Nanofibrils, Lignin And A Polymer Or Co-Polymer Konsumgüterprodukt. U.S. Patent Application 14/854,121, 17 March 2016.
16. Sarazyn, J. Method Of Preparation Of Chitin And Active Principle Complexes And Complexes Obtaned. U.S. Patent Application 14/111,886, 30 January 2014.
17. Priyashantha, H.; Höjer, A.; Saedén, K.H.; Lundh, Å.; Johansson, M.; Bernes, G.; Geladi, P.; Hetta, M. Use of near-infrared hyperspectral (NIR-HS) imaging to visualize and model the maturity of long-ripening hard cheeses. *J. Food Eng.* **2020**, *264*, 109687. [CrossRef]
18. Sacré, P.Y.; Lebrun, P.; Chavez, P.F.; De Bleye, C.; Netchacovitch, L.; Rozet, E.; Klinkenberg, R.; Ziemons, E. A new criterion to assess distributional homogeneity in hyperspectral images of solid pharmaceutical dosage forms. *Anal. Chim. Acta* **2014**, *818*, 7–14. [CrossRef] [PubMed]
19. Yang, H.T.; Jiang, P. Scalable fabrication of superhydrophobic hierarchical colloidal arrays. *J. Colloid Interface Sci.* **2010**, *352*, 558–565. [CrossRef] [PubMed]
20. Mitsutake, H.; Castro, S.R.; de Paula, E.; Poppi, R.J.; Rutledge, D.N.; Breitkreitz, M.C. Comparison of different chemometric methods to extract chemical and physical information from Raman images of homogeneous and heterogeneous semi-solid pharmaceutical formulations. *Int. J. Pharm.* **2018**, *552*, 119–129. [CrossRef]
21. Suhandy, D.; Yulia, M. The Use of Partial Least Square Regression and Spectral Data in UV-Visible Region for Quantification of Adulteration in Indonesian Palm Civet Coffee. *Int. J. Food Sci.* **2017**, *2017*, 6274178. [CrossRef]

22. Li, X.; Sun, C.; Zhou, B.; He, Y. Determination of Hemicellulose, Cellulose and Lignin in Moso Bamboo by Near Infrared Spectroscopy. *Sci. Rep.* **2015**, *5*, 17210. [CrossRef]
23. Biniaś, W.; Biniaś, D. Application of ftnir spectroscopy for evaluation of the degree of deacetylation of chitosan fibres. *Fibres Text. East. Eur.* **2015**, *23*, 10–18.
24. Cloutis, E.; Norman, L.; Cuddy, M.; Mann, P. Spectral reflectance (350-2500 nm) properties of historic artists' pigments. II. Red-orange-yellow chromates, jarosites, organics, lead(-tin) oxides, sulphides, nitrites and antimonates. *J. Near Infrared Spectrosc.* **2016**, *24*, 119–140. [CrossRef]
25. Hayem-Ghez, A.; Ravaud, E.; Boust, C.; Bastian, G.; Menu, M.; Brodie-Linder, N. Characterizing pigments with hyperspectral imaging variable false-color composites. *Appl. Phys. A Mater. Sci. Process.* **2015**, *121*, 939–947. [CrossRef]
26. Elsoud, M.M.A.; el Kady, E.M. Current trends in fungal biosynthesis of chitin and chitosan. *Bull. Natl. Res. Cent.* **2019**, *43*, 1–12. [CrossRef]

© 2020 by the authors. Licensee MDPI, Basel, Switzerland. This article is an open access article distributed under the terms and conditions of the Creative Commons Attribution (CC BY) license (http://creativecommons.org/licenses/by/4.0/).

Article

Electrosprayed Chitin Nanofibril/Electrospun Polyhydroxyalkanoate Fiber Mesh as Functional Nonwoven for Skin Application

Bahareh Azimi [1,2], Lily Thomas [2,3], Alessandra Fusco [1,4], Ozlem Ipek Kalaoglu-Altan [5], Pooja Basnett [6], Patrizia Cinelli [1,2], Karen De Clerck [5], Ipsita Roy [7], Giovanna Donnarumma [1,4], Maria-Beatrice Coltelli [1,2], Serena Danti [2,*] and Andrea Lazzeri [1,2]

[1] Interuniversity National Consortiums of Materials Science and Technology (INSTM), 50121 Firenze FL, Italy; b.azimi@ing.unipi.it (B.A.); alessandra.fusco@unicampania.it (A.F.); patrizia.cinelli@unipi.it (P.C.); giovanna.donnarumma@unicampania.it (G.D.); maria.beatrice.coltelli@unipi.it (M.B.-C.)
[2] Department of Civil and Industrial Engineering, University of Pisa, 56126 Pisa PI, Italy; andrea.lazzeri@unipi.it
[3] Schools of Biosciences, Cardiff University, Cardiff CF10 3AT, UK; thomaslm13@cardiff.ac.uk
[4] Department of Experimental Medicine, University of Campania "Luigi Vanvitelli", 81100 Caserta CE, Italy
[5] Centre for Textile Science and Engineering, Department of Materials, Textiles and Chemical Engineering, 9000 Gent, Belgium; ozlemkalaoglu@gmail.com (O.I.K.-A.); karen.declerck@ugent.be (K.D.C.)
[6] School of Life Sciences, College of Liberal Arts and Sciences, University of Westminster, London W1W 7BY, UK; p.basnett@westminster.ac.uk
[7] Department of Materials Science & Engineering, Kroto Research Institute, University of Sheffield, Sheffield S10 2TG, UK; i.roy@sheffield.ac.uk
* Correspondence: serena.danti@unipi.it

Received: 20 July 2020; Accepted: 24 August 2020; Published: 3 September 2020

Abstract: Polyhydroxyalkanoates (PHAs) are a family of bio-based polyesters that have found different biomedical applications. Chitin and lignin, byproducts of fishery and plant biomass, show antimicrobial and anti-inflammatory activity on the nanoscale. Due to their polarities, chitin nanofibril (CN) and nanolignin (NL) can be assembled into micro-complexes, which can be loaded with bioactive factors, such as the glycyrrhetinic acid (GA) and CN-NL/GA (CLA) complexes, and can be used to decorate polymer surfaces. This study aims to develop completely bio-based and bioactive meshes intended for wound healing. Poly(3-hydroxybutyrate)/ Poly(3-hydroxyoctanoate-co-3-hydroxydecanoate), P(3HB)/P(3HO-co-3HD) was used to produce films and fiber meshes, to be surface-modified via electrospraying of CN or CLA to reach a uniform distribution. P(3HB)/P(3HO-co-3HD) fibers with desirable size and morphology were successfully prepared and functionalized with CN and CLA using electrospinning and tested in vitro with human keratinocytes. The presence of CN and CLA improved the indirect antimicrobial and anti-inflammatory activity of the electrospun fiber meshes by downregulating the expression of the most important pro-inflammatory cytokines and upregulating human defensin 2 expression. This natural and eco-sustainable mesh is promising in wound healing applications.

Keywords: biopolymer; bio-based; surface modification; nanolignin; electrospinning; electrospray; anti-inflammatory

1. Introduction

Polyhydroxyalkonates (PHAs) have recently attracted wide interest in research and product development, as the biomedical market has started to look for better sustainable alternatives to typical petroleum-based products [1]. In nature, PHAs exist as intracellular energy storage and can be produced by various microorganisms. Therefore, these macromolecules demonstrate very good biocompatibility, which has been extensively studied in multiple cell lines, making them ideal polymers for biomedical applications to be produced inside fermenters [2]. Along with low toxicity towards mammalian tissues, PHAs may also show an antimicrobial and bactericidal effect, a key property than can be utilized in the development of wound dressings for tissue regeneration [3]. The immunomodulatory properties of these materials also make them particularly useful in the wound repair process. PHA/starch blend films also showed potential for innovative bio-based beauty masks, wearable after wetting, and releasing starch [4,5]. Different fermentation process conditions result in a variety of PHAs with different properties—for example, they can span from a rigid to an elastomeric behavior [6]. Polyhydroxybutyrate, P(3HB), one of the more studied PHAs holds promise with a lower inflammatory effect in vivo, due to lower acidity of the released hydroxy acids compared to other widely available bioplastics, such as polylactic acid (PLA) [1]. A downfall of P(3HB) is its tendency to low processability and brittleness [1], but through co-polymerization, these limitations can be overcome in order to obtain greater flexibility and improve other processing factors [2]. To date, primarily co-polymerization of P(3HB) with 3-hydroxyvalerate (3HV) has been investigated to increase the flexibility of polymer chains [2]. Other PHAs are under development to achieve tunable mechanical properties, such as the poly(3-hydroxyoctanoate-co-3-hydroxydecanoate) P(3HO-co-3HD) [7]. In tissue engineering, micro/nano fibrous scaffolds have shown the ability to mimic the fibrillar part of the natural extracellular matrix (ECM), which plays a key role in cell migration, adhesion, and colonization [1]. To this purpose, electrospinning is very convenient if compared to other techniques such as spin coating, being a one-step process, providing parameter control to obtain the desired fiber morphology and providing an easy recoverability of the end product [8,9]. Specifically, electrospun fibers have an extremely high surface area to volume ratio, and the produced nonwovens show high porosity, which enable more effective cell attachment and colonization. PHA production on an industrial scale only holds a small proportion of the overall polymer market due to typically higher costs of production, but it holds the potential to reduce waste, decrease emissions, and develop green jobs [10]. Electrospinning could also promote the upscaling and industrialization of PHAs—in particular, in the biomedical sector. Versatility in the final product obtained via electrospinning allows fiber, particle, and fibril production [11]. Electrospray is typically performed via an electrospinning system using a low polymer concentration in the solution, along with other variations in parameters, in order to generate particles and fibrils often in micro or nanometric ranges [11]. The surface functionalization materials are possible by using the electrospray method.

Chitin is a natural polysaccharide found in the shells of crustaceans, cuticles of insects, and cell walls of fungi. It is the second most abundant polymerized carbon found in nature. When fibrillated on the nanoscale, chitin loses its pro-inflammatory and allergenic character and is able to proficiently interact with many cellular compounds in biological tissues [12]. Due to its positive charge, chitin nanofibrils (CN) interacts with negatively charged nanolignin (NL), derived from lignin found in plant biomass, to generate micro-complexes. Such complexes have demonstrated the ability to incorporate biomolecules such as the glycyrrhetinic acid (GA), which is derived from the licorice plant and possesses anti-inflammatory and antimicrobial properties that are interesting for treating skin disorders [13]. CN-NL/GA (CLA) complexes were developed by spray dry technology in previous studies and showed promise for skin contact applications, as CN-NL provides a suitable carrier to promote GA delivery [14,15].

The aim of this study was to set up an electrospray method to decorate the PHA surface with CN and CLA, thus providing a functionalization method for electrospun fiber meshes to be used for wound healing applications. In particular, we used a blend of P(3HB) and P(3HO-co-3HD) to obtain a suitable fiber morphology. CN and CLA were suspended in water-based solutions and preliminarily electrosprayed on the surface of P(3HB)/P(3HO-co-3HD) films to assess the best uniform nano/microparticle distribution on their surface. Thereafter, CN and CLA were electrosprayed upon P(3HB)/P(3HO-co-3HD) electrospun fiber meshes. Finally, we assessed which of the two chitin-based compounds had the highest beneficial effects for the epidermis, in vitro, using a human keratinocyte HaCaT cell line. The expression of a panel of cytokines involved in inflammation and immune response was investigated, including the pro-inflammatory interleukins (ILs) IL-1, IL-6, IL-8, the tumor necrosis factor α (TNF-α), the transforming growth factor β (TGF-β), and the human beta defensin 2 (HBD-2), the latter being an endogenous antimicrobial peptide. The successful fabrication of innovative surface-functionalized PHA fibers would enable the development of novel bio-based products for potential use for skin-related applications, such as wound healing or skin contact.

2. Materials and Methods

2.1. Materials

CN, NL, and GA were supplied by Mavi Sud, Aprilia (LT, Italy). Chloroform (code: 102442), 2-butanol (code: 109630), lithium bromide (LiBr$_2$), acetic acid (code: 33209), ethanol (EtOH), poly(ethylene glycol) (PEG8000), and Dulbecco's phosphate-buffered saline (DPBS) were provided by Sigma-Aldrich (Milan, Italy). Immortalized human keratinocytes, HaCaT cell line, were obtained from ATCC-LGC Standards (Milan, Italy). MgCl$_2$, Dulbecco's Modified Essential Medium (DMEM), L-glutamine, penicillin, streptomycin and fetal calf serum were purchased from Invitrogen, (Carlsbad, CA, USA). Alamar Blue was bought from Thermo Fisher Scientific (Waltham, MA, USA). LC Fast Start DNA Master SYBR Green kit was obtained from Roche Applied Science (Euroclone S.p.A., Pero, Italy).

2.2. P(3HO-co-3HD) and P(3HB) Production

P(3HO-*co*-3HD) was produced using *Pseudomonas mendocina* CH50 using 20 g/L of glucose in 15 L bioreactors, with 10 L working volume (Applikon Biotechnology, Tewkesbury, UK). Batch fermentation was carried out in two stages as described in Basnett et al., 2020 [7]. P(3HB) was produced using *Bacillus subtilis* OK2 using glucose at 20 g/L, using the same protocol as P(3HO-*co*-3HD) production but using a single stage batch fermentation. For simplicity P(3HB)/P(3HO-*co*-3HD) will be further referred to as PHB/PHOHD.

2.3. Preparation of CN and CLA Solutions and Electrospraying Protocols

CN was used at 0.52 w% in aqueous acetic acid and distilled water (50:50 *w/w*). The solution magnetically stirred for 3 h until it appeared uniform. The solution was electrosprayed using an electrospinning bench apparatus (Linari Engineering s.r.l., Pisa, Italy) for 20 min with a static aluminum collector with a ground charge 10 cm from the needle tip. The flow rate of 0.298 mL/h and a voltage of 15 kV were employed. CLA complexes were prepared in powder using a Buchi Mini B-190 spray drier (Flawil, Switzerland) [16] and by adding 2% CN-NL *w/w*% with respect to PEG. CLA complexes were thus prepared in accordance with previous works [17]. The ratio between CN and NL is 2:1 by weight. The content of GA in CLA is 0.2% by weight. CLA powder was dissolved at 0.52 *w/w*% in distilled water and magnetically stirred for 1 h. The solution was electrosprayed using an electrospinning bench apparatus (Linari Engineering s.r.l.) with a distance of 10 cm from the positively charged needle tip to the grounded aluminum static collector at 15 kV at a flow rate of 0.298 mL/h for 20 min.

2.4. PHB/PHOHD Solution Preparation

A solution of 11 *w/w*% PHB/PHOHD (1:10) was produced using chloroform and 2-butanol as solvents with a ratio of 70:30 (*v/v*). This was performed using a stepwise procedure consisting of adding PHOHD to chloroform, under magnetic stirrer for 1 h before the addition of P(3HB) and 2-butanol. The solution was left overnight under stirring to reach homogeneity before adding 0.002 g/mL LiBr$_2$ and then magnetically stirred until it reached a uniform appearance.

2.5. Production of PHB/PHOHD Films and Fiber Meshes

The PHB/PHOHD solution at 11% (the weight of the polymers with respect to the weight of solvent mix) was poured onto a sterile glass petri dish and left for 48 h under the chemical laminar flow hood to allow full solvent evaporation. The electrospinning parameters used to produce PHB/PHOHD fiber meshes were 40 kV, a 0.5 mL/h flow rate, and a 40 cm distance from needle tip to the static aluminum collector. The process ran for 1 h. Humidity around 40% and a temperature of 20 °C was maintained throughout.

2.6. Electrospray of CN and CLA on PHB/PHOHD Film and Fiber Mesh

CLA and CN were electrosprayed onto the previously produced PHB/PHOHD fiber mesh and film using a flow rate of 0.298 mL/h, the voltage of 15 kV with a distance of 10 cm between the positive needle tip and grounded static collector for 60 min.

2.7. Morphological Characterization

Morphological analysis of the samples was performed using field emission electron scanning microscopy (FE-SEM) with FEI FEG-Quanta 450 instrument (Field Electron and Ion Company, Hillsboro, OR, USA) and Inverted optical microscope (Nikon Ti, Nikon Instruments, Amsterdam, The Netherlands). The samples were sputtered with gold or platinum for analysis. Image J software (version 1.52t) was used to evaluate the size of nanofibrils and fibers. The average of 50 measurements has been reported for each sample.

2.8. Chemical Structure Characterization

Infrared spectroscopy using Nicolet T380 instrument (Thermo Scientific, Waltham, MA, USA) equipped with a Smart ITX ATR attachment with a diamond plate was employed for chemical structure characterization of both solid chitin-based substances and electrospun/electrosprayed samples.

2.9. Evaluation of HaCaT Cell Line Viability

Each material sample was sterilized overnight in absolute ethanol, then rinsed three times with PBS. HaCaT cells were cultured in D-MEM supplemented with 1% Penicillin-streptomycin, 1% L-glutamine and 10% fetal calf serum in a humidified incubator set at 37 °C in 95% air and 5% CO_2. The materials were put in contact with HaCaT cells seeded in 12-well plates for 6 h and 24 h. At the endpoint, the Alamar blue test was performed following the manufacturer's protocol after 4 h incubation with the dye. Briefly, Alamar blue incorporates a redox indicator that changes color according to cell metabolic activity. The supernatants were read with a spectrophotometer using a double wavelength reading at 570 nm and 600 nm. Finally, the reduced percentage of the dye (%AB_{RED}) was calculated by correlating the absorbance values and the molar extinction coefficients of the dye at the selected wavelengths.

2.10. Anti-Inflammatory and Immune Responses Evaluation of HaCaT Cells

The immunomodulatory properties of PHB/PHOHD fiber meshes electrospun with CN and CLA were assayed using HaCaT cells. The cells, cultured as described above, were seeded inside 12-well plates until 80% of confluence was reached. At the time-points of the experiment (6 h and 24 h), the mRNA was extracted from the cells and the levels of expression of the proinflammatory cytokines IL-8, IL-6, IL-1β, IL-1 α, and TNF-α anti-inflammatory cytokine TGF-β and antimicrobial peptide HBD-2 were evaluated by real-time reverse transcriptase polymer chain reaction (RT-PCR). Briefly, the total RNA was isolated with TRizol, and 1 μm of RNA was reverse-transcribed into complementary DNA (cDNA) using random hexamer primers at 42 °C for 45 min, according to the manufacturer's instructions. PCR was carried out with the LC Fast Start DNA Master SYBR Green kit using 2 μL of cDNA, corresponding to 10 ng of total RNA in a 20 μL final volume, 3 mM $MgCl_2$, and 0.5 μM sense and antisense primers (Table 1). The results were normalized by the expression of the same cytokine in untreated cells, as a control.

Table 1. Real-time reverse transcriptase polymer chain reaction (RT-PCR) details, including gene, primer sequences, operational conditions, and product size.

Gene	Primers Sequence	Conditions	Product Size (bp)
IL-1α	5′-CATGTCAAATTTCACTGCTTCATCC-3′ 5′-GTCTCTGAATCAGAAATCCTTCTATC-3′	5 s at 95 °C, 8 s at 55 °C, 17 s at 72 °C for 45 cycles	421
IL-1β	5′-GCATCCAGCTACGAATCTCC-3′ 5′-CCACATTCAGCACAGGACTC-3′	5 s at 95 °C, 14 s at 58 °C, 28 s at 72 °C for 40 cycles	708
IL-6	5′-ATGAACTCCTTCTCCACAAGCGC-3′ 5′-GAAGAGCCCTCAGGCTGGACTG-3′	5 s at 95 °C, 13 s at 56 °C, 25 at 72 °C for 40 cycles	628
IL-8	5′-ATGACTTCCAAGCTGGCCGTG-3′ 5′-TGAATTCTCAGCCCTCTTCAAAAACTTCTC-3′	5 s at 94 °C, 6 s at 55 °C, 12 s at 72 °C for 40 cycles	297
TNF-α	5′-CAGAGGGAAGAGTTCCCCAG-3′ 5′-CCTTGGTCTGGTAGGAGACG-3′	5 s at 95 °C, 6 s at 57 °C, 13 s at 72 °C for 40 cycles	324
TGF-β	5′-CCGACTACTACGCCAAGGAGGTCAC-3′ 5′-AGGCCGGTTCATGCCATGAATGGTG-3′	5 s at 94 °C, 9 s at 60 °C, 18 s at 72 °C for 40 cycles	439
HBD-2	5′-GGATCCATGGGTATAGGCGATCCTGTTA-3′ 5′-AAGCTTCTCTGATGAGGGAGCCCTTTCT-3′	5 s at 94 °C, 6 s at 63 °C, 10 s at 72 °C for 50 cycles	198

3. Results

3.1. Morphological Characterization

3.1.1. Morphological Characterization of Electrosprayed CN and CLA

On aluminum foil as a substrate, electrosprayed CN suspensions allowed us to obtain uniform surface decoration (Figure 1a).

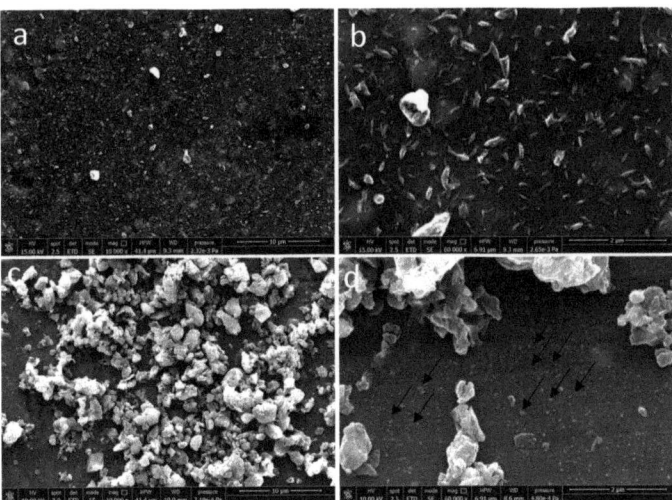

Figure 1. Scanning electron microscopy (SEM) micrographs of (**a**,**b**) electrosprayed CNs at 10,000× and 60,000× magnifications and (**c**,**d**) electrosprayed CN -NL assembled into micro-complexes loaded with (GA), i.e., (CLA) complexes, at 10,000× and 60,000× magnification. Arrows show subpopulation of CLA complexes with low size.

CNs were well dispersed and showed an average size of 180 nm ± 47 nm (Figure 1b). Two main subpopulations of CLA complexes with an average size of 65 ± 20 nm (shown with arrows) and 1239 ± 626 nm were observed (Figure 1c,d). There are also some aggregations on CLA fibrils.

3.1.2. Morphological Analysis of Functionalized PHB/PHOHD Films

Figure 2 shows SEM images of the surface of PHB/PHOHD films plain and functionalized with CNs or CLA complexes at different magnifications. CNs with uniform size and morphology were homogeneously electrosprayed and coated the whole surface of PHB/PHOHD films. CLA complexes distributed locally, and thus non-uniformly, on the surface of the films. The aggregated CLA microparticles can also be observed on the surface of the films.

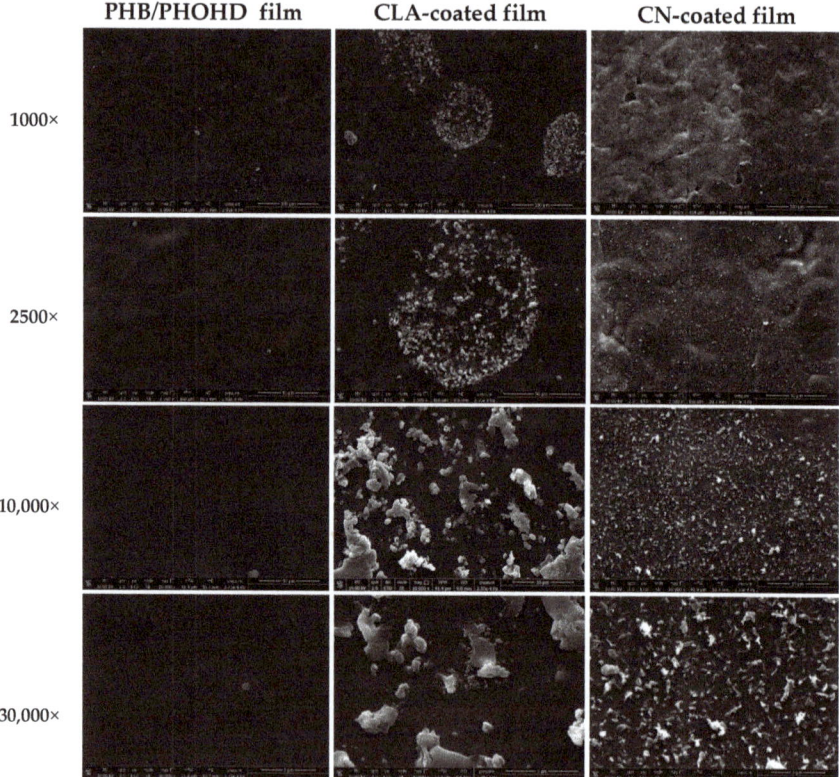

Figure 2. SEM micrographs of PHB/PHOHD films, PHB/PHOHD films functionalized with CNs, and PHB/PHOHD films functionalized with CLA complexes, observed at different magnifications.

3.1.3. Morphological Analysis of Functionalized PHB/PHOHD-Electrospun Fiber Meshes

The electrospun meshes displayed anisotropic fibrous morphology with a homogenous fiber feature (Figure 3a). Ultrafine PHB/PHOHD fibers with an average diameter of 1.28 ± 0.58 µm were successfully produced (Figure 3b). By using PHB/PHOHD fibers as a substrate, CNs (Figure 3c,d) and CLA (Figure 3e,f) were uniformly electrosprayed on the surface of fibers. Figure 4 shows the size distribution of electrosprayed CNs and CLA complexes on the surface of PHB/PHOHD fibers and films. After electrospraying, CNs were found with lower size on the surface of the fibers than on films, although still more aggregated than on the aluminum foil ($p < 0.05$) (Figure 4a). In an opposite manner,

CLA microparticles exhibited lower size and size distribution on the surface of fibers and film than on the aluminum foil (Figure 4b).

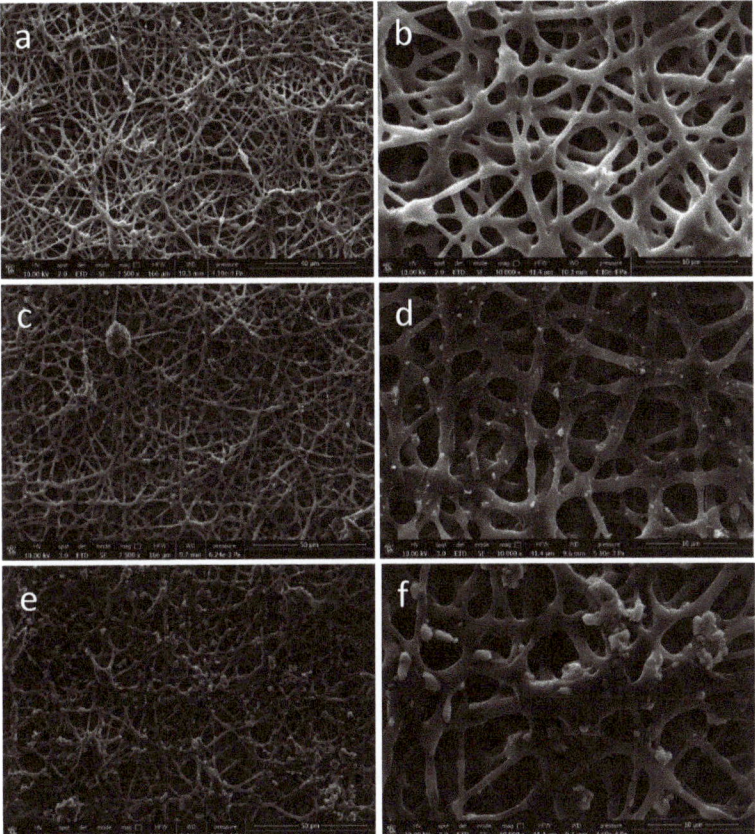

Figure 3. SEM images of (**a**,**b**) PHB/PHOHD electrospun fiber meshes at different magnifications: (**a**) 2500× (**b**) 10,000×; (**c**,**d**) PHB/PHOHD electrospun fiber meshes functionalized with electrosprayed CN at different magnifications: (**c**) 2500×, (**d**) 10,000×; (**e**,**f**) PHB/PHOHD electrospun fiber meshes functionalized with electrosprayed CLA at different magnifications—(**e**) 2500×, (**f**) 10,000×.

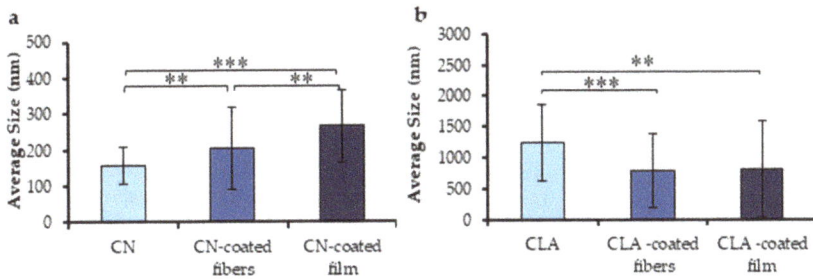

Figure 4. Size of electrosprayed CNs (**a**) and CLA complexes (**b**) on different surfaces—no polymer (aluminum foil), PHB/PHOHD fibers, and PHB/PHOHD film. Data are expressed as ± SD (n = 50 particles), ** $p < 0.001$; *** $p < 0.0001$.

3.2. Chemical Characterization

3.2.1. Chemical Characterization of (PHB/PHOHD)-Electrospun Fiber Mesh and Film

The Fourier-transform infrared spectroscopy (FTIR) spectra of PHB/PHOHD-electrospun fibers and film were compared (Figure 5) by normalizing on the peak at 1730 cm^{-1}, attributable to the C=O stretching of the ester group of PHA.

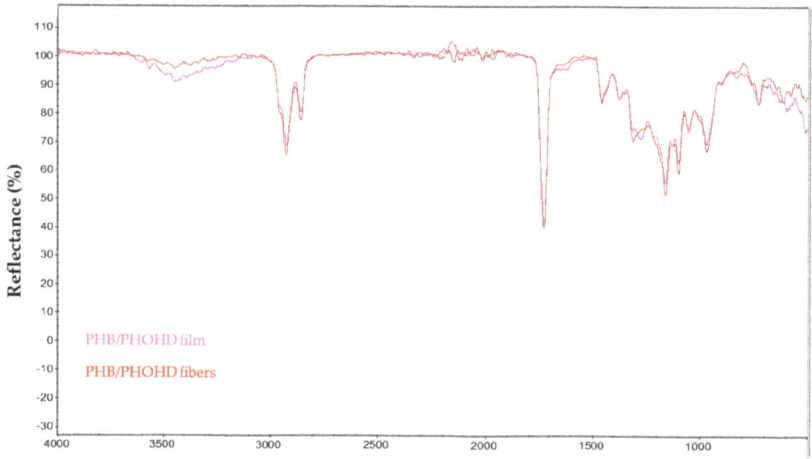

Figure 5. Fourier-transform infrared spectroscopy (FTIR) spectra of PHB/PHOHD film and (PHB/PHOHD)-electrospun fibers.

Weaker peaks at 3500 cm^{-1} and 1600 cm^{-1} in the fiber mesh indicated a reduction in O-H stretching. As this band intensity highly depends on the humidity content of the samples, that can be influenced by ambient conditions, this change cannot be attributed to a change in the –OH concentration of the samples but more reasonably to some water content fluctuations. Overall, there is no strong indication of any detrimental effect on the polymer structural integrity during the electrospinning process.

3.2.2. Chemical Characterization of CN-Coated PHB/PHOHD Fibers

The dried pristine CNs, PHB/PHOHD fibers and CN-coated PHB/PHOHD fibers were characterized by FTIR (Figure 6). The characteristic bands of CN are 1010 cm^{-1} and 1070 cm^{-1}, typical of C–O stretching, 1552 cm^{-1} attributed to amide II, 1619 cm^{-1} and 1656 cm^{-1} attributed to amide I, 2874 cm^{-1} attributed to C–H stretching, 3102 cm^{-1} and 3256 cm^{-1} attributed to N–H stretching of the amide and amine groups, and 3439 cm^{-1} attributable to O–H stretching. All these bands were observed in the FTIR spectrum of CNs.

The main characteristic bands of CN (1552 cm^{-1}, 1619 cm^{-1} and 1656 cm^{-1}) can be also observed on CN-coated PHB/PHOHD fiber spectrum. The increased intensity of the band at 1010 cm^{-1}, 1070 cm^{-1}, and 1385 cm^{-1} was clearly detected in the spectrum of CN-coated fibers. Indeed, the ratio of bands at 1010 cm^{-1} and 1070 cm^{-1}, attributable to C–O linkages (abundant in polysaccharides like CN), to the reference band of 1733 cm^{-1} was higher for the CN-coated PHB/PHOHD fibers than for the plain PHB/PHOHD fibers.

These observations corroborated the presence of CNs on the surface of PHB/PHOHD fibers.

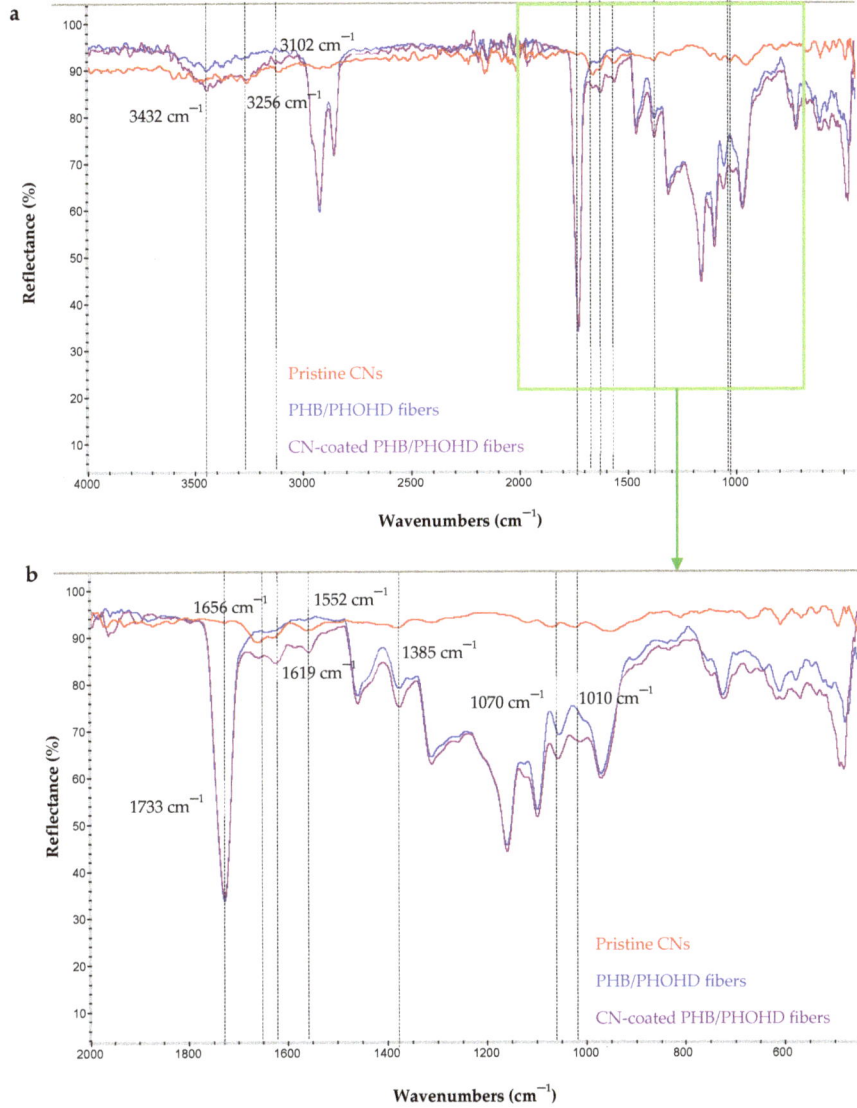

Figure 6. FTIR spectra of pristine CN, PHB/PHOHD-electrospun fibers, both CN-coated and uncoated (plain). (**a**) The whole investigated spectrum and (**b**) Zoomed-in spectrum in the 2000–1100 cm^{-1} range.

3.2.3. Chemical Characterization of CLA-Coated PHB/PHOHD Fibres

CLA complexes, PHB/PHOHD fibers and CLA-coated PHB/PHOHD fibers were characterized by FTIR (Figure 7). The characteristic bands of CN described above, and NL, namely, 1052 cm^{-1} attributed to aromatic C–H deformation [18]; 1222 cm^{-1} associated with C–C plus C–O, 1511 cm^{-1} and 1601 cm^{-1} associated with aromatic skeleton vibrations; and GA, namely 3430 cm^{-1} typical of OH stretching and 2945 cm^{-1} attributable to CH stretching; 1700 cm^{-1} and 1660 cm^{-1} attributed to the C=O stretching of carboxylic and ketone groups, respectively, and 1470 cm^{-1} to CH$_2$ bending, 1025 cm^{-1} to C–O stretching, and 990 cm^{-1} attributed to the rocking of the methyl group, were detected in the

CLA complex spectra. The main characteristic bands of CLA complexes were considered, 1385 cm^{-1}, 1470 cm^{-1}, 1619 cm^{-1}, and 1656 cm^{-1}, which were present in the FTIR spectra of functionalized PHB/PHOHD fibers. In addition, a higher ratio of CLA characteristic bands with respect to 1733 cm^{-1}, as a reference band, were obtained for the coated fibers than uncoated ones, which demonstrated an effective functionalization.

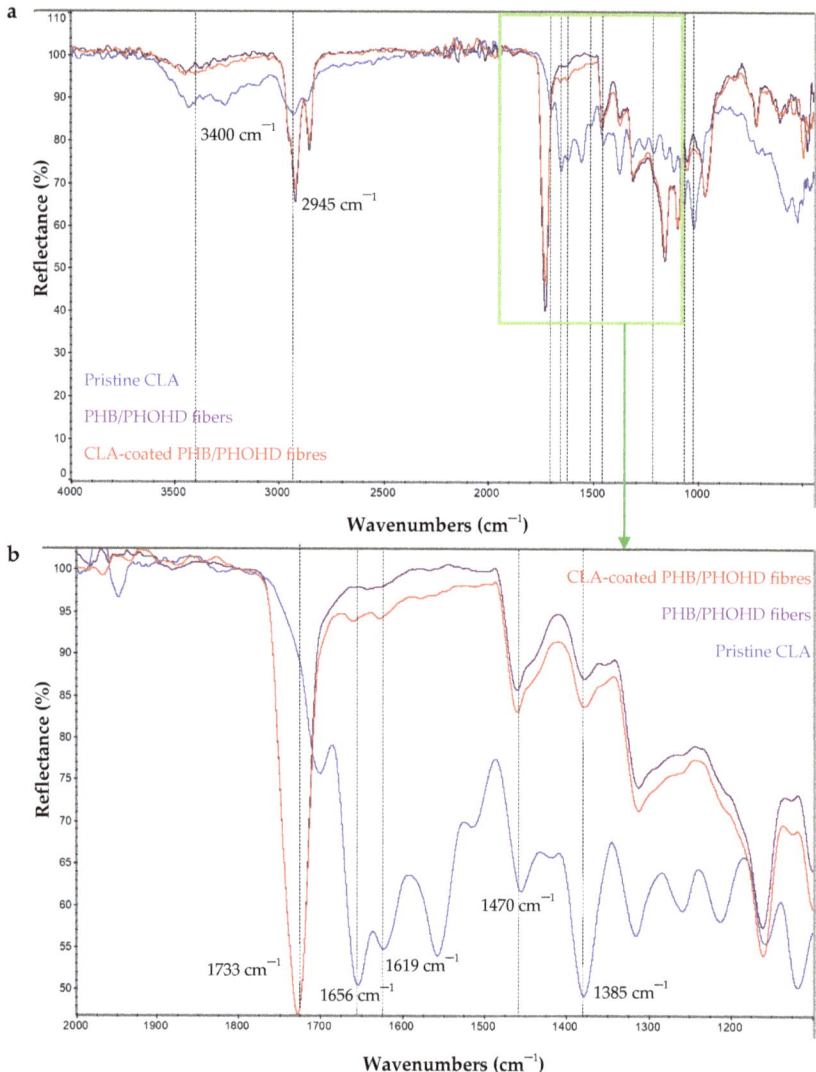

Figure 7. FTIR spectra of CLA complexes, PHB/PHOHD-electrospun fibers both CLA-coated and uncoated (plain). (**a**) The whole investigated spectrum and (**b**) zoomed-in spectrum in 2000–1100 cm^{-1}.

3.3. HaCaT Cell Metabolic Activity

AlamarBlue® test was performed on the HaCaT cells in order to assess the cytocompatibility of the CN-coated fibers and CLA-coated fiber meshes. The results obtained highlighted good cytocompatibility, even though the low concentration of the CN-based particles reduced the metabolic activity as compared to that of the plain polymer (Table 2).

Table 2. AlamarBlue results on PHB/PHOHD fiber meshes, coated with CNs or CLA complexes.

Sample	%AB$_{RED}$
PHB/PHOHD fiber mesh	76
CLA-coated PHB/PHOHD fiber mesh	64
CN-coated PHB/PHOHD fiber mesh	69

3.4. Immunomodulatory Properties

The results show that the samples were able to induce a powerful anti-inflammatory activity in HaCaT cells (Figure 8).

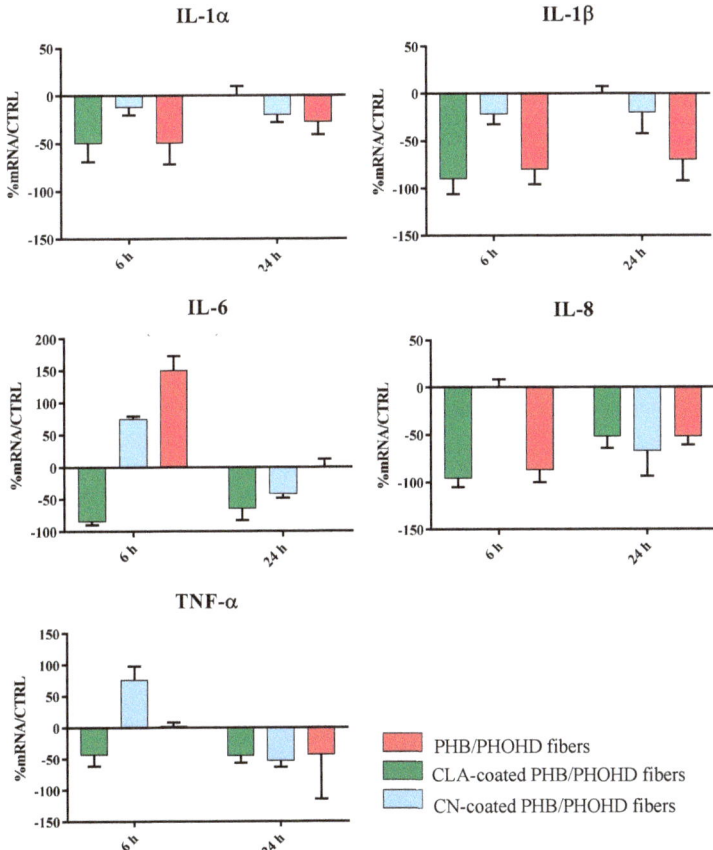

Figure 8. Bar graphs showing the results of real time RT-PCR performed related to different cytokines involved in the inflammatory response of HaCaT cells after being exposed to the PHB/PHOHD fibers (plain, CLA-coated, and CN-coated) for 6 h and 24 h. The results were normalized by the expression in untreated cells as control.

Indeed, all the samples strongly downregulated the expression of the main proinflammatory cytokines IL-1 (α and β), IL-6, IL-8 and TNF-α after 24 h, and most of them within 6 h. The plain fibers and the fibers coated with CNs initially upregulated IL-6; however, this cytokine was subsequently downregulated. TGF-β was not modulated with respect to the untreated cells. In addition, the samples showed the ability to induce the expression of HBD-2 in HaCat cells (Figure 9) after 6 h exposure, thus suggesting a role in stimulating an indirect antibacterial activity.

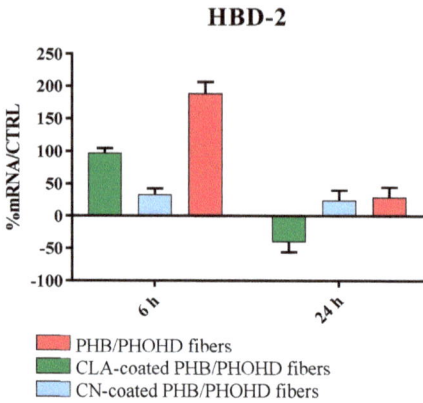

Figure 9. Bar graph showing the results of real time RT-PCR analysis for HBD-2 produced by HaCaT cells exposed to the PHB/PHOHD fibers (plain, CLA-coated, and CN-coated) at 6 h and 24 h. The results are normalized by the expression in untreated cells as control.

4. Discussion

The skin is the first organ to be injured, since it protects the other tissues and organs of the body, so it retains a high capacity for self-repairing. However, the natural wound-healing process of skin may be ineffective in cases of profound lesions or large surface loss, leading to detrimental and painful conditions that require repair adjuvants or tissue substitutes. For example, wounds deriving from diabetes show difficultly when healing and tend to become chronic via self-inflammation. Such a process causes the degradation of the growth factors deputed to healing, thus concurring to generate pathologic states, like recurrent infections and tumor onset. Therefore, the application of specific biomaterial dressings able to stimulate and/or accelerate the wound healing process can be of great importance to restore the normal native tissues [19]. In a lifetime, many other events also induce skin damage and irritation; therefore, anti-inflammatory biomaterials find a number of applications in skin contact products.

In our study, we produced electrospun PHA fibers with improved antibacterial and anti-inflammatory properties as a bioactive nonwoven able to promote skin self-repair and prevent the infection processes. PHAs are highly biocompatible and naturally occurring bacteria-derived polyesters and are produced by bacterial fermentation [20,21]. In particular, we used a blend of two PHAs (i.e., P(3HB) and P(3HO-co-3HD)), which allowed the best fiber production process. The P(3HB)/P(3HO-co-3HD) blend was used in order to achieve a good balance between the good processability of P(3HB) and the desirable elastomeric property of P(3HO-co-3HD). Different techniques such as solvent casting, dip molding, and 3D printing have been used for processing of PHAs in different structures for a variety of biomedical applications [22,23]. Fibrous PHA scaffolds with specific characteristics can promote tissue formation, since their topography is similar to the fibers and fibrils naturally found in the tissue extracellular matrix (ECM), and can be produced by electrospinning as a versatile method [24–26]. Salvatore et al. demonstrated the potential of electrospun PHB/collagen meshes as suitable substrates for wound healing, with the PHB/collagen ratio controlling the morphological, mechanical, and degradation properties of the meshes [27].

The formation of beads during the electrospinning process is a common phenomenon that can affect the quality of the fibers [4]. By using appropriate electrospinning conditions, including solvent mixture and $LiBr_2$, we produced beadless PHB/PHOHD ultrafine fibers with a homogeneous morphology and quite uniform diameter. We showed that electrospinning is a safe technology to produce the ultrafine fibers since no significant difference was observed in the chemical structure of electrospun fibers in comparison to solvent-casted films.

CNs and CLA complexes are bio-based nano- and micro-compounds that can be considered useful bioactive agents for functionalizing skin contact substrates, as they showed proficient interaction in an in vitro skin model [14]. It is important to obtain an easy and effective method for applying such components to biomaterial surfaces to put in contact with skin, encompassing bulk incorporation and surface functionalization [15,28]. In fact, if plain CNs are applied as water suspension, further steps like evaporation must be taken into account. Moreover, the substrate surface may be affected by wetting. On the other hand, when CNs are used to incorporate bioactive molecules, such as GA in CLA complexes, it is important to avoid as much as possible any processes that would lead to possible dispersion or modification of the bioactive molecule. We considered that electrospraying could be a valuable method for CN and CLA complex deposition on PHA surfaces, as it can be performed with the same equipment used for fiber production without any post treatment. At first, PHB/PHOHD films were used to set up electrospraying conditions. Indeed, the chemical nature of the substrate used as a collector membrane can affect the local electric field and cause different deposition outcomes [29]. We observed a different behavior of the CN and CLA solutions electrosprayed on PHB/PHOHD films. Specifically, CNs were even, whereas CLA complexes were locally dispersed. This is a consequence of the diverse interaction of positively charged CNs and the neutral CLA complexes with the electric field. Subsequently, the same technique was used to decorate the surface of the electrospun P(3HB)/P(3HO-co-3HD) fibers in order to improve their antibacterial and anti-inflammatory properties. Due to the porous nature of these nonwovens, both CNs and CLA complexes were homogeneously delivered to the fiber surface, possibly due to a more powerful interaction of the metallic collector with the particles. SEM analysis demonstrated the presence of both CNs and CLA complexes on the surface of P(3HB)/P(3HO-co-3HD) fibers and films. Using different substrates also led to the formation of particles with different size caused by aggregation. This polydispersion effect can be attributed to the interaction of different substrates with the electric field applied during electrospray too [30,31]. However, in all the tested substrates, electrosprayed CN had smaller and more uniform size than electrosprayed CLA. FTIR results also confirmed the presence of CN and CLA nanofibrils on the surface of fibers without any remarkable change in their chemical structures. These results demonstrate that electrospray is a suitable technique for functionalizing the substrates for wound healing applications.

Since the nanosized materials can migrate into tissues and potentially induce unknown reactions, studying the interaction with skin cells is fundamental to demonstrate safety in skin contact applications [25,31,32]. In this study, bio-based and bioresorbable materials were selected to minimize the harmful risks due to organ accumulation, while providing beneficial bioactive properties. The short term interaction of the functionalized substrates with human keratinocytes was investigated in vitro using HaCaT cells. Preliminary investigation of cell metabolic activity suggested good cytocompatibility, as also observed in previous studies [14]. In order to predict the reaction of the epidermis layer, the expression of an array of cytokines involved in the inflammation response were studied, including IL-1, IL-6, IL-8, and TNF-α [14]. Cytokines are multi-functional biological molecules that are involved in autocrine, paracrine, and endocrine signaling as immunomodulating agents and play an important role in biological activities such as tissue repair, growth, and cell development [33]. In particular, IL-1 promotes local inflammation and coagulation, increases the expression of adhesion molecules, and causes the release of chemokines and recruitment of leukocytes to the site of inflammation [34]. IL-6 is involved in the inflammatory acute phase response. IL-8 is a multi-functional chemokine, inducing the activation of polymorphonuclear leukocytes and angiogenesis [35]. Finally, TNF-α is an essential mediator in inflammation with reactive roles in blood coagulation process [36]. All these cytokines were strongly downregulated in all the produced samples, showing that these nano- and micro-compounds in combination with the P(3HB)/P(3HO-co-3HD) fibers inherently support the reduction of inflammatory states without the upregulation of TGF-β.

The PHB/PHOHD fiber meshes, plain or functionalized with CNs and CLA complexes, were also able to upregulate the HaCaT cell expression of an antimicrobial peptide, HBD-2, which acts as an endogenous antibiotic and plays an important role in the innate immune response [37]. Indeed, if the healing process does not proceed properly, a wound may become chronic, thus it is not able to self-repair in a normal, orderly, and timely manner and is highly prone to infections. Since traditional treatment methods are not sufficient for changing the microenvironment of chronic wounds, they do not represent breakthrough approaches [38]. Recently, Zhang et al. comprehensively reviewed the functional biomaterials that can improve chronic wound healing through debridement, anti-infection and antioxidant effects, immunoregulation, angiogenesis, and ECM remodeling [39]. They pointed out that functional biomaterials are expected to improve the patients' quality of life via resolving the treatment dilemma for chronic wounds. Functional biomaterials are also expected to elicit an appropriate immune response, thus orchestrating a cascade of biological events that promote safe and uncomplicated healing. To this purpose, it is important to design scaffold topography that helps both dermal cell organization and epidermal cell migration. Even though many scaffold architectures have been revealed to support fibroblast growth [40], by mimicking the structure of the fibrillar ECM, electrospun scaffolds are expected to allow an optimal dermal regeneration [26]. Moreover, by downregulating inflammatory cytokines and upregulating HBD-2 in keratinocytes, thanks to CN-based particles, such surface functionalized scaffolds should act as highly functional biomaterials [41].

In this view, the development of bio-based and biodegradable functional nonwovens with powerful anti-inflammatory activity and indirect antimicrobial properties opens an interesting scenario in skin repair and regeneration products, including, but not limited to, wound dressings, which can use a sustainable and green route.

5. Conclusions

We developed an easy and effective method to obtain surface-decorated electrospun nonwovens via electrospray of CN and CLA complex solution of P(3HB)/P(3HO-co-3HD) fiber meshes. Such completely bio-based and biodegradable functional nonwovens possessed strong anti-inflammatory activity as they downregulated the main pro-inflammatory cytokines and exhibited the capacity of stimulating HBD-2 by human keratinocytes. Having green and effective substrates for skin contact and repair would allow better treatment of irritated skin and complex skin wounds.

Author Contributions: Conceptualization, M.-B.C. and S.D.; methodology, B.A., L.T., A.F., O.I.K.-A., P.C., and P.B.; investigation, B.A., L.T., and S.D.; writing—original draft preparation, B.A. and S.D.; writing—review and editing, B.A., L.T., M.-B.C., I.R., K.D.C., and S.D.; visualization, S.D.; supervision, S.D., M.-B.C., and G.D.; project administration, S.D.; funding acquisition, M.B.C., A.L., G.D., I.R., and K.D.C. All authors have read and agreed to the published version of the manuscript.

Funding: This research was funded by the Bio-Based Industries Joint Undertaking under the European Union Horizon 2020 research program (BBI-H2020), PolyBioSkin project, grant number G.A. 745839.

Acknowledgments: CISUP—Centre for Instrumentation Sharing—University of Pisa is acknowledged for SEM analysis.

Conflicts of Interest: The authors declare no conflict of interest.

Abbreviations

CLA	Chitin nanofibril/nanolignin-glycyrrhetinic acid
CN	Chitin nanofibril
DMEM	Dulbecco's modified essential medium
DPBS	Dulbecco's phosphate-buffered saline

EtOH	Ethanol
GA	Glycyrrhizin acid
HBD-2	Human beta-defensin 2
IL	Interleukin
NL	Nanolignin
P(3HB)	Poly(3-hydroxybutyrate)
P(3HO-co-3HD)	Poly(3-h ydroxyoctanoate-co-3-hydroxydecanoate)
PEG	Poly(ethylene glycol)
PHAs	Polyhydroxyalkanoates
TGF-β	Transforming growth factor beta
TNF-α	Tumor necrosis factor alpha

References

1. Ying, T.; Ishii, D.; Mahara Murakami, S.; Yamaok, T.; Kumar, S.; Samian, R.; Fujita, M.; Maeda, M.; Iwata, T. Scaffolds from electrospun polyhydroxyalkanoate copolymers: Fabrication, characterization, bioabsorption and tissue response. *Biomaterials* **2008**, *29*, 1307–1317. [CrossRef] [PubMed]
2. Sombatmankhong, K.; Suwantong, O.; Waleetorncheepsawat, S.; Supaphol, P. Electrospun fiber mats of poly(3-hydroxybutyrate), poly(3-hydroxybutyrate-co-3-hydroxyvalerate), and their blends. *J. Polym. Sci. Polym. Phys.* **2006**, *44*, 2923–2933. [CrossRef]
3. Dinjaski, N.; Fernández-Gutiérrez, M.; Selvam, S.; Parra-Ruiz, F.J.; Lehman, S.M.; San Román, J.; García, E.; García, J.L.; García, A.J.; Prieto, M.A. PHACOS, a functionalized bacterial polyester with bactericidal activity against methicillin-resistant Staphylococcus aureus. *Biomaterials* **2014**, *35*, 14–24. [CrossRef] [PubMed]
4. Coltelli, M.B.; Panariello, L.; Morganti, P.; Danti, S.; Baroni, A.; Lazzeri, A.; Fusco, A.; Donnarumma, G. Skin-compatible biobased beauty masks prepared by extrusion. *J. Funct. Biomater* **2020**, *11*, 23. [CrossRef]
5. Coltelli, M.B.; Danti, S.; Trombi, L.; Morganti, P.; Donnarumma, G.; Baroni, A.; Fusco, A.; Lazzeri, A. Preparation of innovative skin compatible films to release polysaccharides for biobased beauty masks. *Cosmetics* **2020**, *5*, 70. [CrossRef]
6. Piarali, S.; Marlinghaus, L.; Viebahn, R.; Lewis, H.; Ryadnov, M.J.; Groll, J.; Salber, J.; Roy, I. Activated Polyhydroxyalkanoate Meshes Prevent Bacterial Adhesion and Biofilm Development in Regenerative Medicine Applications. *Front. Bioeng. Biotech.* **2020**, *8*, 442. [CrossRef]
7. Basnett, P.; Marcello, E.; Lukasiewicz, B.; Nigmatullin, R.; Paxinou, A.; Haseeb Ahmad, M.; Gurumayum, B.; Roy, I. Antimicrobial Materials with Lime Oil and a Poly(3-hydroxyalkanoate) Produced via Valorisation of Sugar Cane Molasses. *J. Funct. Biomater.* **2020**, *11*, 24. [CrossRef]
8. Ramos Avilez, H.; Castilla Casadiego, D.; Vega Avila, A.; Perales Perez, O.; Almodovar, J. Production of chitosan coatings on metal and ceramic biomaterials. *Chitosan Based Biomater.* **2017**, *1*, 255–293. [CrossRef]
9. Azimi, B.; Millazo, M.; Lazzeri, A.; Berrettini, S.; Uddin, Z.J.; Qin, M.; Buehler, M.J.; Danti, S. Electrospinning Piezoelectric Fibers for Biocompatible Devices. *Adv. Healthcare Mater.* **2020**, *9*, 1901287. [CrossRef]
10. Dietrich, K.; Dumont, M.; Del Rio, L.; Orsat, V. Producing PHAs in the bioeconomy—Towards a sustainable bioplastic. *SPAC* **2017**, *9*, 58–70. [CrossRef]
11. Kavadiya, S.; Biswas, P. Electrospray deposition of biomolecules: Applications, challenges, and recommendations. *J. Aerosol Sci.* **2018**, *125*, 182–207. [CrossRef]
12. Morganti, P.; Morganti, G. Chitin nanofibrils for advanced cosmeceuticals. *Clin. Dermatol.* **2008**, *26*, 334–340. [CrossRef]
13. Kowalska, A.; Kalinowska-Lis, U. Beta-Glycyrrhetinic acid: Its core biological properties and dermatological applications. *Int. J. Cosmet. Sci.* **2019**, *41*, 325–331.
14. Danti, S.; Trombi, L.; Fusco, A.; Azimi, B.; Lazzeri, A.; Morganti, P.; Coltelli, M.; Donnarumma, G. Chitin Nanofibrils and Nanolignin as Functional Agents in Skin Regeneration. *Int. J. Mol. Sci.* **2019**, *20*, 2669. [CrossRef]
15. Obisesan, K.A.; Neri, S.; Bugnicourt, E.; Campos, I.; Rodriguez-Turienzo, L. Determination and Quantification of the Distribution of CN-NL Nanoparticles Encapsulating Glycyrrhetic Acid on Novel Textile Surfaces with Hyperspectral Imaging. *J. Funct. Biomater.* **2020**, *11*, 32. [CrossRef]

16. Morganti, P.; Muzzarelli, C. Spray-Dried Chitin Nanofibrils, Method for Production and Uses. Thereof. Patent WO2007060628, 5 May 2007.
17. Morganti, P. Method of Preparation of Chitin and Active Principles Complexes and the so Obtained. Complexes. Patent WO2012143875, 26 October 2012.
18. Boeriu, C.G.; Bravo, D.; Gosselink, R.J.A.; Van Dam, J.E.G. Characterisation of structure-dependent functional properties of lignin with infrared spectroscopy. *Ind. Crops Prod.* **2004**, *20*, 205–218. [CrossRef]
19. Danti, S.; D'Alessandro, D.; Mota, C.; Bruschini, L.; Berrettini, S. Applications of bioresorbable polymers in skin and eardrum. In *Bioresorbable Polymers for Biomedical Applications: From Fundamentals to Translational Medicine*; Perale, G., Hilborn, J., Eds.; Woodhead Publishing, Elsevier: Cambridge, UK, 2017; pp. 423–444.
20. Lebeaux, D.; Ghigo, J.M.; Beloin, C. Biofilm-related infections: Bridging the gap between clinical management and fundamental aspects of recalcitrance toward antibiotics. *Microbiol. Mol. Biol. Rev.* **2014**, *78*, 510–543. [CrossRef]
21. Li, Z.; Yang, J.; Loh, X.J. Polyhydroxyalkanoates: Opening doors for a sustainable future. *NPG Asia Mater.* **2016**, *8*, e265. [CrossRef]
22. Reis, R.L.; Neves, N.M.; Mano, J.F.; Gomes, M.E.; Marques, A.P.; Azevedo, H.S. *Natural-Based Polymers for Biomedical Applications*; Elsevier: Amsterdam, The Netherland, 2008.
23. Roy, I.; Visakh, P.M. (Eds.) *Polyhydroxyalkanoate (PHA) Based Blends, Composites and Nanocomposites*; Royal Society of Chemistry: London, UK, 2014.
24. Ding, J.; Zhang, J.; Li, J.; Li, D.; Xiao, C.; Xiao, H.; Yang, H.; Zhuang, X.; Chen, X. Electrospun polymer biomaterials. *Prog. Polym. Sci.* **2019**, *90*, 1–34. [CrossRef]
25. Danti, S.; Azimi, B.; Candito, M.; Fusco, A.; Sorayani Bafqi, M.S.; Ricci, C.; Milazzo, M.; Cristallini, C.; Latifi, M.; Donnarumma, G.; et al. Lithium niobate nanoparticles as biofunctional interface material for inner ear devices. *Biointerphases* **2020**, *15*, 031004. [CrossRef]
26. Repanas, A.; Andriopoulou, S.; Glasmacher, B. The significance of electrospinning as a method to create fibrous scaffolds for biomedical engineering and drug delivery applications. *J. Drug Deliv. Sci. Technol.* **2016**, *31*, 137–146. [CrossRef]
27. Salvatore, L.; Carofiglio, V.E.; Stufano, P.; Bonfrate, V.; Calò, E.; Scarlino, S.; Nitti, P.; Centrone, D.; Cascione, M.; Leporatti, S.; et al. Potential of Electrospun Poly(3-hydroxybutyrate)/Collagen Blends for Tissue Engineering Applications. *J. Healthcare Eng.* **2018**, 6573947. [CrossRef]
28. Coltelli, M.-B.; Aliotta, L.; Vannozzi, A.; Morganti, P.; Panariello, L.; Danti, S.; Neri, S.; Fernandez-Avila, C.; Fusco, A.; Donnarumma, G.; et al. Properties and Skin Compatibility of Films Based on Poly(Lactic Acid) (PLA) Bionanocomposites Incorporating Chitin Nanofibrils (CN). *J. Funct. Biomater.* **2020**, *11*, 21. [CrossRef]
29. Mirjalili, M.; Zohoori, S. Review for application of electrospinning and electrospun nanofibers technology in textile industry. *J. Nanostructure Chem.* **2016**, *6*, 207–213. [CrossRef]
30. Merrill, M.H.; Pogue, W.R.; Baucom, J.N. Electrospray Ionization of Polymers: Evaporation, Drop Fission, and Deposited Particle Morphology. *ASME. J. Micro Nano-Manuf.* **2015**, *3*, 011003. [CrossRef]
31. Zamani, M.; Prabhakaran, M.P.; Ramakrishna, S. Advances in drug delivery via electrospun and electrosprayed nanomaterials. *Int. J. Nanomed.* **2013**, *8*, 2997–3017. [CrossRef]
32. Morganti, P.; Palombo, M.; Tishchenko, G.; Yudin, V.E.; Guarneri, F.; Cardillo, M.; Del Ciotto, P.; Carezzi, F.; Morganti, G.; Fabrizi, G. Chitin-hyaluronan nanoparticles: A multifunctional carrier to deliver anti-aging active ingredients through the skin. *Cosmetics* **2014**, *1*, 140–158. [CrossRef]
33. Mantovani, A.; Sica, A.; Sozzani, S.; Allavena, P.; Vecchi, A.; Locati, M. The chemokine system in diverse forms of macrophage activation and polarization. *Trends Immunol.* **2004**, *25*, 677–686. [CrossRef]
34. Dinarello, C.A. Interleukin-1. *Cytokine Growth Factor Rev.* **1997**, *8*, 253–265. [CrossRef]
35. Koch, A.E.; Polverini, P.J.; Kunkel, S.L.; Harlow, L.A.; DiPietro, L.A.; Elner, V.M.; Elner, S.G.; Strieter, R.M. Interleukin-8 as a macrophage-derived mediator of angiogenesis. *Science* **1992**, *258*, 1798–1801. [CrossRef]
36. Esposito, E.; Cuzzocrea, S. TNF-alpha as a therapeutic target in inflammatory diseases, ischemia-reperfusion injury and trauma. *Curr. Med. Chem.* **2009**, *16*, 3152–3167. [CrossRef]
37. Donnarumma, G.; Paoletti, I.; Fusco, A.; Perfetto, B.; Buommino, E.; de Gregorio, V.; Baroni, A. β-Defensins: Work in Progress. *Adv. Exp. Med. Biol.* **2016**, *901*, 59–76. [CrossRef]
38. Nunan, R.; Harding, K.G.; Martin, P. Clinical challenges of chronic wounds: Searching for an optimal animal model to recapitulate their complexity. *Dis. Models Mech.* **2014**, *7*, 1205–1213. [CrossRef]

39. Zhang, X.; Shu, W.; Yu, Q.; Qu, W.; Wang, Y.; Li, R. Functional Biomaterials for Treatment of Chronic Wound. *Front. Bioeng. Biotechnol.* **2020**, *8*, 516. [CrossRef]
40. Lazzeri, L.; Cascone, M.G.; Danti, S.; Serino, L.P.; Moscato, S.; Bernardini, N. Gelatine/PLLA sponge-like scaffolds: Morphological and biological characterization. *J. Mater. Sci. Mater Med.* **2007**, *18*, 1399–1405. [CrossRef]
41. Milazzo, M.; Gallone, G.; Marcello, E.; Mariniello, M.D.; Bruschini, L.; Roy, I.; Danti, S. Biodegradable polymeric micro/nano-structures with intrinsic antifouling/antimicrobial properties: Relevance in damaged skin and other biomedical applications. *J. Funct. Biomater.* **2020**, *11*, 60. [CrossRef]

© 2020 by the authors. Licensee MDPI, Basel, Switzerland. This article is an open access article distributed under the terms and conditions of the Creative Commons Attribution (CC BY) license (http://creativecommons.org/licenses/by/4.0/).

 Journal of
Functional Biomaterials

Article

Development of Bionanocomposites Based on Poly(3-Hydroxybutyrate-co-3-Hydroxyvalerate)/PolylActide Blends Reinforced with Cloisite 30B

Clément Lacoste [1,*], Benjamin Gallard [1], José-Marie Lopez-Cuesta [1], Ozlem Ipek Kalaoglu-Altan [2] and Karen De Clerck [2]

1. Polymers Composites and Hybrids (PCH), IMT Mines Ales, 6 avenue de Clavières, 30319 Ales CEDEX, France; benjamin.gallard@mines-ales.fr (B.G.); jose-marie.lopez-cuesta@mines-ales.fr (J.-M.L.-C.)
2. Centre for Textile Science and Engineering, Department of Materials, Textiles and Chemical Engineering, Faculty of Engineering and Architecture, Ghent University (UGent), Tech Lane Science Park 70A, 9052 Ghent, Belgium; ozlemkalaoglu@gmail.com (O.I.K.-A.); karen.declerck@ugent.be (K.D.C.)
* Correspondence: clement.lacoste@mines-ales.fr; Tel.: +3-34-6678-5655

Received: 29 July 2020; Accepted: 11 September 2020; Published: 16 September 2020

Abstract: In the present study, poly(3-hydroxybuturate-co-3-hydroxyvalerate) (PHBV) and plasticized polylactide acid (PLA) blends were processed by melt extrusion with different weight ratio (up to 20 wt.% of PHBV). Bionanocomposites were obtained through the incorporation of an organomodified montmorillonite (C30B) at 3 wt.%. The main features of the processing and physico-chemical characterization of films and injected samples were assessed and the influence of the components on the chemical, thermal and mechanical properties of the bionanocomposites was investigated. The results indicated that plasticized PLA/PHBV/C30B bionanocomposites present optimal mechanical properties for sanitary applications. Moreover, plasticized PLA/PHBV could lead to finely tuned biomaterials able to form electrospun nanofibers.

Keywords: biopolymer; blends; PLA; PHBV; nanocomposite

1. Introduction

Nowadays, due to environmental concerns, the substitution of non-renewable polymers by bio-based and biodegradable polymers has attracted considerable attention, especially for short term applications. In medical applications such as wound healing, the biocompatible character of the polymer is also required. In this context, biopolyesters seem to be the most promising materials gathering the '3B' criteria: bio-based, biodegradable, and biocompatible.

Among them, poly(lactic acid) (PLA) is an aliphatic linear thermoplastic derived from the fermentation of starch from renewable crops (corn, potato) or from cellulose that can be degraded by micro-organisms within few weeks under environmental conditions. PLA exhibits high transparency, thermal plasticity and good mechanical properties, making it currently the biopolymer most used for short-term applications such as packaging (films, rigid cups, cutlery, etc), sutures and medical devices.

Poly(3-hydroxybutyrate-co-3-hydroxyvalerate) (PHBV) is also an aliphatic biopolyester synthesized by bacterial fermentation from renewable resources like sugar and vegetable oil or other microorganisms [1]. Due to its non-toxicity and biocompatible character, PHBV has been extensively used in pharmaceutical (encapsulation, drug delivery), biomedical (sutures, implant, surgical materials), packaging (containers, films) and coating applications [2]. Moreover, compared to other biodegradable polymers, PHBV shows low thermal resistance ability due to its high crystallinity.

In the recent past, market demand of those two promising candidates has seriously increased. Nevertheless, none of these two polymers can fulfill the requirement of almost all structural materials when used alone. Indeed, PLA and PHBV exhibit brittleness and efforts have been made recently to improve their ductility. Different methods can be used to increase the mechanical properties of biopolymers and to extend their range of industrial applications like random and block copolymerization, chemical modification or filler addition [3]. Among them, the blending of polymers is a simple and cost-effective approach to prepare composites of different morphologies and physical characteristics. Despite the immiscibility of PLA and PHBV [4–6], the physical and mechanical properties of their blends can be tailored by varying the composition. Ferreira et al. [4] have reported that adding PHBV to PLLA (poly-L-lactic acid) led to tougher materials compared to the neat samples, and the blend films showed yield behavior. Zembouai et al. [3,5] have studied the thermo-mechanical properties of PLA/PHBV blends of several ratios prepared by melt compounding and observed that the tensile stress and elongation at break decreased with increasing PHBV content, whereas the modulus increased. The less brittle blend was obtained for PLA/PHBV 75/25 with an elongation of 7% (±1.3) due to the fine spherulitic structure operated by the distribution of PHBV particles into the PLA matrix. It is then necessary to finely control the morphology of the blend to extend their properties.

Moreover, the combined association with nanoclays is an effective strategy to improve polymer mechanical strength since the addition of nanofillers is able to widely extend the final properties of the reinforced material compared to the pristine polymers. Among them, the addition of organomodified layered silicates (OMLS) such as montmorillonites has given rise to an increasing interest [7–9]. As an example of reinforcement using montmorillonite, Paul et al. [7] attested the intercalation of the clays in PLA matrix, and noticed that the intercalation could also be supported by interlayer migration phenomenon for plasticized PLA matrix. Ozdemir at al. [8] also observed the good dispersion of montmorillonite (3 wt.%) into PLA/polyethylene glycol, enhancing the rigidity and the tensile properties of the nanocomposites. The preparation of OMLS-based nanocomposites with enhanced mechanical properties could lead to numerous competitive biomaterials like electrospun fibers for wound dressing.

Although biopolyester polymers can be considerably improved with the above mentioned strategies, difficulties in improving films' production through cost-effective processes still exist. In this sense, the electrospinning process has gained considerable attention for the production of effective biopolymer films. Electrospinning is indeed a simple, viable, and attractive method that can be used for industrial applications as it can be effectively up-scaled [10]. Concerning biomedical materials, studies have reported that electrospun nanofibers are the most effective wound dressing materials compared to sponges, hydrocolloids, and hydrogel materials. The nanoporous structure enables liquid evaporation and has excellent oxygen permeability, preventing bacteria contamination [11].

Therefore, the aim of this study was to develop biopolyester nanocomposites for the production of electrospun nanofibers with appropriate strength/stiffness balance. Plasticized PLA was used as matrix, whereas PHBV was used as minor phase, and organomodified montmorillonite nanofillers were used to control the mechanical strength. The miscibility, crystallization and melting behavior, as well as the mechanical properties of the bionanocomposite blends and films, were investigated.

2. Materials and Methods

2.1. Materials

Plasticized PLA was supplied by NatureWorks in pellet form under the trade name NP SF 141, extrusion grade. This polymer is a semicrystalline one having the following properties according to the supplier: density = 1.22 g.cm^{-3} and T_m = 150–160 °C, clarity = transparent.

PHBV was supplied by NaturePlast in pellet form under the trade name PHI002, injection molding grade. The average molecular weight is Mw = 203,000 g.mol^{-1} and the polydispersity index is PDI = 3.2. According to the supplier, the main properties of this polyhydroxy alkanoate are the following: density = 1.23 g.cm^{-3} and T_m = 170–176 °C, clarity = opaque.

Cloisite 30B (C30B) was supplied by Southern Clay Products (Gonzales, TX, USA). This powder (d_{001} = 1.9 nm) is a montmorillonite modified with bis-(2-hydroxyethyl) methyl tallow alkyl ammonium cations. C30B was dried under vacuum at 60 °C overnight before use.

2.2. Preparation of the Blends, Films and Nanofibers

The polymers were pre-dried at 60 °C at 24 h before molding in order to avoid hydrolytic degradation of the polymers in the extruder. The formulations of the samples are reported in Table 1. As reported in previous works [3,7,8], incorporation of 3% of montmorillonite increased the mechanical and thermal stability of plasticized PLA and was selected in this work.

Table 1. Bio-nanocomposites' formulations.

Sample	PLA NF 141 (wt.%)	PHBV (wt.%)	C30B (wt.%)
PLA	100	-	-
PLA/5P	95	5	-
PLA/5P/3C	92	5	3
PLA/10P	90	10	-
PLA/10P/3C	87	10	3
PLA/15P	85	15	-
PLA/20P	80	20	-

Firstly, the biopolymer blends were prepared by melt mixing in a twin screw extruder BC21 from CLEXTRAL (CLEXTRAL SAS, Firminy, France), with the screw profile reported in Figure 1. The diameter of the screws was 25 mm and the length of the extruder was 900 mm. The blend was extruded into a single hole die of 5 mm diameter.

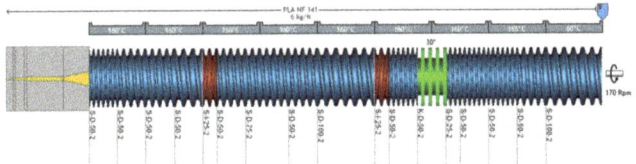

Figure 1. Screw profile developed from LUDOVIC®software (2016, Sciences Computers Consultants, Saint-Etienne, France).

Beforehand the processing the polymers were dried by PIOVAN dryer at 80 °C for 12 h and in an oven at 100 °C for 12 h for C30B. The main processing parameters were fixed for all formulations as follows: the temperatures were fixed at 60 °C in the first zone (Z1), 155 °C in the second zone (Z2), and finally 160 °C from Z3 to Z12. The screw speed was fixed at 170 rpm, the flow rate was 6 kg/h, the torque was around 65 N.m-1 and the pressure was 30 bars.

Secondly, the films were prepared by a single screw extruder from Thermofisher Electron: Polylab OS7 with a Rheomex 19/25. The screw profile was a standard one 3:1 L = 25D (Diameter = 19 mm). The width of the flat die was 100 mm and the thickness was adjustable. The temperatures were Z1 = 170 °C, Z2 = 170 °C, Z3 = 165 °C, Z4 = 160 °C. The screw speed was 120 rpm. The temperature of the roll was 15 °C and the speed was adjusted manually depending of the thickness of the film. The film thickness measured by a Palmer average was 180µm along the width.

Electrospinning solutions were prepared by dissolving PLA/5P, PLA/10P and PLA/10P/3C in chloroform/DMF (N,N-Dimethylformamide) (5/1, v/v) solvent mixture with a concentration of 8 wt.%. Mass concentrations are expressed by weight percentages (wt.%) defined by the ratio of the polymer mass and the sum of the polymer and solvent mass. All electrospinning experiments were performed on a mononozzle setup using the solvent electrospinning technique with an 18 gauge Terumo mixing needle without bevel. A stable Taylor cone was obtained with a flow rate of 1 mL/h, a tip to collector distance of 20 cm and an applied voltage of 17 kV at room conditions ((25 ± 5) °C and (35 ± 5)% relative humidity).

2.3. Spectral and Thermal Characterizations

The thermal properties of the products were studied through differential scanning calorimetry (DSC), thermogravimetric analysis (TGA), and Fourier Transform infrared (FTIR) analysis. Each measurement was done in duplicate.

DSC measurements were performed on a Perkin Elmer Diamond (Perkin Elmer SAS, Villebon sur Yvette, Evry, France) operating under a constant flow of nitrogen (30 mL·min^{-1}). Samples were weighed (circa 10 mg) into aluminum crucibles. Analysis in comparison to a blank crucible was performed using a temperature profile (ramp rate of 10 °C·min^{-1}) from 15 °C to 200 °C. Thermal transitions were calculated from Perkin software through the tangent method.

Measurements of mass loss versus temperature were performed using a Perkin Elmer thermogravimetric analyzer (TGA) module Pyris 1 (Perkin Elmer SAS, Villebon sur Yvette, Evry, France), under N_2 purge (flow rate of 50 mL·min^{-1}). Typically, 8–10 mg of sample were placed on an aluminum oxide pan, and heated from 30 °C to 900 °C at 10 °C·min^{-1}.

FT-IR spectra of PHBV/PLA/C30B blends were recorded with a Bruker Vertex 70 spectrometer (Bruker France SAS, Champs sur Marne, France) in transmission mode using a resolution of 4 cm^{-1} and 32 scans per sample. The spectra were recorded in the range of 4000–400 cm^{-1}.

2.4. Contact Angle

The wettability of the blends was determined using a Krüss Wettability Meter (KRUSS GmbH France, Villebon sur Yvette, France) equipped with a 60 picture/s camera. First, a 2 µL drop of water was put at the sample's surface. Then, contact angle was determined as the mean of 10 values measured each second after drop deposition.

2.5. Tensile Tests

The static tensile tests were performed at room temperature according to ISO 527 on a ZWICK TH010 universal testing machine (ZWICKROELL, Ars-Laquenexy, France) equipped with a 10 kN load head and using a loading speed of 1 mm/min. An extensometer was used at low elongation to measure the elastic moduli. The dumbbell-shape samples (75 × 4 × 1 mm^3) were tested at least 5 times for each sample.

3. Results

3.1. Thermal Degradation

The thermal stability of the pristine polymers and their blends was determined by TGA and the curves are presented in Figure 2. The decomposition temperatures at different stages (5%, 10% and 50% of mass loss) as well as residues at 600 °C are reported in Table 2.

Table 2. Decomposition temperatures at 5%, 10%, 50% weight loss and residues at 600 °C of the blends.

Sample	$T_{5\%}$ (°C)	$T_{10\%}$ (°C)	$T_{50\%}$ (°C)	$Char_{600\ °C}$ (%)
PLA	301	323	365	3.6
PLA/5P	302	314	364	3.4
PLA/5P/3C	300	317	365	5.8
PLA/10P	296	306	356	3.5
PLA/10P/3C	300	309	356	5.0
PLA/15P	291	303	358	2.9
PLA/20P	293	302	355	3.3
PHBV	287	291	306	1.3

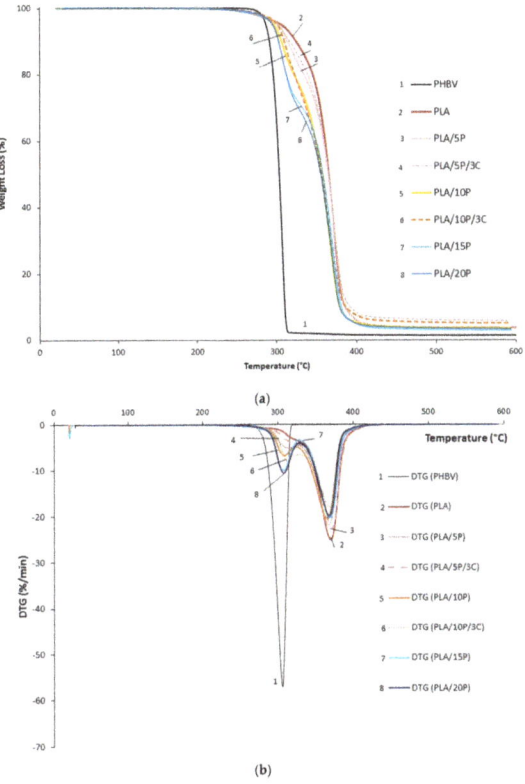

Figure 2. (a) Weight loss curves and (b) Derivative thermogram (DTG) of PHBV, PLA, and their blends from TGA.

As expected, PLA ($T_{50\%}$ = 365 °C) is more thermally stable than PHBV ($T_{50\%}$ = 306 °C). It can be observed that the degradation of PHBV occurred in a single step separately from PLA, which presented a double peak (first peak circa 320 °C) attributed to the presence of plasticizer (Figure 2b). It was previously reported in the literature that the $T_{50\%}$ of pristine PLA is around 360 °C [3,8,12,13], which is in accordance with the measured value $T_{50\%}$ = 365 °C. However, it could be noticed that for this plasticized grade of PLA, the onset temperature $T_{5\%}$ was shifted to 301 °C compared to 330 °C reported in the literature for non-plasticized grades [3]. All the decomposition temperatures of the blends are between those of the PLA and PHBV and occurred in a two-step process. The addition of PHBV in the PLA matrix shifted the onset degradation temperature to lower values (Table 1). The presence of 3 wt.% of cloisite 30B did not significantly modify the thermal stability of the blends as similar degradation temperatures at 5%, 10% and 50% were reported for PLA/5P and PLA/5P/3C, as well as for PLA/10P and PLA/10P/3C. A very similar thermal behavior has been reported in previous studies focused on EVA-based (Ethylene-vinyl acetate) [14] and plasticized PLA-based nanocomposites [7] with an optimal thermal stability noticed at 3 wt.% of montmorillonite content. Meanwhile, the residue was increased with the presence of C30B at 5.0% for PLA/10P/3C against 3.5% for PLA/10P/3C, and at 5.8% for PLA/5P/3C against 3.4% for PLA/5P/3C. This is in agreement to the organic fraction of C30B (around 35 wt.%). Note that this grade of PLA has much more residue (3.6%) than PHBV (1.3%), suggesting the presence of additives able to produce a charred material.

3.2. Thermal and Crystallization Behavior of PLA/PHBV Blends

PLA and PHBV are typical semicrystalline polymers and their properties are highly related to their solid-state morphology and crystallinity. It is then important to study the influence of the minor phase on the crystallization of the matrix. Figure 3a shows the DSC cooling thermogram after being melted at 200 °C. A single crystallization peak was found at 94.7 (±0.3) °C and 89.4 (±0.5) °C for neat PHBV and PLA, respectively. The cold crystallization temperatures (T_c) of PLA/PHBV blends were shifted to lower temperatures of 86.3 °C, 85.7 °C, 85.4 °C and 85.3 °C when the PHBV content increased from 5%, 10%, 15% and 20%, respectively. The presence of PLA restricted the crystallization of PHBV by suppressing the nucleation of PHBV in the blend. Similar results were also reported on the crystallization behavior of PHBV/PLA blends as the degree of crystallinity of PLA generally increased with PHBV [15,16]. For instance, the presence of 3 wt.% of C30B slightly affected the crystallization with an increase in the enthalpy from 15.5 J/g to 17.9 J/g for PLA/10/3C in comparison with PLA/10P.

Figure 3b shows the DSC heating thermogram (second run) after cooling at 10 °C/min. In the case of PLA, the melting peak centered at 157.5 °C with a small shoulder at 145.9 °C, which corresponds to the melt of small and imperfect crystals of lower thermal stability [13]. PHBV presented also a single melting peak at 175.2 °C. Concerning the PLA/PHBV blends, a multi-step process was observed with up to three melting temperatures (T_{m1}, T_{m2}, and T_{m3}) reported in Table 3, suggesting a lack of miscibility. The first double peak around 145–155 °C could be attributed to the formation of different crystal structures due to the melting of PLA, as reported in other PLA blends and composites [13,17]. The third peak could be the consequence of a recrystallization during melting attributed to possible PHBV degradation. As expected [8], the incorporation of C30B into the PHBV/PLA composite did not modify the melting temperatures of the blends.

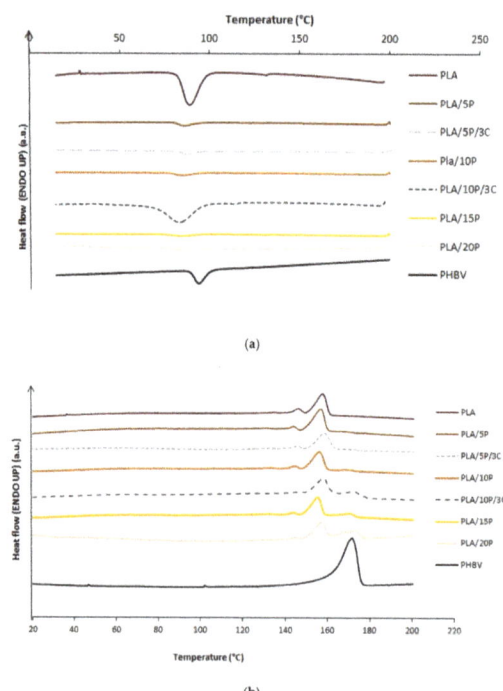

Figure 3. DSC curves of PLA, PHBV and their blends at (a) the cooling stage and (b) the second heating stage.

Table 3. DSC properties of PLA, PHVB and their blends.

Sample	T_c (°C)	ΔH_c (J·g^{-1})	T_{m1} (°C)	ΔH_{m1} (J·g^{-1})	T_{m2} (°C)	ΔH_{m2} (J·g^{-1})	T_{m3} (°C)	ΔH_{m3} (J·g^{-1})
PLA	89.4	21.0	145.9	2.5	156.7	19.4	-	-
PLA/5P	86.3	20.3	143.7	1.6	156.3	22.6	-	-
PLA/5P/3C	87.4	18.2	145.3	1.5	158.2	18.4	171.8	0.50
PLA/10P	85.7	15.5	144.1	1.6	158.1	17.2	172.1	0.60
PLA/10P/3C	84.0	17.8	145.5	0.3	157.9	15.9	172.3	3.90
PLA/15P	85.4	15.9	144.0	1.1	155.4	16.7	171.6	3.40
PLA/20P	85.3	16.1	145.3	0.5	157.2	14.6	173.5	2.50
PHBV	94.7	76.9	138.1	9.8	172.5	80.4	-	-

3.3. FTIR

FTIR spectra of some PLA/PHBV films are shown in Figure 4 with an arbitrary offset for comparison. All spectra displayed the characteristic bands of PLA-based materials. An intense peak was observed at 1749 cm^{-1}, attributed to the carbonyl vibration in polyesters. The bands at 1180 cm^{-1} and 1083 cm^{-1} belong to asymmetric and symmetric C-O-C vibration, respectively. Two bands related to the amorphous and crystalline phases of PLA are found at 867 cm^{-1} and 755 cm^{-1}. A peak located at 1724 cm^{-1} (Figure 4b, grey narrows) is increasingly intense, as the PHBV content increased as it belongs to stretching vibrations of the crystalline carbonyl group of PHBV [17], confirming the low miscibility of the two biopolyester. Regarding PLA/5P/3C, small intense peaks were observed at 519 cm^{-1} and 1035 cm^{-1}, corresponding to Si-O bending and stretching, respectively, confirming the presence of C30B.

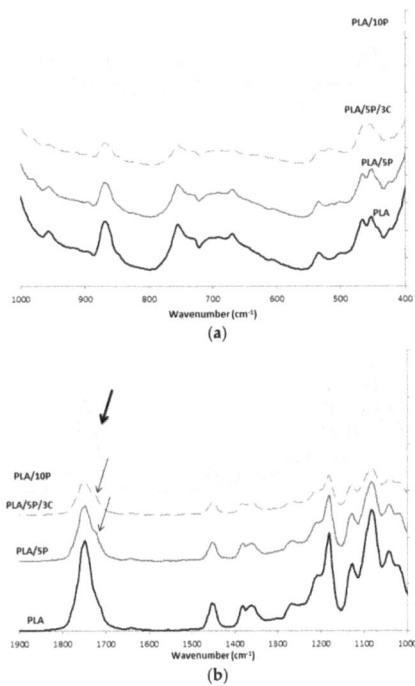

Figure 4. FT-IR spectra of PLA, PLA/5P, PLA/5P/3C and PLA/10P film samples in the range of (a) 1000–400 cm^{-1}; (b) 1900–1000 cm^{-1}.

3.4. Contact Angle

The water contact angle (WCA) measurements were investigated and results are shown in Figure 5. No significant differences of WCA were observed for neat PLA samples, either for the film or for the dumbbell shape. All values were in range of 76.1° ± 0.5. The wettability of the polymers is not changed by the addition of cloisite 30B.

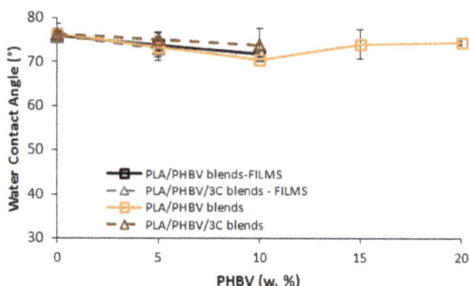

Figure 5. Water contact angle (WCA) of the bio-nanocomposites on films and dumbbell shape surfaces.

3.5. Tensile Properties

Figure 6 shows the tensile modulus, strength, and elongation at break of neat PLA, PHBV, and their blends (with and without C30B) according to PHBV content. As expected [17,18], the presence of plasticizer in this grade of PLA decreases both tensile strength and modulus compared to usual grades with elongation close to 5% [5]. Thereby, the high elongation at break (circa 350% for the films and 285% for dumbbell-shape samples) confirms the very good plasticizer efficiency, making it a good candidate when ductile behavior is required. However, it is obvious that the shape of the materials will highly influence their mechanical properties (Figure 6 and Table 4). Regarding dumbbell-shape PLA, the strength of the polymer is slightly higher (+11%) and its rigidity is clearly stronger (+686%) in comparison to the film, but the elongation at break is reduced (−18%). To a lesser extent, similar observations can be done for PLA/PHBV blends.

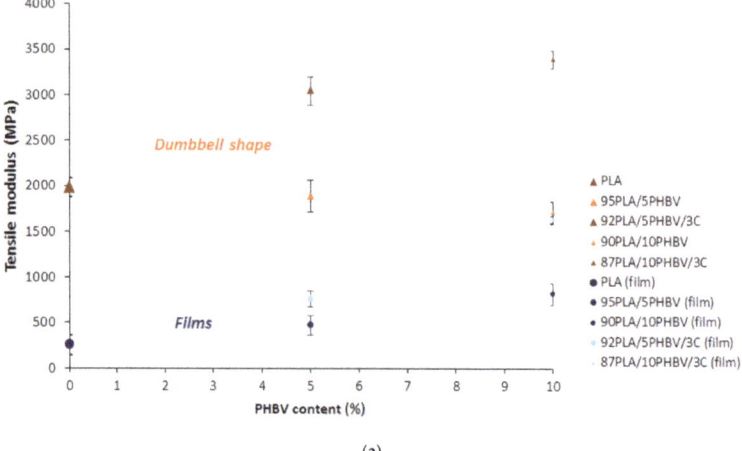

(a)

Figure 6. *Cont.*

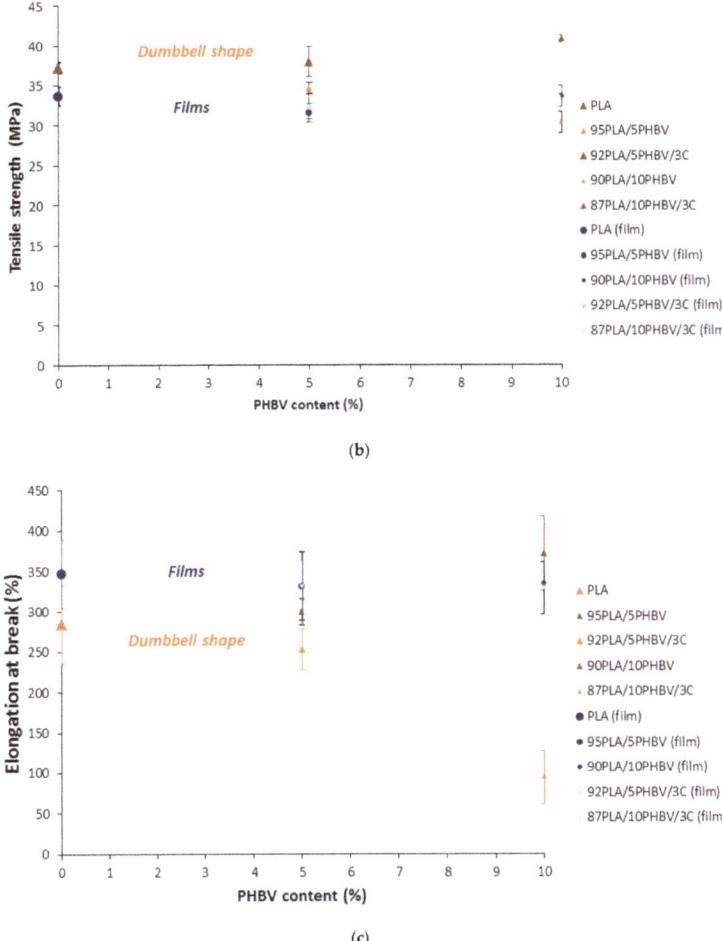

Figure 6. (a) Tensile modulus, (b) strength, and (c) elongation at break of neat PLA, PHBV, and their blends.

The addition of 5% and 10% of PHBV into the PLA matrix induced a diminution of the tensile properties of the blends (elongation and stress at break) whereas an increase in the Young modulus was observed. Similar results were reported regarding the diminution of the stress at break and the modulus. It is believed that the formation of spherulitic PHBV particles finely dispersed in the PLA matrix act as fillers [3,19].

Nevertheless, the addition of nanoclays helped to offset the loss of mechanical properties of the blends. The addition of 3% of C30B significantly increased the Young modulus of the blend measured at 3043 ± 157 MPa and 3386 ± 97 MPa for PLA/5PHBV and PLA/10PHBV, respectively, when the modulus of neat PLA was found to be 1981 ± 107 MPa for dumbbell-shape samples. The rigidity of the films also significantly increased with Young's moduli of 252 ± 111 MPa, 762 ± 81 MPa, and 1630 ± 41 MPa for PLA, PLA/5P and PLA/10P, respectively. It might be due to a strong interfacial interaction between the polymers and the silicates layers leading to supramolecular assemblies. To a lesser extent, the maximal stress at break was slightly improved at the expense of a limited reduction in elongation with the addition of 3% of nanoclays.

Table 4. Effect of shape and nanoclay C30B on the tensile properties of PLA/PHBV blends.

Sample	Shape	C30B (%)	Tensile Strength (MPa)	Young modulus (MPa)	Elongation (%)
PLA	film	-	33.7 ± 0.7	252 ± 11	347 ± 49
	dumbbell	-	37.3 ± 1.2	1981 ± 107	285 ± 42
95PLA/5PHBV	film	-	31.5 ± 1.5	468 ± 66	331 ± 27
	film	3	32.3 ± 1.0	762 ± 81	333 ± 33
	dumbbell	-	34.7 ± 0.7	1888 ± 174	299 ± 16
	dumbbell	3	37.9 ± 1.9	3043 ± 157	253 ± 35
90PLA/10PHBV	film	-	33.6 ± 2.2	811 ± 87	334 ± 18
	film	3	34.4 ± 0.5	1630 ± 41	329 ± 32
	dumbbell	-	30.4 ± 1.3	1703 ± 123	371 ± 46
	dumbbell	3	40.8 ± 0.4	3386 ± 97	95 ± 33

The formulation of the bio-nanocomposite PLA/10P/3C appears then as a promising candidate for sanitary products like wound dressing or face masks, which are required to fit the body shape. The mechanical properties of the films offered excellent rigidity (E = 1630 ± 41 MPa), good resistance (σ = 1630 ± 41 MPa), as well as considerable stretchability (ε = 329 ± 32%) in comparison with the limited ductility of the neat PLA (Table 4).

3.6. Electrospun Nanofibers

The morphological aspects and the diameter of the electrospun fibers corresponding to the PLA/5P, PLA/10P blends and PLA/10P/3C nanocomposite compositions were investigated by SEM (Figure 7). PLA/PHBV blends have shown good spinnability with the formation of uniform, randomly oriented fibers, although some defects like beads are observed for the nanocomposite. The average fiber diameters were calculated as 324 ± 89 nm, 402 ± 113 nm and 433 ± 97 nm for PLA/10P/3C, PLA/10P and PLA/5P, respectively. The presence of the nanoclay provided narrower fibers. It is believed that the presence of modified montmorillonite with organic alkyl ammonium cations can improve the electrical conductivity, which may lead to smaller average fiber diameter. However, the SEM image in Figure 7a involving 3 wt.% of cloisite 30B indicates a non-homogeneous surface along the fiber, probably due to clay aggregation.

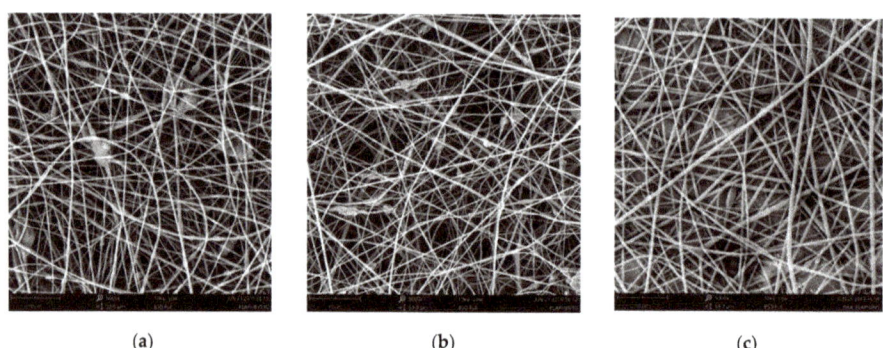

Figure 7. Electrospun nanofibers of (**a**) PLA/10P/3C; (**b**) PLA/10P and (**c**) PLA/5P observed by SEM.

4. Conclusions

Bionanocomposites based on plasticized PLA matrix, PHBV in a minor phase, and C30B as nanofiller were prepared. As expected, poor miscibility between PLA and PHBV was observed. The addition of PHBV up to 20 wt.% provoked a decrease in the thermal stability and has shifted crystallization and melting temperatures to lower values for all PLA/PHBV blends in comparison to the neat PLA matrix. Then, an increase in the material rigidity was observed with a significant reduction in the tensile strength. However, the large efficiency of the plasticized PLA matrix allowed elongation at break values close to 300% to be maintained. Meanwhile, the addition of 3 wt.% of clay C30B, well dispersed into the polymeric phase, did not significantly damage the thermal behavior of the blends but enabled a considerable increase in the mechanical properties of the blends. The formulation with moderate PHBV content (up to 10%) and reinforced with C30B appeared to be particularly interesting with balanced strength/elongation properties to provide stretchable and resistant films. Thus, the bionanocomposite could be a promising candidate to extend the functional properties of the biodegradable polymers, as well as for the production of electrospun nanofibers.

Author Contributions: Conceptualization, C.L., J.-M.L.-C. and K.D.C.; methodology, B.G. and O.I.K.-A.; formal analysis, C.L.; investigation, C.L., O.I.K.-A., and B.G.; data curation, C.L. and O.I.K.-A.; writing—original draft preparation, C.L.; writing—review and editing, all authors.; supervision and funding acquisition, J.-M.L.-C. and K.D.C. All authors have read and agreed to the published version of the manuscript.

Funding: This research was funded by of the European project POLYBIOSKIN (H2020), grant number 745839, and all the authors acknowledge the funding support.

Conflicts of Interest: The authors declare no conflict of interest. The funders had no role in the design of the study; in the collection, analyses, or interpretation of data; in the writing of the manuscript, or in the decision to publish the results.

References

1. Laycock, B.; Halley, P.; Pratt, S.; Werker, A.; Lant, P. The chemomechanical properties of microbial polyhydroxyalkanoates. *Prog. Polym. Sci.* **2014**, *39*, 397–442. [CrossRef]
2. Tarrahi, R.; Fathi, Z.; Seydibeyoğlu, M.Ö.; Doustkhah, E.; Khataee, A. Polyhydroxyalkanoates (PHA): From production to nanoarchitecture. *Int. J. Biol. Macromol.* **2020**, *146*, 596–619. [PubMed]
3. Zembouai, I.; Bruzaud, S.; Kaci, M.; Benhamida, A.; Corre, Y.M.; Grohens, Y.; Taguet, A.; Lopez-Cuesta, J.M. Poly(3-Hydroxybutyrate-co-3-Hydroxyvalerate)/Polylactide Blends: Thermal Stability, Flammability and Thermo-Mechanical Behavior. *J. Polym. Environ.* **2014**, *22*, 131–139. [CrossRef]
4. Ferreira, B.M.P.; Zavaglia, C.A.C.; Duek, E.A.R. Films of PLLA/PHBV: Thermal, morphological, and mechanical characterization. *J. Appl. Polym. Sci.* **2002**, *86*, 2898–2906.
5. Zembouai, I.; Kaci, M.; Bruzaud, S.; Benhamida, A.; Corre, Y.M.; Grohens, Y. A study of morphological, thermal, rheological and barrier properties of Poly(3-hydroxybutyrate-Co-3-Hydroxyvalerate)/polylactide blends prepared by melt mixing. *Polym. Test.* **2013**, *32*, 842–851.
6. Yang, J.; Zhu, H.; Zhang, C.; Jiang, Q.; Zhao, Y.; Chen, P.; Wang, D. Transesterification induced mechanical properties enhancement of PLLA/PHBV bio-alloy. *Polymer* **2016**, *83*, 230–238. [CrossRef]
7. Paul, M.-A.; Alexandre, M.; Degée, P.; Henrist, C.; Rulmont, A.; Dubois, P. New nanocomposite materials based on plasticized poly(L-lactide) and organo-modified montmorillonites: Thermal and morphological study. *Polymer* **2003**, *44*, 443–450. [CrossRef]
8. Ozdemir, E.; Hacaloglu, J. Characterizations of PLA-PEG blends involving organically modified montmorillonite. *J. Anal. Appl. Pyrolysis* **2017**, *127*, 343–349. [CrossRef]
9. Zembouai, I.; Kaci, M.; Zaidi, L.; Bruzaud, S. Combined effects of Sepiolite and Cloisite 30B on morphology and properties of poly(3-hydroxybutyrate-co-3-hydroxyvalerate)/polylactide blends. *Polym. Degrad. Stab.* **2018**, *153*, 47–52. [CrossRef]
10. Sanhueza, C.; Acevedo, F.; Rocha, S.; Villegas, P.; Seeger, M.; Navia, R. Polyhydroxyalkanoates as biomaterial for electrospun scaffolds. *Int. J. Biol. Macromol.* **2019**, *124*, 102–110. [PubMed]
11. Mutlu, G.; Calamak, S.; Ulubayram, K.; Guven, E. Curcumin-Loaded electrospun PHBV nanofibers as potential wound-dressing material. *J. Drug Deliv. Sci. Technol.* **2018**, *43*, 185–193. [CrossRef]
12. Wang, S.; Ma, P.; Wang, R.; Wang, S.; Zhang, Y.; Zhang, Y. Mechanical, thermal and degradation properties of poly(d,l-lactide)/poly(hydroxybutyrate-co-hydroxyvalerate)/poly(ethylene glycol) blend. *Polym. Degrad. Stab.* **2008**, *93*, 1364–1369. [CrossRef]
13. Arrieta, M.P.; López, J.; López, D.; Kenny, J.M.; Peponi, L. Development of flexible materials based on plasticized electrospun PLA-PHB blends: Structural, thermal, mechanical and disintegration properties. *Eur. Polym. J.* **2015**, *73*, 433–446.
14. Lim, S.T.; Hyun, Y.H.; Choi, H.J.; Jhon, M.S. Synthetic Biodegradable aliphatic polyester/montmorillonite nanocomposites. *Chem. Mater.* **2002**, *14*, 1839–1844. [CrossRef]

15. Qiu, Z.; Ikehara, T.; Nishi, T. Miscibility and crystallization behaviour of biodegradable blends of two aliphatic polyesters. Poly(3-hydroxybutyrate-co-hydroxyvalerate) and poly(butylene succinate) blends. *Polymer* **2003**, *44*, 7519–7527. [CrossRef]
16. Jiang, N.; Abe, H. Morphological changes in poly(L-lactide)/poly(3-hydroxybutyrate- co-3-hydroxyvalerate) blends induced by different miscibility. *Polymer* **2015**, *66*, 259–267. [CrossRef]
17. Armentano, I.; Fortunati, E.; Burgos, N.; Dominici, F.; Luzi, F.; Fiori, S.; Jiménez, A.; Yoon, K.; Ahn, J.; Kang, S.; et al. Bio-Based PLA_PHB plasticized blend films: Processing and structural characterization. *LWT Food Sci. Technol.* **2015**, *64*, 980–988.
18. Arrieta, M.P.; Fortunati, E.; Dominici, F.; López, J.; Kenny, J.M. Bionanocomposite films based on plasticized PLA-PHB/cellulose nanocrystal blends. *Carbohydr. Polym.* **2015**, *121*, 265–275. [PubMed]
19. Noda, I.; Satkowski, M.M.; Dowrey, A.E.; Marcott, C. Polymer Alloys of Nodax Copolymers and Poly(lactic acid). *Macromol. Biosci.* **2004**, *4*, 269–275. [PubMed]

© 2020 by the authors. Licensee MDPI, Basel, Switzerland. This article is an open access article distributed under the terms and conditions of the Creative Commons Attribution (CC BY) license (http://creativecommons.org/licenses/by/4.0/).

Journal of
Functional
Biomaterials

Review

Pullulan for Advanced Sustainable Body- and Skin-Contact Applications

Maria-Beatrice Coltelli [1,2,*], Serena Danti [1], Karen De Clerck [3], Andrea Lazzeri [1,2] and Pierfrancesco Morganti [4,5,*]

1. Department of Civil and Industrial Engineering, University of Pisa, 56122 Pisa, Italy; serena.danti@unipi.it (S.D.); andrea.lazzeri@unipi.it (A.L.)
2. Consorzio Interuniversitario Nazionale per la Scienza e Tecnologia dei Materiali (INSTM), 50121 Florence, Italy
3. Department of Materials, Textiles and Chemical Engineering, Faculty of Engineering and Architecture, Ghent University, Technologiepark 70A, 9052 Ghent, Belgium; Karen.DeClerck@ugent.be
4. Department of Mental Health and Physics and Preventive Medicine, Unit of Dermatology, University of Campania "Luigi Vanvitelli", 80138 Naples, Italy
5. Academy of History of Health Care Art, 00193 Rome, Italy
* Correspondence: maria.beatrice.coltelli@unipi.it (M.-B.C.); pierfrancesco.morganti@iscd.it (P.M.)

Received: 31 January 2020; Accepted: 9 March 2020; Published: 18 March 2020

Abstract: The present review had the aim of describing the methodologies of synthesis and properties of biobased pullulan, a microbial polysaccharide investigated in the last decade because of its interesting potentialities in several applications. After describing the implications of pullulan in nano-technology, biodegradation, compatibility with body and skin, and sustainability, the current applications of pullulan are described, with the aim of assessing the potentialities of this biopolymer in the biomedical, personal care, and cosmetic sector, especially in applications in contact with skin.

Keywords: pullulan; biopolymers; exopolysaccharides; biodegradation; biocompatibility

1. Introduction

Pullulan (PUL) is a biopolymer produced by strains of the polymorphic fungus *Aureobasidium pullulans* as an extracellular, water soluble polysaccharide [1–3]. This natural fibrous polymer, which is produced using various substrates such as starch, distilled by-products, bakery waste, or agro-industrial residues is a linear, unbranched, odorless, tasteless, and neutral exopolysaccharide resistant to changes in temperature and pH. PUL [4], produced from the fungus as an elongated branched septate and large chlamydospores during its life-cycle, consists of α-(1,6)-repeated maltotriose units via an α-(1,4)glycosidic bond with chemical formula $(C_6H_{10}O_5)_n$ and the structure reported in Figure 1.

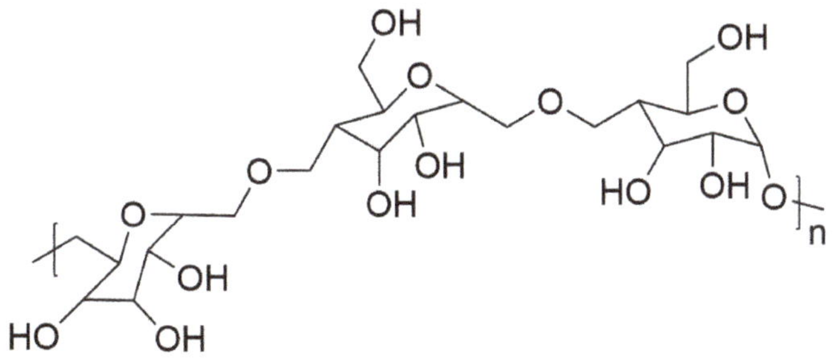

Figure 1. Structure of pullulan.

The alpha linkages give PUL physical properties such as water solubility and fiber flexibility, enabling its capacity of forming fibers and films, which is similar to certain petrol-derived plastics [5]. However, it is a non-ionic, non-hygroscopic, non-immunogenic, non-toxic, non-carcinogenic, non-mutagenic polymer that can be processed into a transparent and edible film, thanks to its excellent filming and casting properties [6,7]. Moreover PUL was declared safe with a Generally Recognized as Safe (GRAS) status by the Food and Drug Administration in USA and certified harmless for food usage by regulations in many countries. Thus, it has potential industrial applications in food, pharmaceutical, cosmetic, and biomedical fields [8] and many reviews have been recently dedicated to its production and applications [9–14]. In fact, this polymer is a flocculant, foaming, and adhesive agent and for its antioxidant and prebiotic properties, it is used in low calorie dietary fibers in healthy and functional food [6,7]. Moreover, PUL also permits cross linking and delivery of genetic material and some therapeutic cytokines, thanks to its fibrous characteristic and chemical structure in each repeating unit multiple –OH groups, suitable for scaffolds compatible with physiological conditions [15].

Thus, the structure results show that it is ideal for the hydrogel-based delivery of both cells and biomolecules, due also its own physicochemical property, which exhibits water retention capability. If used on the submicron-to-nanoscale dimension or electrospun, PUL fibers possess several advantages: huge surface area to volume ratios, accessibility and flexibility in surface modifications as well as excellent mechanical properties such as tensile strength and modulus, suggesting a huge potential for biomedical, cosmetic, and engineering applications [6].

Thus, PUL electrospun fibers used in combination with other natural-origin biomaterials contribute to make a tissue as structured as the natural extracellular matrix (ECM). Therefore, PUL shows good capacity to support and regulate the cell types involved in skin homeostasis [16]. Indeed, ECM provides a microenvironment that allows for nutrient diffusion and modulates the biochemical and physical stimuli among the different cells, inducing their favorable interactions, increasing the proliferation rate, maintaining their phenotype, supporting the differentiation of stem cells, and activating cell-signaling [17].

On one hand, biomaterials and their relative combinations (e.g., polymers and composites in their different forms: non-woven tissues, films, emulsions) enable support and correct structural architecture for tissue regeneration, ultimately improving quality of life, when used in medical and cosmetic applications [18]. In addition, scaffolding materials play a critical role in tissue engineering by serving as an artificial ECM to provide temporary mechanical support for the cells and mimic their natural nanofibrous environment [19]. Additionally, thanks to the electrospinning technique, the PUL fibers can be used to produce non-woven tissues and films to be further physically coated or chemically bound by various active ingredients. These ingredients may help to produce advanced medications and innovative cosmetic products as well as coatings for medical, food, and cosmetic packaging [20,21]. However, the non-toxicity and easy biodegradability of this natural compound enables the production of innovative biopolymer-based nanocomposites, both skin-friendly and environmentally friendly. Indeed, these products, based on renewable biowaste sources, may be utilized as scaffolds for medical, food, and cosmetic use [22,23]. Therefore, for the above-mentioned interesting characteristics, PUL is highly useful to fabricate innovative cosmetic products free of preservatives and other chemicals, which are often the cause of allergic and sensitizing phenomena. These safe and smart cosmeceuticals could modulate the aging phenomena as they are positively involved in the orchestra of signaling, regulating the reproduction and survival of different skin cells. Moreover, being made from agroforestry waste, this new category of biomedical/personal care/cosmetic products could help to reduce the CO_2 production and save the environment, ultimately preserving natural raw materials and biodiversity of our planet for the incoming generations [18,19].

2. Polysaccharides and Waste

Over the last decades, many natural biopolymers have shown to be of interest for many different industrial fields including pharmaceutical, biomedical, and cosmetics. Among them, the extracellular polysaccharides (EPS), obtainable from agri-food waste such as the fungal polysaccharides including glucans, mannans, and chitinous polymers are considered interesting. Thus, bioconversion of the waste by-products using biotechnological ways offers two important advantages: (a) a reduction in environmental pollution by sustainable development ways; and (b) to provide value-added biomaterials with a wide spectrum of activities in many industrial sectors such as pharmaceutical, food, cosmetics, and other industrial sectors [24]. These biomaterials can be classified into several categories: neutral (beta-glucan, dextrans, cellulose, and pullulan); acid (alginic acid, hyaluronic acid); basic (chitin, chitosan), or sulfated (heparin sulfate and chondroitin sulfate). However, there is a need to obtain biomaterials by cost-effective processes that are able to ensure the total utilization of processing wastes through green eco-friendly technologies. Among all of these biomolecules, polysaccharides are the more inexpensive and easily degradable bio-products, representing, by ~75%, the majority of all organic materials on earth [25]. They are, therefore, the most abundant natural polymers on the planet, serving as major element in plants (i.e., cellulose) and animals (i.e., chitin in arthropods and hyaluronic acid in mammals) or as a food storage mechanism (i.e., starch or glycogen). These biopolymers, which include plant exudates such as gum arabic, Karaya gum, and tragacanthin gum; seed gums such as guar gum; and algal or microbial polysaccharides such as xanthan gum and pullulan are the simplest building blocks of natural carbohydrate molecules that comprise repeating units joined together by glycosidic bonds, are often used as chemical markers, particularly in cell recognition [26,27]. They are found, therefore, as components of glycoproteins and glycolipids showing, among others, immunostimulation and antioxidant effects. However, apart from the natural raw material found in nature from which it is possible to obtain polysaccharides, a great quantity of them can be extracted from agri-food waste, accounting for about 300 billion tons/year [28,29]. By utilizing this waste, it will be possible to preserve natural materials for the incoming generations, eliminating part of the pollutants and reducing the production of greenhouse gas (GHG) emissions.

3. Exopolysaccharide Synthesis and Applications

As previously reported, PUL is an extracellular, linear, unbranched, and water-soluble microbial polysaccharide that is produced by microorganism fermentation similar to dextran and xanthan gum. It belongs to EPS and shows film forming properties [3]. It is interesting to underline that EPS, synthesized from microorganisms to produce the biofilm, are responsible for the physicochemical and biological properties of this microbial structure, which involves, among others, surface adhesion, cell–cell communication, protective barrier, and water retention as well as protective activity against predation by protozoa, oxidizing biocides, UV, and environmental stress. For their interesting properties, these EPSs can be used in many industrial applications. Thus, the different EPS are used in the pharmaceutical and cosmetic industry as bio-polymeric carriers for controlled release and in tissue engineering as well as in agriculture to improve the fertility and productivity of soil, or in the food industry to produce lactic acid and in the wastewater for color removal [30,31]. Naturally, the production of exopolysaccharides depends on the type of microorganism, the system, and substrate adopted with the operating condition and the biomass utilized. However, substrate composition and environmental conditions seem to be the main factors affecting the production of EPS [32]; especially if there exists no single set of culture that may assure their high productivity due to the different requirements requested from microorganisms such as temperature, pH, aeration rates, and fermentation conditions [33]. Thus, these and other biodegradable polysaccharides have been used and continue to be used in many different medical and cosmetic applications, opening new opportunities in future delivery technology.

4. Pullulan Characteristics, Applications, and Market

PUL is one of the emerging biopolymers synthesized and elaborated by the polymorphic fungus *A. pullulans*, characterized by its non-hygroscopic nature, and water-solubility, showing a relatively low viscosity compared to other polysaccharides [34]. Similar to chitin, it possesses good thermal stability, showing decomposition temperatures in the range of 250–280 °C. PUL can be produced by utilizing the waste generated by many agri-food as the substrate. In fact, they are rich in carbon and nitrogen sources and other nutrients, so they are essential for the adequate growth of *A. pullulans*. However, although the large-scale production of this polymer has been developed, the major problem remains its discoloration from the pigment melanin present in the fermentation broth of the fungus. Although it has the facility to be processed into a transparent and edible film with interesting gas barrier and anti-static properties, the obtained film becomes sticky by absorbing atmospheric moisture at 80% relative humidity (RH) [35]. Therefore, the necessity to blend and laminate it with other polymers occurs such as agar, alginate, carrageenan, or other polymers. Interestingly, PUL shows good adhesive and binder properties, is non-irritant, non-toxic, edible, biodegradable, skin- and environment-friendly, thus having properties that are extremely useful in different industrial applications, as previously reported [34,36]. Thus PUL, being resistant to mammalian amylases and providing low calories, is used to produce low-calorie dietary products as starch replacers and functions as a prebiotic to promote the growth of beneficial bifidum bacteria. Moreover, its adhesive properties may be used as a binder and stabilizer in food pastes, denture adhesive, and tobacco and as an adhesive to stick nuts to cookies. Additionally, for its water retention and film forming properties, it may result in an interesting moisturizing and protective ingredient for cosmetic lotions, powders, facial packs, and protective agent for hair shampoos, hair dressings, and tooth powders [20,34–40]. PUL, in fact, is produced by microorganisms because they like to surround themselves with a highly hydrated EPS layer to protect their organisms from desiccation and against predation by protozoa [33,37]. Additionally, it is interesting to underline that fungal polysaccharides such as PUL have shown to stimulate the immune response, have antimicrobial activity, and lowers cholesterol and triglycerides. Moreover, it can develop peculiar extended structures including hydrogels or micro/nanoparticles that may be used as biomaterials for other specific medical and cosmetic applications [3]. Pullulan has also been used in composite materials for biomedical applications (e.g., bone substitutes) [41–44].

The applications of pullulan are shown in a schematic in Figure 2. However, it must be pointed out that many research groups are continuing their investigations into its interesting properties. Hence, it is reasonable that in the future, the number of applications will grow extensively.

Regarding its physicochemical characteristics, the molecular weight of PUL ranges from 362 to 480 kDa and is highly influenced by: (a) types of strain used; (b) initial medium pH; (c) nature and composition of substrate used; (d) media composition; and (e) time of harvesting [36]. Regarding the productive process, its highest molecular weight is obtained by the fermentation of agri-food waste by different steps, whereas that obtained from a synthetic medium such as glucose generally have a lower molecular weight. Thus, the microbial fermentation of PUL is carried out in five steps: (a) the harvesting of microbes; (b) removal of undesired by-products such as melanin and cellular proteins, indispensable for its purity; (c) precipitation of polysaccharide; (d) ultracentrifugation/dialysis; and (e) freeze drying. Sometimes, its purity can be further improved by the use of ultrafiltration membranes [36]. For this purpose, agro-based industries such as potato, sugar, rice, grape, and coconut processing generate a huge amount of solid/liquid waste that can cause severe environmental issues, if not discarded [25]. However, aeration, the appropriate pH range between 5.5 and 7.5, the presence of specific enzymes, glucose, and a nitrogen source are considered fundamental for optimizing both fermentation and yield in pullulan [20,36]. Moreover, different mathematical models can be used to facilitate the control and optimization of pullulan production while several analysis techniques are available during and after the fermentation period to control its purity, especially when used for medical and cosmetic purposes. In conclusion, while PUL—compared to the other exopolysaccharides—shows a high potentiality of applications for its interesting bioactive properties, the production cost represents the critical challenge for its wide diffusion in the market. Hence, the costs must be further reduced to become competitive with other microbially-produced polysaccharides [20]. Thus, the pullulan market is expected to grow at a Compound Annual Growth Rate (CAGR) of roughly 2.2% over the next five years and will reach US$ 130 million in 2023 from US$ 129 million in 2017 [37]. Currently, the pharmaceutical industry accounted for the largest markets with about 40.74% of the global consumption in 2015, showing Japan as the biggest market with 667 million tons, China ranked second with a share of 20.65%, and the USA was third with a share of 29.65% [37].

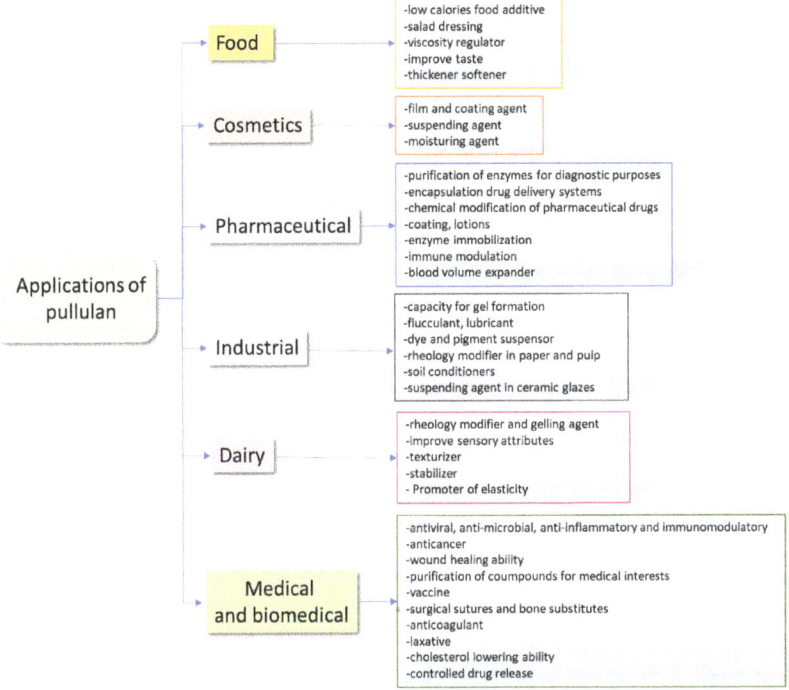

Figure 2. Applications of pullulan.

However, in our opinion, this sugar-like polymer could have a greater future in the medical and cosmetic sectors, if innovative carriers for new and smart products are realized in the coming years [38–41].

5. Production, Biodegradation, and Biomedical Applications of Pullulan

PUL is the major exopolysaccharide synthesized intracellularly by the polymorphic fungus *A. pullulans* and further secreted out to the cell surface. First isolated and characterized by Bernier (1958) from culture broths of *A. pullulans*, PUL has become the object of an ever increasing research effort. So far, different microbial sources of PUL have been found: *Tremella mesenterica*, *Cytaria harioti*, *Cytaria darwinii*, *Cryphonectria parasitica*, *Teloschistes flavicans*, and *Rhodototula bacarum*; however, *A. Pullulans* is still the main microorganism used to produce PUL [33]. Elemental analysis revealed PUL to have the chemical formula $C_6H_{10}O_5$. Treatment with the extracellular enzyme pullulanase, derived from *Enterobacter aerogenes*, led to maltotriose subunits as the main product. Infrared spectroscopy proved the presence of both α and β-glycoside bonds along the backbone of the polysaccharide and the co-existence of α-(1→4)- and α-(1→6)-glycosidic linkages in the PUL structure [33]. To obtain PUL, the microorganism is cultured in a fermenter, and different fermentation parameters have been used [20] Glucose units needed the presence of three key enzymes to be converted into pullulan, namely: α-glucose mutase, uridine diphosphoglucose pyrophosphorylase (UDPG) pyrophosphorylase, and glucosyltransferase. Aside from glucose, *A. pullulans* also consumes sucrose, mannose, galactose, maltose, fructose, and even agricultural wastes; the presence of hexokinase and isomerase are necessary for *A. Pullulans* to convert different carbon sources into PUL. It can be synthesized by cell-free enzymes of *A. Pullulans* when both UDPG and adenosine triphosphate (ATP) are present in the reaction mixture. Indeed, ATP is essential for biosynthesis, and UDPG cannot be replaced by ADPG [45]. Isotopic labelling experiments revealed that lipid-linked oligosaccharides were produced during PUL biosynthesis. Since 1976, Hayashibara Company Limited (Okama, Japan) has been the main commercial producer of PUL. A continuum fermentation process, with acidic pH, nitrogen, and carbon sources is used to obtain PUL. The main problems related to PUL production are a decrease in molecular weight, the high viscosity of the fermentation broth, and the discoloration of the polysaccharide resulting from the simultaneous synthesis of melanin pigment.

Polymer degradation occurs through scission of the main chains or side chains of its macromolecules. There are different factors that can induce polymer degradation such as thermal activation, hydrolysis, biological activity (due to enzymes action), oxidation, photolysis, and radiolysis [46]. Polymer biodegradation is thus a consequence of its chemical and physical conditions and interaction with the environment it is in contact with, which can be the human body (e.g., implanted device) or water or soil, if the manufact is disposed of. Biodegradation can occur due to biological activity, in which microorganisms identify the polymers as a source of organic building blocks and a source of energy they need for life. Environmental factors both regulate the polymer to be degraded and have an impact on the microbial population, therefore parameters such as humidity, temperature, pH, and others are really important in the biodegradation process. PUL can thus be degraded by enzymatic action [44]. These enzymes can be divided into four groups depending on what bond they act on: (1) Pullulanase enzymes perform hydrolysis on the $\alpha(1\rightarrow 6)$ glycosidic bonds of pullulan, leading to the formation of maltotriose; (2) Isopullulanase hydrolyzes the $\alpha(1\rightarrow 4)$ glycosidic bonds of pullulan, leading to isopanose; (3) Neopullulanase hydrolyzes the $\alpha(1\rightarrow 4)$ glycosidic bonds of pullulan, leading to panose; and (4) glucoamylase enzymes, which lead to the hydrolysis of pullulan, giving glucose as the major degradation product. Among these groups, we can find the α-amylase enzyme, which is mainly secreted by salivary glands and the pancreas and acts on the $\alpha(1\rightarrow 4)/\alpha(1\rightarrow 6)$ glycosidic bonds [44]. There are other enzymes that are able to degrade the pullulan, but are not found in the human body such as β-amylase. PUL degradation follows this reaction:

$$\text{Biodegradable polymer (PUL)} + \text{Microorganisms/Enzymes} \rightarrow CO_2 + H_2O + \text{Biomass}$$

Reaction products (the biomass) are biocompatible, therefore the immune system is able to interact with them. Moreover, a material such as PUL complies with second (biodegradable/bioactive) and third generation (induce tissue regeneration and function) biomaterials, thanks to its biocompatibility. After its approval by the FDA, PUL and its many derivatives are widely exploited in the food industry, pharma and cosmetic products, and biomedical applications including tissue engineering [47–50]. Pullulan derivatives are obtained by reactions of esterification, sulfation, oxidation, and others, or by the grafting of chemical structures to the main backbone. For example, PUL/nano-hydroxyapatite composites proposed for bone tissue engineering have shown potential to regenerate bone defects both in vivo and in vitro [42].

Methacrylated pullulan (PULMA) hydrogels were printed via multiscale light-assisted 3D printing techniques by using visible stereolithography apparatus (SL) and two-photon lithography (TPL), thus enabling 3D patterns from the millimeter down to the micron range. Mechanical properties, in particular rigidity, were controlled by adding a bifunctional crosslinker. PULMA structures were cytocompatible with mesenchymal stem cells, finally confirming the ability of this polysaccharide to be used in the biomedical field [51]

PUL and its derivatives such as nanogels, nanoparticles, and microspheres, can act as efficient carriers in drug delivery systems (DDS), reducing the toxicity of drugs and improving their activity and stability.

Moreover, the nano/micro size ensures the persistence in blood circulation of DDS for a prolonged time so that they can achieve their therapeutic role. Some examples include tumor, gene, and liver targeting. The efficiency of antitumor drugs like doxorubicin (which is toxic for the heart and stomach) is enhanced using a PUL-based system that is able to target cancer cells. PUL-based DDS for anticancer drugs include pH-sensitive PUL nanoparticles [52]. Usually, the PUL is modified to have a spacer acting as an acid-sensitive bond that is stable at physiological pH, but hydrolysable under acidic conditions, in order to conjugate the drug to the PUL backbone. To obtain gene targeting, cationized PUL (mixed in aqueous solution rich of Zn^{2+} with plasmid DNA) was shown to favor drug release; the PUL-PEI (low Mw polyethyleneimine) conjugate supports non-viral systems due to the high DNA affinity and blood compatibility. Finally, high affinity between liver exosomes and cationized PUL, which easily accumulates in the hepatic tissue, improves the activity of its specific macrophages to better suppress the inflammation.

6. Pullulan and Nanotechnology

Nanotechnology aims at understanding and exploiting the techniques used by nature to make its ingredients, which range in the nano domain. It consists of the construction and characterization of nanoscale materials and structures exhibiting enhanced physicochemical properties compared to the bulk materials. A nanometer, in fact, is one billionth of a meter and typical atoms are about one third of a nanometer. On the other hand, the so called bionanotechnology is focused on the ways the nanotechnology is used to make devices, bio-macromolecules, and products that are able to copy and deliver biological machines that are similar to nature, which are necessary to regenerate biological tissues [53]. Transmucosal delivery is the first-line option for the systemic delivery of many drugs and nanoparticles have been generally demonstrated to improve protein pharmacokinetic profiles, not only providing increased stabilization, but also allowing for controlled release and enhanced drug absorption. Pullulan-based nanoparticles have been reported to adhere to the nasal epithelium in a study regarding nasal vaccination [54], and the adherence of pullulan nanoparticles to respiratory epithelial cells, although to a limited extent, was verified [55].

Hydrophobic derivatives of pullulan, namely cholesteryl-pullulan, form nanoparticles by self-aggregation. These nanoparticles have been demonstrated to form stable complexes with both hydrophobic and hydrophilic drugs such as proteins, reinforcing their flexibility [56]. Reversibly disulfide-crosslinked pullulan nanoparticles with folic acid decoration were fabricated for dual-targeted and reduction-responsive anti-tumor drug delivery due to the specific affinity of pullulan and folic acid to overexpressed ASGPR (asialoglycoprotein receptor) and FR (folate receptor) on the surface of tumor cells, respectively [57]. Phthalyl pullulan nanoparticle (PPN)-treated *Lactobacillus plantarum* was synthesized and characterized to develop a new type of prebiotic for *Lactobacillus plantarum* [58]. These polymeric nanoparticles, as prebiotics, can exert substantial effects on probiotics, which lead to the increased production of an antimicrobial peptide that is powerful against Gram-positive and Gram-negative pathogens.

The interactions at the interfaces between nanomaterials and biological systems, therefore, is of significant interest in discovering the material activity at the nano level. Nanoparticles, in fact, interact with small organic ligands, therapeutic molecules, proteins, DNA, and cell membranes, establishing a series of nanoparticle/biological interfaces that depend on colloidal forces as well as dynamic bio-physicochemical interactions [59]. Thus, it is possible to establish a direct connection and functionality between the biomolecules of the skin-cell structures and the nanoparticles and nanocomposites produced and used, depending on their composition, size, shape, surface charge, and physicochemical functions. These interactions could impart unique physical properties to the nanomaterials, simultaneously regulating the biological responses of the bio-nanoconjugates [59]. However, it should be underlined that all nanomaterials have been distinguished and are acting in dependence of their natural origin, and are incidental and engineered nanoparticles by which they are made. It should also be remembered that the efficiency of a specific controlled-release of any nano-system is determined by its physicochemical properties and biodegradation rate [60]. Controlling and mimicking nano-biomolecules and nano devices, therefore, represents the greatest and fundamental challenge of nanotechnology. This nano dimension may be obtained, for example, by the electrospinning technology as reported below. The goal of the medical and cosmetic sectors in tissue engineering is to use biomaterials to produce bioengineered scaffolds and cell therapies to act as "smart band aids" to replace senescent and/or diseased resident cells, reestablishing their anatomy and physiology [61]. A better knowledge of the longer-term effectiveness and safeness of these novel materials is important in order to realize innovative drugs for preventing and trying to solve many diseases in regenerative medicine as well as to formulate smart cosmetic products that are able to prevent and slow down the signs of aging and photo-aging. The production of nanostructured tissues is possible thanks to electrospinning techniques. More specifically, electrospun PUL could be used to improve tissue engineering scaffolds by increasing their cytocompatibility by acting on the surface nano-rugosity and nano-porosity of these materials. Its biofunctionality and biocompatibility, in fact, may help to produce tissues that are able to mimic the native extracellular matrix (ECM) [62]. To improve the final product strength and further tune the properties, our group has worked for some years on the use of nanochitin, nanolignin, and its complexes [20–24,38–41,53]. Multiscale structures of PULMA enabled the control of mechanical properties by the virtue of cross-linking [51]. Moreover, different nanoparticles can be obtained by this polysaccharide and its derivatives [63].

7. Pullulan and Extra Cellular Matrix

Thus, PUL and other polymers seem able to provide not only support, but also the correct structural architecture essential for tissue regeneration [18]. Moreover, a combination of different polymers such as chitin nanoparticles and nanolignin and their complexes that are used to deliver selected active ingredients can enhance the efficacy of the final formulation [38–41]. They can, in fact, play a critical role, serving as a synthetic ECM for cells, mimicking the native microenvironment that lead to tissue formation. These biopolymers, therefore, allow nutrient diffusion, also modulating the biochemical and physical stimuli necessary to guide cell proliferation, differentiation, migration, and growth [64]. Nutrition, reproduction, and communications are the three vital functions of living beings. Cells communicate with each other and with connective tissues via signals that are sent by surface molecules or membrane proteins. Thus, it seems possible that PUL effectiveness, especially when the polymer is nanostructured, could be connected to a better and gradual comprehension of the finely orchestrated nature of intercellular communication, enhancing the innate surveillance mechanism of the skin and the continuous turnover of its cells, altered, for example, during premature aging. Thus, as previously reported, there is the necessity to formulate products through a biomimetic approach; adopting problem-solving methods inspired by nature's functions to make structures from the molecular level as nature does. An interesting technique is the electrospinning method to make non-woven tissue through the use of biopolymers. Through this technology, it is possible to produce scaffolds that are able to directly influence cell viability, migration, proliferation, and differentiation, thanks to their specific surface and structural properties [64]. Electrospinning is therefore a particularly promising cost-efficient technique for tissue engineering to fabricate biomimetic nano matrices, characterized by their surface to volume ratio and interconnecting pores, which is indispensable for cell adhesion and survival. This technique, useful to produce continuous ultra-thin, nano-scaled, non-woven dry fibers collected as non-woven tissues, involves the use of a high-voltage electrostatic field applied to create electrically charged jets of polymer solutions [65,66]. This stems from the principle in which a liquid droplet could be charged by electrostatic charge to form a fiber. In its setup, therefore, there are three basic components: the material extrusion system equipped with microscale spinneret, high-voltage power supply, and counter electrode for collecting fibers. Thus, electrospinning is a highly versatile technique that allows for the processing of different biopolymers into nanofibers that are covered or bound by various nanoparticles, and embedded by selected active ingredients. Through this technology, it is possible to produce a great variety of non-woven tissues, which having a particular morphology mimicking the native ECM and containing proper ingredients, that may be used in the medical and cosmetic field. Naturally, the formation and structure of the fibers is dependent on their chain entanglements and concentration as well as on the chemical structure of the polymers used. Compact globular-like polymer chains produce fewer entanglements than random walk-coil chains at the same concentration [66]. The concentration of the polymer solution—used as the main parameter to control the fiber diameter—is critical and needs to be controlled in order to obtain spun fibers, and also because it modulates viscosity and surface tension of the final solution. Regarding PUL manufacturing, it has already been electrospun with success from an aqueous solution, and parameter mapping including environmental parameters such as humidity and temperature allows for reproducible electrospun mats with tuned fiber diameters [67]. Sun et al. [68] obtained pullulan nanofibers with a diameter of 100–700 nm using redistilled water as the solvent through electrospinning technology. The result was achieved by the control of the processing parameters such as polymer concentration, applied voltage, distance between the capillary and collector, and flow rate. Karim et al. [69] prepared electrospun mats based on pullulan/montmorillonite clay nanocomposites. The study showed that the introduction of clay resulted in the improvement in tensile strength and thermal stability of the PUL matrix. The coexistence of intercalated montmorillonite layers and an increase in the crystallinity of the blended nanofiber mats with the addition of clay filler were also observed.

Interestingly, Qian et al. [70] incorporated up to 7.41% of rutin as a UVA and UVB absorber in electrospun pullulan/poly(vinyl alcohol). Rutin (3′,4′,5,7-tetrahydroxyflavone-3b-D-rutinoside) is one of the most abundant flavonoids from natural sources and has many biological properties, being anti-inflammatory, antiallergenic, and antimicrobial. Wang and Ziegler [71] prepared electrospun nanofiber mats from aqueous starch-pullulan dispersions. Moreover, Wang et al. [72,73] prepared electrospun gelatin/PUL fibers mimicking ECM. Hence, PUL was electrospun in combination with specific polymer, nanofillers, or compounds for obtaining functional nanostructured mats. In fact, as previously reported, among the several features that can affect a scaffold's performance are fiber surface properties such as roughness, topography, and chemistry as well as porosity, pore size, inter connectivity, mechanical properties, and degradation rates of the non-woven-tissues, aside from the atmospheric conditions (mainly the humidity and the temperature of the environment).

8. Skin Structure and Skin Contact Applications

Skin is a biological barrier that is generally less than 2 mm thick, prevents water loss, and protects the body's internal organs from the external environment and pathogen aggression (Figure 3) [74]. It is composed of two layers: the outer part, the epidermis, which is in contact with the environment, and the inner part, the dermis.

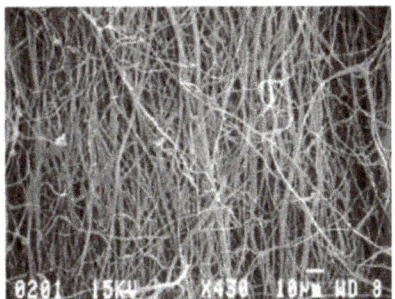

Electrospun pullulan/gelatin tissue   Extracellular Matrix (ECM)

Figure 3. Comparison between electrospun pullulan/gelatin tissue [73] (**left**) with extracellular matrix [73] (**right**).

The first layer, which can be further divided into the stratum corneum (SC), stratum granulosum, stratum spinosum, and stratum basalis, is composed of cellular layers of keratinocytes that self-renew their complex structure in 2–3 weeks. Keratinocytes, which produce keratin and make up more than 85% of the epithelial mass, are generated by mitosis in the basal layer of the epidermis. They pass outward through successive stages of differentiation to be shed at the skin surface together with a thin protective hydrolipidic film [75]. The dermis has a sponge-like structure composed of various ECM fibrous proteins and fibroblasts that synthesize collagen and elastin. While the skin outermost layer (i.e., stratum corneum made of corneocytes) prevents the skin delivery of any substance, the interior layers perform the task of keeping the skin hydrated and glossy. It is pointed out that keratinocytes, thanks to keratin, provide a covering tissue that is chemically unreactive, hard, waterproof, elastic, and resistant to any abrasion and physical insult. Moreover, these specialized cells secrete low-molecular weight mediators by which cells communicate and influence the activities of each other. On the other hand, the dead cell corneocytes, filled with keratin, water and enzymes, represent not only the main protective function as a barrier, but are also involved in the maintenance of skin flexibility and hydration by natural moisturizing factors (NMF), which is also necessary to retain water, reinforcing and supporting the collagen function [62,63]. NMF, which comprise between 20% and 30% of the corneocytes dry weight, is principally composed of amino acids or derivatives [i.e., pyrrolidine carboxylic acid (PCA)] together with lactic acid, urea, citrate and sugars [75]. Moreover, corneocytes are surrounded by a protein envelope layer linked to lipids (skin lamellae) connected with the lamellar bodies of the stratum granulosum that are rich in profilaggrin. The process that converts profilaggrin in filaggrin is fundamental for the organization and production of lipid lamellae and NMF (Figure 4).

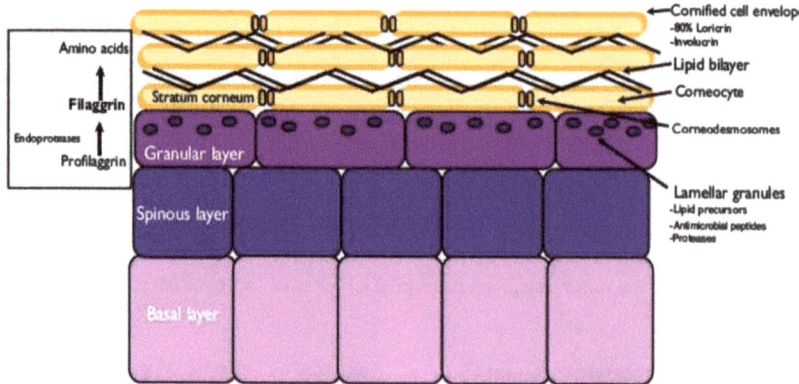

Figure 4. Production of amino acids and natural moisturizing factors (NMF) by filaggrin at the stratum corneum level.

In any way, intercellular lipids and NMF are pivotal for the rate-limiting efficiency of the skin's hydrophobic and hydrating barrier, forming the only continuous domain in the SC. The skin barrier function remains efficient when the presence of lipid lamellae is correctly balanced by an intact superficial hydrolipidic film enriched by NMF [75]. Thus, the necessity of formulating innovative cosmetic products that, thanks to efficient carriers and effective active ingredients, are able to pass through this barrier. As a result, the selected ingredients, released at the right skin layers, may have the possibility of communicating with the cells, with the aim to re-establish the intercellular communication, keep the stratum corneum hydrated, and renew the native ECM structure of the dermis when diseased or aged [76]. This activity requires the use of the right active ingredients and carriers such as pullulan and its composites, which are capable of regenerating the structural architecture not only of healthy skin, but also for wounded or aged skin when altered or destroyed by different injuries or diseases [77,78]. For instance, Li et al. [79] prepared hyaluronic acid grafted pullulan polymers and used them in the formation of novel biocompatible wound healing films. Moreover, Vora et al. [80] studied pullulan-based dissolving microneedle arrays for enhanced transdermal delivery of small and large biomolecules. In good agreement, the cosmetic industry is continually searching for new ingredients and functions in order to respond to the diversification of consumer needs.

9. Pullulan and Personal Care/Cosmetic Market

Carriers play a critical role in the creation of a cosmetic product, helping to preserve the efficacy, safety, and stability of the selected active ingredients. They have to load, transport, and deliver the active ingredients at the programmed level of the skin to give the promised benefits to the customers. The detergent or cosmetic formulation is made of active ingredients and other compounds that form the physical structure of the vehicle/carrier that are necessary for delivering the "actives" at the level of different skin layers. These specialized compound-systems can affect the load and delivery of the active components by a number of different means [75]: interacting with the active agents, controlling the rate of their release from the carrier, altering the SC resistance, or enhancing its hydration state [77]. Optimal selection and use of the right carriers can assist cosmetic manufacturers in developing new and innovative products characterized for their enhanced functionality and cost-saving formulations. Due to increasing requests from consumers of nature-oriented products, natural polysaccharides and their derivatives represent a group of interesting polymers to be used as carriers to make the usual and classic emulsions as well as innovative cosmeceutical-tissues [22–24,38–41]. Specifically, they may be used to make not only the new micro/nano emulsions, but also the smart non-woven tissues made by natural fibers covered on their surface or bound into their structure by selected active ingredients [78,79]. Moreover, polysaccharides such as PUL, its derived compounds, and complexes may be used in the formulation and manufacture of new matrix systems, scaffolds, films micro/nano particles that resemble the natural ECM. Furthermore, as previously reported polysaccharides are not only biodegradable and toxicologically harmless, but also of low cost and relative abundance compared to their synthetic counterparts [59]. Thus PUL, given its particular physicochemical properties due to the multiple functional groups characterizing its biomolecules that are capable of different complex active compounds, can be processed into hydrogels, transparent, and edible films and tissues [78–83]. In addition, it shows an ideal capability to make advanced medications and innovative cosmetic products due to its safeness, effectiveness, water retention capacity, interesting gas barrier (i.e., the skin *perspiratio insensibilis*), and regenerative property to rejuvenate or reconstruct the ECM structural architecture of an aged or injured skin [76]. The aging process, in fact, slows down the skin mechanisms attributed to innate surveillance, reproduction, and cell growth, influencing their daily regeneration [82–85]. For this purpose, regenerative medicine aims to recuperate lost tissues by guiding cell growth and restoring the original tissue-scaffold architecture. Thus, the necessity of using biomaterials such as PUL or other polysaccharides, chitin nanofibrils, nanolignin, and their derivatives, as interesting and innovative polymers, is useful to reconstruct the aged skin structure [23].

On one hand, chitin nanofibrils, which keep balanced cellular metabolism and intercellular communication, seems able to stimulate collagen and elastin production and neutralize the oxidation phenomena caused by UV and pollution particles [22–24]. On the other hand, a composite consisting of polysaccharides containing chitin nanofibrils and its complex with lignin, mimicking the skin ECM structure, represents a biomimetic matrix that is able to simulate the effects of a natural tissue and favor cell growth and differentiation [39–41]. Scaffolds, in fact, as previously reported, should serve as a platform for cellular localization, adhesion, and differentiation as well as a guide for the development of new functional tissues. Therefore, the importance of the safe and effective reconstruction of these scaffolding-architectures, made by the incorporation of the right bioactive ingredients, is evident. They have to positively influence the skin cellular interactions via the creation of the same micro-nano, beneficial, biomimetic environmental structures, and should also be capable of stimulating biochemical communications [85]. Currently, the available cosmetic treatment options are lacking in establishing both functional and cosmetic satisfaction. Both women and men, are looking for effective and miraculous products because they do not want to appear aged and would like to obtain a durable juvenile aspect. For this purpose, and according to the Euromonitor International research study, the global and personal care market was worth US$ 454 billion in 2013, US$ 25 billion of which is represented by the anti-aging niche market [86–88]. Within this global market, 28% is represented by the Asia Pacific, achieving the highest value sales at US$ 128 billion in 2013, while the anti-aging market accounted for the largest share and the highest value growth of 76% from 2008 to 2013. This underlines that the aging population, projected to grow to ~1.5 billion in 2050, has become the major driver of the cosmetic market due to the strong desire to retain a youthful appearance, just as the younger population is looking to maintain their youthful look. For this reason, there appears to be a robust demand for anti-aging cosmetics in future years [89].

10. Final Remarks

Due to the interesting activity and effectiveness shown from PUL and polysaccharides, especially in the medical sector, it is necessary to evaluate them for a more intensive use of this polymer in applications related to personal care and cosmetics, not only as an innovative active ingredient, but also as a safe compound for biodegradable items and packaging. For all these reasons, the development of biocomposites produced using biopolymer matrices, natural fibers, and reinforcement materials from renewable resources such as polysaccharides and PUL, has become a current and extensive research area because of the recyclability and biodegradability of these raw materials. However, the final aim of nanoparticle/nanocomposite products for the skin has to enhance the permeation of the active ingredients through the skin, which is necessary to make an innovative smart product. The aim of the industry is to go toward a net positive and green direction with the goal to produce safe and effective products, thus reducing the environmental impact of the entire cosmetic supply chain from production to the distribution to the final consumer. Naturally, the major dream of the modern personal care and cosmetic industry is to have sufficient intellectual, scientific, and productive capacity for creating safe and effective products that are capable of reducing the appearance of fine lines, wrinkles, and black spots caused by the aging phenomena, while respecting the environment. To obtain these results, it is necessary to act at the level of the key biological target of senescence boosting and ameliorating the cutaneous architecture of SC and the dermis skin scaffold [90]. Targeting senescent skin, in fact, not only has a positive influence on the cell's life, but also gives emotional benefits on the entire human body, ameliorating the quality of life. Last but not least the future personal care and cosmetic products must be packed in biodegradable containers to really respect the environment [91]. In-progress research studies are moving in this direction to obtain a deeper comprehension of the likeness and difference between the skin and natural polymer based supports, produced by both electrospinning and/or casting/filming technology. Innovative, biobased, and biodegradable tissues and films as well as soft and hard packaging, for their specific structure and the natural ingredients selected, may be capable of slowing down, partially or globally, the skin aging phenomena, and also in reducing or eliminating the great waste invading the planet. The amount of food lost or wasted every year is equivalent to approximately 1.3 billion tons/year [92,93], accounting for GHG emissions estimated at 3.3 Gtons [93]. Regarding plastic waste, it has been verified that the beauty industry annually produces around 120 billion units of packaging with a CAGR of 6%. It has been estimated, therefore, that in 2050, 12 billion tons of plastic in landfills will be generated from the cosmetic sector only [94]. Thus, for example, the use of products made by PUL, which are oil resistant, printable, and primarily all biodegradable and compostable, can help to improve the sustainability of waste management [95]. In conclusion, a more intensive production and use of PUL, obtained from agro-industrial waste [96] such as carob pod, molasses, grape skin pulp, starch, olive oil and so on, may be not only useful for many commercial applications, but also of great help in protecting the environment from pollution.

Author Contributions: Conceptualization, M.-B.C. and P.M.; Writing—original draft preparation, P.M., M.-B.C., and S.D.; Writing—review and editing, M.-B.C. and K.D.C.; Supervision, A.L.; Funding acquisition, M.-B.C and S.D. All authors have read and agreed to the published version of the manuscript.

Funding: This research was funded by the Bio-Based Industries Joint Undertaking under the European Union Horizon 2020 research program (BBI-H2020), PolyBioSkin project, grant number G.A 745839.

Acknowledgments: We thank Alice Bertolini, Luca Bosi, Claudia Lo Nigro, and Letizia Maria Sciara, students in the Master's in Materials and Nanotechnology a.y. 2019–2020 (University of Pisa) for their help in the pullulan bibliography search in Section 5.

Conflicts of Interest: The authors declare no conflicts of interest.

References

1. Singhal, R.S.; Kulkarni, P.R. Production of food Additives by fermentation. In *Biotechnology Food Fermentation*; Joshi, V.K., Pandey, A., Eds.; Asiatech Publishers Inc.: New Delhi, India, 1999; pp. 144–200.
2. Youssef, F.; Roukas, T.; Biliaderis, G.G. Pullulan production by a non-pigmented strain of Aureobasidium pullulans using batch and fedbatch culture. *Process. Biochem.* **1999**, *34*, 355–366. [CrossRef]
3. Shukla, A.; Mehta, K.; Parmar, J.; Pandya, J.; Saraf, M. Depicting the Exemplary Knowledge of Microbial Exopolysaccharides in a Nutshell. *Eur. Polym. J.* **2019**, *119*, 298–310. [CrossRef]
4. Kumar, D.; Saini, N.; Pandit, V.; Ali, S. An Insight to Pullulan: A Biopolymer in Pharmaceutical Approaches. *Int. J. Basic Appl. Sci.* **2012**, *1*, 202–219. [CrossRef]
5. Haniffa, M.; Cader, M.A.; Ching, Y.C.; Abdullah, L.C.; Poh, S.C.; Chuah, C.H. Review of Bionanocomposite Coating Films and Their Applications. *Polymers* **2016**, *8*, 246. [CrossRef] [PubMed]
6. Ran, L.; Tomasula, P.; Moreira de Sousa, A.M.; Kevin, L. Electrospinning Pullulan Fibers from Solutions. *Polymers* **2017**, *9*, 32. [CrossRef]
7. Leathers, T.D. Biotechnological production and applications of pullulan. *Appl. Microbiol. Biotechnol.* **2003**, *62*, 468–473. [CrossRef] [PubMed]
8. FDA: US Food and Drug Administration, Center for Food and Applied Nutrition, Office of Food Safety, Agency Response Letter, GRAS Notice No. GRN 000099. Available online: https://www.ams.usda.gov/sites/default/files/media/PullulanPetition18131.pdf (accessed on 13 March 2020).
9. Alhaique, F.; Matricardi, P.; Di Meo, C.; Coviello, T.; Montanari, E. Polysaccharide-based self-assembling nanohydrogels: An overview on 25-years research on pullulan. *J. Drug Deliv. Sci. Technol. Part B* **2015**, *30*, 300–309. [CrossRef]
10. Tabasum, S.; Noreen, A.; Farzam Maqsood, M.; Umar, H.; Akram, N.; Nazli, Z.; Chatha, S.A.S.; Zia, K.M. A review on versatile applications of blends and composites of pullulan with natural and synthetic polymers. *Int. J. Biol. Macromol. Part A* **2018**, *120*, 603–632. [CrossRef]
11. Badwhar, P.; Dubey, K.K. Insights of Microbial Pullulan Production: A Bioprocess Engineer Assessment. *Curr. Biotechnol.* **2018**, *7*, 262–272. [CrossRef]
12. Grigoras, A.G. Drug delivery systems using pullulan, a biocompatible polysaccharide produced by fungal fermentation of starch. *Environ. Chem. Lett.* **2019**, *17*, 1209–1223. [CrossRef]
13. Zhang, T.; Yang, R.; Yang, S.; Guan, J.; Zhang, D.; Ma, Y.; Liu, H. Research progress of self-assembled nanogel and hybrid hydrogel systems based on pullulan derivatives. *Drug Deliv.* **2018**, *25*, 278–292. [CrossRef] [PubMed]
14. Arora, A.; Sharma, P.; Katti, D.S. Pullulan-based composite scaffolds for bone tissue engineering: Improved osteoconductivity by pore wall mineralization. *Carbohydr. Polym.* **2015**, *123*, 180–189.
15. San Juan, A.; Hlawaty, H.; Chaubet, F.; Letourneur, D. Cationized pullulan 3D as new materials for gene transfer. *J. Biomed. Mat. Res. Part A* **2007**, *82*, 354–362. [CrossRef] [PubMed]
16. Malafaya, P.B.; Silva, G.A.; Reis, R.L. Natural-origin polymers as carriers and scaffolds for biomolecules and cell delivery in tissue enginnering applications. *Adv. Drug Deliv.* **2007**, *59*, 207–233. [CrossRef]
17. Hanson, S.; Hematti, P. Clinical Applications of Mesenchymal Stem Cell-Bionanomaterial Constructs for Tissue Reconstruction. In *Biomaterials in Regenerative Medicine*; Ramalingam, M., Ramakrishna, S., Eds.; CRC Press: Boca Raton, FL, USA, 2012; pp. 479–491.
18. Nair, L.S.; Laurencin, C. Biodegradable polymers as Biomaterials. *Prog. Polym. Sci.* **2007**, *32*, 762–798. [CrossRef]
19. Seidi, A.; Sampathkumar, K.; Srivastava, A.; Ramakrishna, S.; Ramalingam, M. Gradient Nanofiber Scaffolds for Tissue Engineering. *J. Nanosci. Nanotechnol.* **2013**, *13*, 4647–4655. [CrossRef]

20. Cheng, K.C.; Demirci, A.; Catchmark, J.M. Pullulan: Biosynthesis, production, and applications. *Appl. Microbiol. Biotechnol.* **2011**, *92*, 29–44. [CrossRef]
21. Farris, S.; Unalan, I.U.; Introzzi, L.; Fuentes-Alventosa, J.M.; Cozzolino, C.A. Pullulan-Based Films and Castings for Food Packagings: Present Applications, Emergjng Opportunities and Future Challenges. *J. Appl. Polym. Sci.* **2014**. [CrossRef]
22. Morganti, P.; Coltelli, M.B. A new Carrier for Advanced Cosmeceuticals. *Cosmetics* **2019**, *6*, 10. [CrossRef]
23. Danti, S.; Trombi, L.; Fusco, A.; Azimi, B.; Lazzeri, A.; Morganti, P.; Coltelli, M.B.; Donnarumma, G. Chitin Nanofibrils and Nanolignin as Functional Agents in Skin Regeneration. *Int. J. Mol. Sci.* **2019**, *20*, 2669. [CrossRef]
24. Padmaja, G.; Jyoth, A.N. Roots and Tubers. In *Valorization of Food Processing by-Products*; Chandrasekaran, M., Ed.; CRC Press: Boca Raton, FL, USA, 2016; pp. 377–414.
25. Heinze, T. (Ed.) *Polysaccharydes: Structure, Characterization and Use*; Springer: New York, NY, USA, 2005.
26. Hou, Y.; Ding, X.; Hou, W. Composition and antioxidant activity of water-soluble oligosaccharides from Hericium erinaceus. *Mol. Med. Rep.* **2015**, *11*, 3794–3799. [CrossRef] [PubMed]
27. Desnpande, M.S.; Rale, V.B.; Lynch, J. Aureobasidium pullulans in applied microbiology: A status report. *Enzym. Microbial Technol.* **1992**, *14*, 514–527. [CrossRef]
28. Morganti, P. Use of Chitin Nanofibrils from Biomass for an Innovative Bioeconomy. In *Nanofabrication Using Nanomaterials*; Ebothe, J., Ahmed, W., Eds.; One Central Press: Manchester, UK, 2016; pp. 1–22.
29. Singh, R.S.; Kaur, N.; Kennedy, J.F. Pullulan production from agro-industrial waste and its applications in food industry: A review. *Carbohydr. Polym.* **2019**, *217*, 46–57. [CrossRef]
30. Bajpai, V.K.; Rather, I.A.; Majumder, R.; Shunkla, S.; Aeron, A.; Kim, K.; Kang, S.C.; Dubey, R.C.; Maheshwari, D.K.; Lim, J.; et al. Exopolisaccharyde and lactic acid bacteria: Perception, functionality and prospects. *Bangladesh J. Pharmacol.* **2020**, *1*, 1–23. [CrossRef]
31. Rabha, B.; Nadra, R.S.; Amhed, B. Effect of some fermentation substrates and growth temperature on Exopolysaccharide production by Streptococcus thermophilus BN1. *Int. J. Biosci. Biochem. Bioinform.* **2012**, *2*, 44. [CrossRef]
32. Kumar, A.S.; Mody, K.; Jha, B. Bacterial exopolysaccharides-a perception. *J. Basic Microbiol.* **2007**, *47*, 103–117. [CrossRef]
33. Singh, R.S.; Saini, G.K.; Kennedy, J.F. Pullulan: Microbial sources, production and Applications. *Carbohydr. Polym.* **2008**, *73*, 515–531. [CrossRef]
34. Oguzhan, P.; Yangilar, F. Pullulan: Production and usage in food industry. *Afr. J. Food Sci. Technol.* **2013**, *4*, 57–63.
35. Rhim, J.W. Characteristics of Pullulan based Edible films. *Food Sci. Bitechnol.* **2003**, *12*, 161–165.
36. Sugumaran, K.R.; Ponnusami, V. Review on production, downstream processing and characterization of microbial Pullulan. *Carbohydr. Polym.* **2017**, *173*, 573–591.
37. Pullulan Market 2019 Industry Research, Share, Trend, Global Industry Side, Price, Future Analysis, Regional Outlook to 2024. Available online: https://www.marketwatch.com/press-release/pullulan-market-2019-industry-size-and-share-evolution-to-2024-by-growth-insight-key-development-trends-and-forecast-by-market-reports-world-2019-07-24 (accessed on 11 March 2020).
38. Morganti, P.; Coltelli, M.B.; Danti, S. Biobased Tissues for Innovative Cosmetic Products: Polybioskin as an EU Reesrch Project. *Glob. J. Nanomed.* **2018**, *3*, 555620. [CrossRef]
39. Morganti, P.; Morganti, G.; Coltelli, M.B. Chitin Nanomaterials and Nanocompositrs for Tissue Repair. In *Marine-Derived Biomateriaks for Tissue Engineeing Applications*; Choi, A.H., Ben-Nissan, B., Eds.; Springer: Singapore, 2019; pp. 523–544.
40. Morganti, P.; Danti, S.; Coltelli, M.B. Chitin and lignin to produce biocompatible tissues. *Res. Clin. Dermatol.* **2018**, *1*, 5–11. [CrossRef]
41. Schlaubitz, S.; Derkaoui, S.M.; Marosa, L.; Miraux, S.; Renard, M.; Catros, S.; Le Visage, C.; Letourneur, D.; Amédée, J.; Fricain, J.C. Pullulan/dextran/nHA Macroporous Composite Beads for Bone Repair in a Femoral Condyle Defect in Rats. *PLoS ONE* **2014**, *9*, e110251. [CrossRef] [PubMed]
42. Fricain, J.C.; Schlaubitz, S.; Le Visage, C.; Arnault, I.; Derkaoui, S.M.; Siadous, R.; Catros, S.; Lalande, C.; Bareille, R.; Renard, M.; et al. A nano-hydroxyapatite—Pullulan/dextran polysaccharide composite macroporous material for bone tissue engineering. *Biomaterials* **2013**, *34*, 2947–2959. [CrossRef] [PubMed]

43. Takahata, T.; Okihara, T.; Yoshida, Y.; Yoshihara, K.; Shiozaki, Y.; Yoshida, A.; Yamane, K.; Watanabe, N.; Yoshimura, M.; Nakamura, M.; et al. Bone engineering by phosphorylated-pullulan and b-TCP composite. *Biomed. Mater.* **2015**, *10*, 65009. [CrossRef] [PubMed]
44. Aydogdu, H.; Keskin, D.; Baran, E.T.; Tezcaner, A. Pullulan microcarriers for bone tissue regeneration. *Mater. Sci. Eng. C* **2016**, *63*, 439–449. [CrossRef]
45. Donota, A.; Fontana, J.C.; Schorr-Galindo, B.S. Microbial exopolysaccharides: Main examples of synthesis, excretion, genetics and extraction. *Carbohydr. Polym.* **2012**, *87*, 951–962. [CrossRef]
46. Nair, N.R.; Sekhar, V.C.; Nampoothiri, K.M.; Pandey, A. Biodegradation of Biopolymers. In *Current Developments in Biotechnology and Bioengineering*; Pandey, A., Negi, S., Soccol, C.R., Eds.; Elsevier: Amsterdam, The Netherlands, 2017; pp. 739–755.
47. Morganti, P.; Morganti, G.; Colao, C. Biofunctioal Textiles for Aged Skin. *Biomedicines* **2019**, *7*, 51. [CrossRef]
48. Bruneel, D.; Schacht, E. Enzymatic Degradation of Pullulan and Pullulan Derivatives. *J. Bioact. Compat. Polym.* **1995**, *10*, 299–312. [CrossRef]
49. Mishra, B.; Vuppu, S.; Rath, K. The role of microbial pullulan, a biopolymer in pharmaceutical approaches: A review. *J. Appl. Pharm. Sci.* **2011**, *1*, 45–50.
50. Singh, R.S.; Kaur, N.; Rana, V.; Kennedy, J.F. Recent insights on applications of pullulan in tissue engineering. *Carbohydr. Polym.* **2016**, *153*, 455–462. [CrossRef] [PubMed]
51. Della Giustina, G.; Gandin, A.; Brigo, L.; Panciera, T.; Giulitti, S.; Sgarbossa, P.; D'Alessandro, D.; Trombi, L.; Danti, S.; Brusatin, G. Polysaccharide hydrogels for multiscale 3D printing of pullulan scaffolds. *Mater. Des.* **2019**, *165*, 107566. [CrossRef]
52. Singh, R.M.; Kaur, N.; Rana, V.; Kennedy, J.K. Pullulan: A novel molecule for biomedical applications. *Carbohydr. Polym.* **2017**, *171*, 102–121. [CrossRef]
53. Alma Khan, F. (Ed.) *Biotechnology Fundamentals*; CRC Press: Boca Raton, FL, USA, 2012.
54. Nochi, T.; Yuki, Y.; Takahashi, H. Nanogel antigenic protein-delivery system for adjuvant-free intranasal vaccines. *Nat. Mater.* **2010**, *9*, 572–578. [CrossRef] [PubMed]
55. Grenha, A.; Rodrigues, S. Pullulan-based nanoparticles: Future therapeutic applications in transmucosal protein delivery. *Ther. Deliv.* **2013**, *4*, 1339–1341. [CrossRef] [PubMed]
56. Akiyoshi, K.; Kobayashi, S.; Shichibe, S. Self-assembled hydrogel nanoparticle of cholesterol-bearing pullulan as a carrier of protein drugs: Complexation and stabilizationof insulin. *J. Control. Release* **1998**, *54*, 313–320. [CrossRef]
57. Huang, L. Versatile redox-sensitive pullulan nanoparticles for enhanced liver targeting and efficient cancer therapy. *Nanomed. Nanotechnol. Biol. Med.* **2018**, *14*, 1005–1017. [CrossRef]
58. Hong, L.; Kim, W.-S.; Lee, S.-M.; Kang, S.-K.; Choi, Y.-J.; Cho, C.-S. Pullulan Nanoparticles as Prebiotics Enhance the Antibacterial Properties of Lactobacillus plantarum Through the Induction of Mild Stress in Probiotics. *Front. Microbiol.* **2019**, *10*, 142. [CrossRef]
59. Giri, A.; Goswami, N.; Sarkar, S.; Pal, S.K. Bio-Nanomaterials: Understanding Key Biophysics and their Applications. *Nanotechnology* **2013**, *11*, 41–110. [CrossRef]
60. Abedini, F.; Ebrahimi, M.; Roozbehani, A.H.; Domb, A.J.; Hosseinkhani, H. Overview on natural hydrophilic polysaccharide polymers in drug delivery. *Polym. Adv. Technol.* **2018**, *29*, 2564–2573. [CrossRef]
61. Lanza, R.; Langer, R.; Vacanti, J.P. *Principles of Tissue Engineering*; Academic Press: New York, NY, USA, 2007.
62. Li, Y.; Ma, X.; Ju, J.; Sun, X.; Deng, N.; Li, Z.; Kang, W.; Cheng, B. Preparation and characterization of crisslimked electrospun nanofiber membrane as a potential for biomaterial. *J. Text. Inst.* **2018**, *109*, 756–759. [CrossRef]
63. Akiyoshi, K.; Deguchi, S.; Moriguchi, N.; Yamaguchi, S.; Sunamoto, J. Self-aggregates of hydrophobized polysaccharides in water: Formation and characteristic of nanoparticles. *Macromolecules* **1993**, *26*, 3062–3068. [CrossRef]
64. Khan, F.; Ahmed, S.R. Fabrication of 3D Scaffolds on Organ Printing for Tisshe Regeneration. In *Biomaterials and Stem Cells in Regenerative Medicine*; Ramalingam, M., Ramakrishna, S., Eds.; CRC Press: Boca Raton, FL, USA, 2012; pp. 101–122.
65. Kong, L.; Ziegler, G.R. Rheological aspects in fabricating Pululan fibers by electro-wet-spinning. *Food Hydrocoll.* **2014**, *38*, 226–229. [CrossRef]
66. Mendes, A.C.; Stephansen, K.; Chronakis, I.S. Electrospinning of food proteins and polysaccharides. *Food Hydrocoll.* **2017**, *68*, 53–68. [CrossRef]

67. Torre-Muruzabal, A.; Daelemans, L.; Van Assche, G.; De Clerck, K.; Rahier, H. Creation of a nanovascular network by electrospun sacrificial nanofibers fors elf-healing applications and its effect on the flexural properties of the bulk material. *Polym. Test.* **2016**, *54*, 78–83. [CrossRef]
68. Sun, X.; Jia, D.; Kang, W.; Cheng, B.; Li, Y. Research on electrospinning process of pullulan nanofibers. *Appl. Mech. Mater.* **2013**, *268–270*, 198–201. [CrossRef]
69. Karim, M.R.; Lee, H.W.; Kim, R.; Ji, B.C.; Cho, J.W.; Son, T.W.; Oh, W.; Yeum, J.H. Preparation and characterization of electrospun pullulan/montmorillonite nanofiber mats in aqueous solution. *Carbohydr. Polym.* **2009**, *78*, 336–342. [CrossRef]
70. Qian, Y.; Qi, M.; Zheng, L.; King, M.W.; Lyu, L.; Ye, F. Incorporation of Rutin in Electrospun Pullulan/PVA Nanofibers for Novel UV-Resistant Properties. *Materials* **2016**, *9*, 504. [CrossRef]
71. Wang, H.; Ziegler, G.R. Electrospun nanofiber mats from aqueous starch-pullulan dispersions: Optimizing dispersion properties for electrospinning. *Int. J. Biol. Macromol.* **2019**, *133*, 1168–1174. [CrossRef]
72. Wang, Y.; Guo, Z.; Qian, Y.; Zhang, Z.; Lyu, L.; Wang, Y.; Ye, F. Study on the Electrospinning of Gelatin/Pullulan Composite Nanofibers. *Polymers* **2019**, *11*, 1424. [CrossRef]
73. Morganti, P.; Fusco, A.; Paoletti, I.; Perfetto, B.; Del Ciotto, P.; Palombo, M.; Chianese, A.; Baroni, A.; Donnarumma, G. Anti-Inflammatory, Immunomodulatory, and Tissue Repair Activity on Human Keratinocytes by Green Innovative Nanocomposites. *Materials* **2017**, *10*, 843. [CrossRef]
74. Priestley, G.C. (Ed.) *Molecular Aspects of Dermatology*; John Wiley & Sons Ltd.: Chichester, UK, 1993.
75. Rawlings, A.V.; Leyden, J.J. (Eds.) *Skin Moisturizing*, 2nd ed.; Informa Healthcare: New York, NY, USA, 2009.
76. Prausnitz, M.R.; Mitragotri, S.; Lager, R. Current status and future potential of transdermal drug delivery. *Nat. Rev. Drug Discov.* **2008**, *3*, 115–124. [CrossRef] [PubMed]
77. Priya, S.G.; Jungvid, H.; Kumar, A. Skin tissue engineering for tissue repair and regeneration. *Tissue Eng. Part B Rev.* **2008**, *14*, 105–118. [CrossRef] [PubMed]
78. Wong, V.W.; Rustad, K.C.; Gálvez, M.G.; Neofytou, E.; Glotzbach, J.P.; Hanuszy, K.; Major, M.R. Engineered Pullulan Collagen Composite Dermatol Hydrogels Improve Early Cutaneous Wound Healing. *Tissue Eng. Part A* **2011**, *17*, 631–644. [CrossRef] [PubMed]
79. Li, H.; Xue, Y.; Jia, B.; Bai, Y.; Zuo, Y.; Wang, S.; Zhao, Y.; Yang, W.; Tang, H. The preparation of hyaluronic acid grafted pullulan polymers and their use in the formation of novel biocompatible wound healing film. *Carbohydr. Polym.* **2018**, *188*, 92–100. [CrossRef]
80. Vora, L.K.; Courtenay, A.J.; Tekko, I.A.; Larrañeta, E.; Donnelly, R.F. Pullulan-based dissolving microneedle arrays for enhanced transdermal delivery of small and large biomolecules. *Int. J. Biol. Macromol.* **2020**, *146*, 290–298. [CrossRef]
81. Magdassi, S.; Touitou, E. (Eds.) *Novel Cosmetic Delivery Systems*; Marcel Dekker Inc.: New York, NY, USA, 1999.
82. Ogai, I.J.; Nep, E.U.; Audu-Peter, J.D. Advanced in Natural Polymers as Pharmaceutical Excipients. *Pharm. Anal. Acta* **2011**, *3*, 146. [CrossRef]
83. Morganti, P.; Palombo, M.; Tishchenko, G.; Yudin, V.E.; Guarneri, F.; Cardillo, A.; Del Ciotto, P.; Carezzi, F.; Morganti, G.; Fabrizi, G. Chitin-Hyaluronan Nanoparticles to Deliver Anti-Aging Active Ingredients Through the Skin. *Cosmetics* **2014**, *1*, 140–158. [CrossRef]
84. Yildizimer, L.; Thanh, N.T.K.; Seifalian, A.M. Skin regeneration Scaffolds: A multimodal bottom-up approach. *Trends Biotechnol.* **2012**, *30*, 12. [CrossRef]
85. Beneke, C.E.; Viljoen, A.M.; Hamman, J.H. Polymerjc Plant-derived Excipients in Drug Delivery. *Molecules* **2009**, *14*, 2602–2629. [CrossRef]
86. Loh, P.Y. Beauty and Personal Care Trends, In Cosmetics Asia, Euromonotor International, Bangkok. 12 February 2014. Available online: https://www.euromonitor.com/beauty-and-personal-care (accessed on 3 October 2019).
87. *Mordor Intelligence: Beauty and Personal Care Products Market-Growth, Trends, Forecast (2019–2024)*; Industry Report; Sachibouli: Hyderabad, India, 2019; Available online: https://www.mordorintelligence.com/ (accessed on 6 October 2019).
88. Loh, P.Y. Global Trends in Beauty and Personal Care: Opportunities for Future, Euromonitor International. 2014. Available online: https://www.slideshare.net/Euromonitor/global-trends-in-beauty-and-personal-care (accessed on 31 January 2020).

89. Loh, P.Y. Global Trends in Anti-Agers, Euromonitor International. 2014. Available online: https://www.slideshare.net/Euromonitor/global-trends-in-antiagers (accessed on 31 January 2020).
90. Kim, B.S.; Mooney, D.Y. Development of biocompatible synthetic extracellular matrices for tissue engineering. *Trends Biotechnol.* **1998**, *16*, 224–230. [CrossRef]
91. Gigante, V.; Canesi, I.; Cinelli, P.; Coltelli, M.B.; Lazzeri, A. Rubber Toughening of Polylactic Acid (PLA) with Poly(butylene adipate-co-terephthalate) (PBAT): Mechanical Properties, Fracture Mechanics and Analysis of Ductile-to-Brittle Behavior while Varying Temperature and Test Speed. *Eur. Polym. J.* **2019**, *115*, 125–137. [CrossRef]
92. Scott, C.R. Naturals in Cosmetic Science: The Bigger Picture. *Pers. Care Eur.* **2019**, *13*, 55–56.
93. Gustavsson, J.; Cederberg, C.; Sonesson, U.; Van Otterdijk, R.; Meybeck, A. *Save Food: Global Food Losses and Food Waste*; Food and Agriculture Organization of the United Nations (FAO): Rome, Italy, 2011; Available online: http://www.fao.org./save-food/resources/keyfindings/en/ (accessed on 3 October 2019).
94. FAO. *Food Wastage Footprint. Impact on Natural Resources*; Food and Agriculture Organization of the United Nations: Rome, Italy, 2013; Available online: http://www.fao.org/nt/sustainability (accessed on 3 October 2019).
95. Morganti, P.; Morganti, G.; Chen, H.D.; Gagliardini, A. Beauty Mask: Market and Environment. *J. Clin. Cosmet. Dermatol.* **2019**, *3*. [CrossRef]
96. Israilides, C.; Smith, A.; Scanlon, B.; Barnett, C. Pullulan from Agro-Industrial Wastes. *Biotechnol. Genet. Eng. Rev.* **1999**, *16*, 309–324. [CrossRef]

© 2020 by the authors. Licensee MDPI, Basel, Switzerland. This article is an open access article distributed under the terms and conditions of the Creative Commons Attribution (CC BY) license (http://creativecommons.org/licenses/by/4.0/).

Review

Autologous Matrix of Platelet-Rich Fibrin in Wound Care Settings: A Systematic Review of Randomized Clinical Trials

Chayane Karla Lucena de Carvalho, Beatriz Luci Fernandes * and Mauren Abreu de Souza

Graduate Program on Health Technology, Pontifical Catholic University of Paraná, Curitiba, PR 80215-901, Brazil; chaycarvalho7@gmail.com (C.K.L.d.C.); mauren.souza@pucpr.br (M.A.d.S.)
* Correspondence: beatriz.fernandes@pucpr.br

Received: 1 March 2020; Accepted: 6 May 2020; Published: 14 May 2020

Abstract: Platelet-rich fibrin (PRF) consists of a matrix that provides the necessary elements for wound healing, acting as a biodegradable scaffold for cell migration, proliferation, and differentiation, in addition to the delivery of growth factors and angiogenesis. This study aims to determine the effectiveness of the autologous PRF in the treatment of wounds of different etiologies. We carried out a systematic review of randomized clinical trials, guided by the recommendations of the Cochrane Collaboration using the following databases: Pubmed/MEDLINE, EMBASE, Web of Science, and CENTRAL. The search strategy resulted in the inclusion of ten studies that evaluated the use of PRF dressings for the healing of acute or chronic wounds of multiple etiologies. Among the 172 participants treated with PRF in wounds of varying etiologies and different segment times, 130 presented favorable events with the use of the intervention. Among the 10 studies included, only two of them did not demonstrate better results than the control group. The studies showed clinical heterogeneity, making it impossible to perform a meta-analysis. The findings do not provide enough evidence to support the routine use of PRF dressings as the first line of treatment for the healing of acute or chronic wounds of different etiologies. There was great variability in the application of the various protocols and the ways to prepare the PRF, resulting in clinical heterogeneity. Therefore, it makes it impossible to synthesize and to collect evidence from different types of studies in the meta-analysis, which affects the results and their proper discussion.

Keywords: platelet-rich fibrin; wound healing; skin wounds; wound dressing

1. Introduction

Human skin is an organ structured by many tissues, designed to develop multiple functions such as thermoregulation, vitamin D metabolism, detection of sensory stimuli, as well as reacting to mechanical trauma, chemical reagents, and pathogens [1,2]. However, to maintain its functionality, it is necessary to preserve its structural integrity.

Ruptures of the skin layers or adjacent tissues, called wounds, bring anatomical and functional changes, resulting in increased morbidity with a high impact on the public health sectors [3]. Wounds have a huge financial burden on health systems around the world. For example, they account for more than US$ 25 billion per year in the USA, due to the expenses with therapies, which are sometimes ineffective [4].

The establishment of wounds signals to the body the immediate need to correct the lesions through the self-regenerative process known as healing [5]. Wound healing is a physiological process including a cascade of complex, orderly and interconnected events, involving many types of cells interacting in a highly sophisticated temporal sequence, guided by the release of soluble mediators and signals that can influence the direction of the circulating cells to the damaged tissues [6,7].

Over the years, traditional and well-established medicine has been improved, giving rise to regenerative types of medicines. This new approach involves the use of therapeutic alternatives based on the human body itself (i.e., autologous or heterologous). In the context of wound healing, the regenerative technologies are available based on a variety of mechanisms, such as vascular fraction, the stroma of adipose tissue, platelet-rich plasmas, and bone marrow concentration, among others. In this scenario, these options help to reduce the economic and psychosocial burden, usually generated by chronic and highly complex wounds, including ulcerations, surgical, and necrotic wounds as well [8,9].

In this scenario, autologous platelet biomaterials represent an important source of cytokines and growth factors widely used for clinical and surgical applications, including tissue regeneration and wound healing [10]. Among these approaches, platelet-rich fibrin (PRF) is considered the second-generation of platelet concentrate, developed in France in 2001 [11].

The use of emerging cellular therapeutic technologies, such as platelet biomaterials, covers many pathologies as a therapeutic agent and/or coadjuvant treatment, resulting in improvements in the quality of life of the patients [12].

The first platelet concentrates were introduced in 1998. The platelet-rich plasma (PRP) was primarily used in odontology and later in otorhinolaryngology and orthopedics, to accelerate tissue regeneration, especially for soft ones. However, there is a lack of uniformity in the methods of obtaining PRP and the use, or non-use, of bovine thrombin for its activation [13].

Platelet-rich fibrin (PRF), considered a second-generation platelet concentrate, was developed in France in 2001 [11]. It does not need biochemical additives, such as anticoagulants or bovine thrombin. It is possible to obtain the PRF from the controlled centrifugation of the venous blood itself, mainly because of the soluble fibrinogen found in the fibrin, which is responsible to polymerize it in a tridimensional structure [14].

PRF consists of a matrix in which cytokines, growth factors, and platelet cells are retained and can be constantly released, offering the necessary elements for wound healing, acting as a biodegradable scaffold for the delivery of growth factors, collagen synthesis, and angiogenesis. Its applicability is observed in various health fields [15–17]. Its autologous origin and immediate availability are also noteworthy, as well as the factors related to safety, costs, and practical aspects, such as short manufacturing and implementation time [18]. Some growth factors and cytokines, mainly generated from platelets and leukocytes in the fibrin clot during wound healing, regulate some important biological processes like cell migration and differentiation, angiogenesis, and extracellular matrix synthesis [19].

Fibrin membrane (FM) is a natural biopolymer with an important capacity for the regeneration of several injured tissues [20]. The adhesive FM, when arranged on the wound bed, changes its configuration and mechanical properties over time because of the fibrin matrix retraction and expression of the secretome mostly containing the vital signaling molecules. The FM can be combined with secondary dressings, including alginates, hydrocolloids, and gauze, enhancing the action of the biomaterial [21]. Therefore, the PRF-based dressings accelerate the healing of hard and soft tissues and can be used in the treatment of different types of lesions [20,22,23].

The preparation protocol for the FM consists of blood collection through the venous puncture and subsequent centrifugation in tubes without anticoagulant, in order to form a strong polymerized fibrin clot [24]. The fibrin clot, a natural polymer, is one of the three layers resulting from the centrifugation process presenting about 97% of platelets and 50% of leukocytes from the initial blood volume, which is incorporated and distributed in a tridimensional way [25].

Regenerative medicine has especially evolved, aimed at skin healing, with different approaches: (1) individual application of skin cells, (2) biopolymer scaffolding, and (3) with the combination of both—which is classified as acellular scaffolding (i.e., temporary skin substitutes with allogeneic or autologous epithelial cells). Their final purpose is for the treatment and healing of both acute and chronic skin wounds, contributing to a reduction in the morbidity and mortality of the affected population [21,26].

Although there have been technological advances in wound treatments, innovative approaches using natural biopolymers with higher effectiveness and lower costs need further clinical studies. Therefore, the main purpose of this study is to provide information from available literature to analyze the effectiveness of autologous FM in the treatment of wounds of varying etiologies.

2. Materials and Methods

This paper presents a systematic review of randomized clinical trials, guided by the Cochrane Collaboration recommendations contained in the Handbook for Systematic Reviews of Interventions, version 6.0 [27], and described by the Preferred Reporting Items for Systematic Reviews and Meta-Analyses (PRISMA) [28].

The systematic review is a type of secondary study conducted from a defined research question to identify, evaluate, select, and synthesize pieces of evidence from primary studies that meet the predefined eligibility criteria [27].

We used the acronym PICO (Patient, Intervention, Comparison and Outcomes) to elaborate the research question, in which we related "P" to wounds of any etiology; "I" to the autologous matrix of platelet-rich fibrin (in different formulations, concentrations, and modes of obtention); "C" to different dressings technologies; and "O" to healing, reduction of the wound area and adverse events. Therefore, the research question was: what is the effectiveness of dressing based on the autologous matrix of platelet-rich fibrin for the treatment of wounds of different etiologies, when compared to other dressings technologies for the healing outcomes and reduction of the wound area?

In this study, we included randomized clinical trials with any size, where the autologous matrix of PRF was adopted, including at least one of the groups treated to achieve the proposed outcome. We excluded studies involving periodontal proceduresm, and interventions not limited to the use of PRF and the use of heterologous materials.

We recovered the relevant studies through a search strategy in databases: Medical Literature Analysis and Retrieval System Online/Pubmed (MEDLINE); Excerpta Medica Database (EMBASE); Web of Science; and the Cochrane Central Register of Controlled Trials (CENTRAL). The keywords and the searching strategies used in each database are presented in Table 1.

Table 1. Literature searching strategies.

Database	Keywords and Searching Strategies
Pubmed/MEDLINE	((("platelet-rich fibrin" (MeSH Terms) OR ("platelet-rich" (All Fields) AND "wounds" (All Fields) OR "wounds" (MeSH Terms) OR ("wound healing" (MeSH Terms) OR "wound healing" (All Fields)) AND Clinical Trial (ptyp)
EMBASE	('platelet-rich fibrin'/exp OR 'platelet-rich fibrin') AND 'wound healing'/exp
Web of Science	TS = (Platelet-rich Fibrin * AND Wounds * OR Wound Healing *)
CENTRAL	Platelet-rich Fibrin * AND wounds * OR Wound Healing

Additionally, a manual search was conducted by gray literature, which consists of studies not controlled by scientific editors, such as government reports, thesis, dissertations, and abstracts published in conference proceedings. We evaluated the reference list of clinical trials in order to identify not eligible studies, i.e., the ones not included in the searching strategies.

We selected descriptors and their synonyms for the search of primary studies in previously established databases: Medical Subject Headings (MeSH)—platelet-rich fibrin OR autologous platelet-rich fibrin OR platelet-rich fibrin matrix; AND wound AND randomized controlled trial.

All the recovered citations were screened and evaluated for their eligibility according to the inclusion criteria by two independent reviewers. The screening and selection process included two phases: the evaluation of the titles and abstracts of all identified studies, fully reading the selected studies, and making a justification for the exclusions.

We performed a critical analysis of the included studies using the Cochrane Risk of Bias Tool to assess the risk of bias, available in the Review Manager, version 5.3. The two reviewers judged the studies according to three categories: low risk of bias, high risks of bias and undetermined risk of bias for the generation of the domain of the randomization of the samples by allocation sequence (selected bias); blinding of participants and researchers (performance bias); blinding of outcomes evaluators (detection bias); systematic differences in segment losses (frictional bias); incomplete outcomes or selective report of outcomes (report bias). The Kappa coefficient determined the level of agreement between the reviewers on the inclusion or exclusion of the analyzed studies. A third reviewer re-evaluated the divergences.

We organized the data from these studies in a narrative synthesis presentation form, including authorship and year of publication, country of origin, the title of the manuscript and the journal, clinical information such as the number of participants, intervention groups and comparison between them, intervention time, and main outcomes.

We categorized all included studies through allocation confidentiality, according to the Cochrane Handbook, as described: category A—the allocation process was adequately described; category B—although the allocation process has not been described, the study points out to randomization; category C—allocation confidentiality was conducted improperly (for example, arrival order, medical record number, and date of birth); category D—the randomization of the participants was not demonstrated.

3. Results

The search strategy resulted in the recovery of 500 studies (Figure 1). After the first screening, 44 studies remained, of which eight were duplicated, and 26 did not meet the eligibility criteria. Through full reading, only ten clinical trials comprised the final sample of this systematic review. The Kappa agreement index was 0.729 ($p \leq 0.001$).

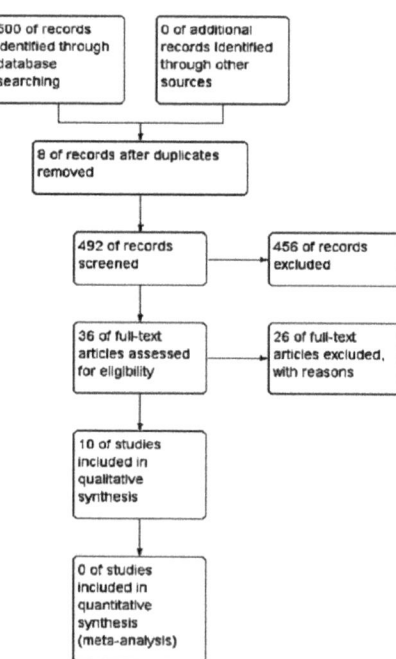

Figure 1. Flow diagram showing the preferred reporting items for systematic reviews and meta-analyses (PRISMA). Review Manager 5.3.

We reinforce that some of the reasons for the exclusion of some studies were wounds of periodontal and maxillary etiology, biopolymers other than platelet-rich fibrin, as well as the methodologic design, which were different from the inclusion criteria.

Therefore, Table 2 presents the selected studies with their respective references, year of publication, title, journal, and the database. Table 3 shows a summary of the clinical findings in these ten studies.

Table 2. Distribution of the studies according to the authors, year of publication, title, journal, and database.

Authors/Year	Title	Journal/Database
Danielsen et al., (2008) [29]	Effect of Topical Autologous Platelet-Rich Fibrin versus No Intervention on Epithelialization of Donor Sites and Meshed Split-Thickness Skin Autografts: A Randomized Clinical Trial	Plast. Reconstr. Surg./Pubmed
Weber et al., (2012) [30]	Platelet-Rich Fibrin Matrix in the Management of Arthroscopic Repair of the Rotator Cuff A Prospective, Randomized, Double-Blinded Study	Am. J. Sports. Med./EMBASE
Gür et al., (2016) [31]	Use of a platelet-rich fibrin membrane to repair traumatic tympanic membrane perforations: a comparative study	Acta Otolaryngol./EMBASE
Pravin et al., (2016) [32]	Autologous platelet-rich plasma (PRP) versus leucocyte-platelet rich fibrin (l-PRF) in chronic non-healing leg ulcers—a randomized, open-labeled, comparative study	J. Evol. Med. Dent. Sci./EMBASE
Somani et al., (2017) [33]	Comparison of Efficacy of Autologous Platelet-rich Fibrin versus Saline Dressing in Chronic Venous Leg Ulcers: A Randomised Controlled Trial	J. Cutan. Aesthet. Surg./Web of Science
Goda (2018) [34]	Autogenous leucocyte-rich and platelet-rich fibrin for the treatment of venous leg ulcer: a randomized control study	Egypt J. Surg./Web of Science
Garrido-Castells et al., (2019) [35]	Effectiveness of Leukocyte and Platelet-Rich Fibrin versus Nitrofurazone on Nail Post-Surgery Bleeding and Wound Cicatrization Period Reductions: Randomized Single Blinded Clinical Trial	J. Clin. Med./Web of Science
Zhang et al., (2019) [36]	Platelet-rich fibrin as an alternative adjunct to tendon-exposed wound healing: A randomized controlled clinical trial	Burns/CENTRAL
Elkahwagi et al., (2019) [37]	Role of autologous platelet-rich fibrin in relocation pharyngoplasty for obstructive sleep apnoea	Int. J. Oral Maxillofac. Surg./CENTRAL
Vaheb et al., (2020) [38]	Evaluation of the Effect of Platelet-Rich Fibrin on Wound Healing at Split-Thickness Skin Graft Donor Sites: A Randomized, Placebo-Controlled, Triple-Blind Study	Int. J. Low Extrem. Wounds/CENTRAL

About 130 patients presented favorable events using the PRF, according to the clinical studies analyzed herein, involving 172 participants with various etiologies of wounds. Among the ten studies included, only in two clinical trials, the PRF was not superior to the outcome evaluated.

We carried out the methodological quality assessment of the ten studies included in this review using the Cochrane Collaboration tool that assesses the risk of bias in randomized clinical trials. Through the judgment of the reviewers for each domain, it was possible to infer the overall quality of the studies presented in Figure 2. The description of the results for the methodological evaluative categorization of each study is shown in Figure 3, as well as the individual judgment of the five domains. The agreement between evaluators of the risk of bias in each study was measured using the Kappa index. The value was 0.832 ($p \leq 0.01$), which points out the credibility of the interpretation.

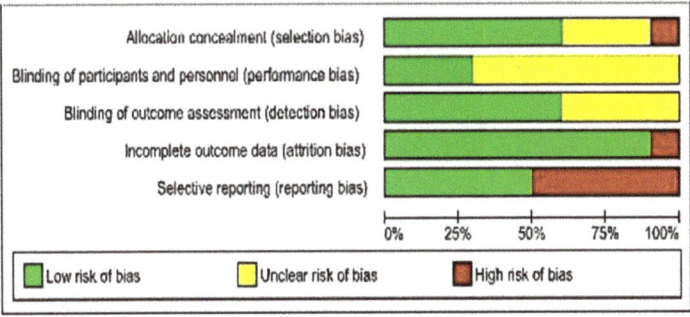

Figure 2. Bias Risk Summary: reviewers' judgment for each domain and their percentages on the overall quality of the studies. (Review Manager, version 5.3).

Table 3. Distribution of studies according to the number of participants, intervention group, control group, and main outcomes.

Study	Groups	Main Outcomes
[29]	Intervention group (n = 51): PRF in the incisional acute wound of laparoscopic cholecystectomy Control group (n = 51): human albumin and subcutaneous collagen deposition	The PRF in acute surgical wounds did not promote significant repairs but suppressed the synthesis and subcutaneous deposition of collagen. The study does not support the use of PRF to accelerate wound healing after surgery. However, it suggests that the PRF should be explored in the treatment of chronic wounds.
[30]	Intervention group (n = 30): PRF in acute wounds from the rotator cuff surgery Control group (n = 30): without PRF	There were no significant differences in perioperative pain, functional recovery, or structural outcomes with the use of PRF in arthroscopic repairing surgeries of the rotator cuff.
[31]	Intervention group (n = 30): PRF on the repair of the tympanic membrane perforations Control group (n = 30): paper patch, moist with polyvinylpyrrolidone 10%	The total closure of the perforations was observed in 24 (80%) patients from the PRF group and 16 (53%) from the control group ($p < 0.05$). The average improvement was 14.1 dB in the PRF group and 12.4 dB in the control group 45 days after the medical procedure ($p < 0.05$). The PRF provided faster healing than the polyvinylpyrrolidone.
[32]	Intervention group (n = 15): Platelet-rich fibrin and leukocytes (L-PRF) in chronic unhealed leg ulcers Control group (n = 15): Platelet-rich (PRP)	L-PRF had a better effect on the cure outcome of the lesion when compared to PRP. L-PRF has great anti-inflammatory effects and protects the wound against infections. At the end of the sixth application, 100% of healing was seen in 11 ulcers treated with L-PRF and eight ulcers treated with PRP (73.3% vs. 53.3%, respectively). More than 90% of improvement in the area and volume of the wounds was observed in 13 PRF cases and 10 PRP cases (86.6% vs. 66.6%).
[33]	Intervention group (n = 9): PRF in the treatment of chronic venous ulcers in legs Control group (n = 6): saline dressing	The mean reduction in the ulcer area in the PRF group was 85.51%, while in the saline group was 42.74% ($p < 0.001$). The PRF is effective, inexpensive, safe, and an outpatient procedure.
[34]	Intervention group (n = 18): PRF in the treatment of chronic venous ulcers in legs Control group (n = 18): conventional dressing	The closing rate of the wounds with initial area > 10 cm^2 was 50% in the sixth week and 100% in the seventh week of treatment with PRF, while in the control group was only 14.3% in the sixth week and 42.6% in the seventh one.
[35]	Intervention group (n = 20): L-PRF in post-surgical bleeding and acute wound healing in patients with bilateral onychocryptosis Control group (n = 20): use of Nitrofurazone	Statistically significant differences ($p < 0.001$) were observed between the groups showing reduction of wound healing period and post-surgical bleeding for L-PRF intervention concerning nitrofurazone treatment. L-PRF can be considered first-line supporting intervention after the surgical procedure for patients suffering from nail problems such as onychocryptosis.
[36]	Intervention group (n = 18): PRF to treat lower limb acute injury after with exposed tendons before skin grafts Control group (n = 18): treatment with dermal regeneration matrix Integra®	The graft acceptance rate was 92.3% in the Integra® group compared to 97.83% in the PRF one ($p < 0.001$). The changes in the texture of the scar tissue were superior in the Integra® group at all times in the three-months postoperative period.
[37]	Intervention group (n = 15): PRF in the postoperative acute wound, before suture, in Pharyngoplasty for treatment of obstructive sleep apnea Control group (n = 15): conventional suture	There was lower dehiscence of the wounds in the PRF group ($p = 0.013$) than in the control group. The patients from the PRF group related less pain in days 3, 5, and 10 after the surgery than those from the control group ($p < 0.001$). Additionally, the time taken to return to a normal diet was shorter in the PRF group ($p = 0.001$).
[38]	Intervention group (n = 17): PRF in burns that require a divided thickness skin graft Control group (n = 17): treatment with vaseline petrolatum gauze	The wound healing time in the PRF and control group was 11.80 ± 3.51 and 16.30 ± 4.32 days, respectively ($p < 0.001$). The PRF group presented higher rates of wound healing in days 8 and 15 compared with the control group ($p < 0.001$). There was a significant difference in average pain levels between the two groups (lower in the PRF group) ($p < 0.001$).

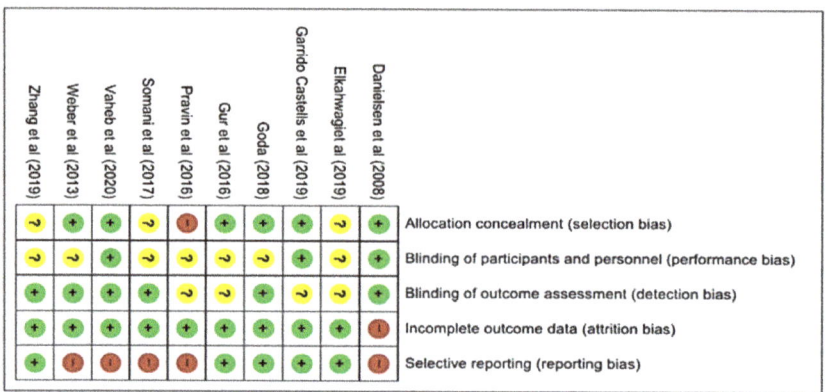

Figure 3. Bias Risk Summary: authors' judgements on each item, risk bias for each included study. Review Manager, version 5.3. Legend: "+" = low risk of bias; "−" = high risk of bias; "?" = uncertain risk of bias.

In the individual internal validation of the included studies, for the domain selection of bias, seven studies were classified as low risk of bias [29–31,34,35,38], three studies as the uncertain risk of bias [33,36,37], and one study as high risk of bias [32].

In the performance bias domain, three studies were judged as low risk of bias [26,31,35], seven studies were judged as the uncertain risk of bias [30–34,36,37], and no study with a high-risk rating of bias.

Regarding the detection bias domain, seven studies were classified as low risk of bias [29,30,33,34,36,38], three as the uncertain risk of bias [31,32,35,37], and no study with a high-risk rating of bias.

For the fourth evaluated domain, frictional bias, all studies were classified as low risk of bias. Finally, for the domain reporting bias, five studies were classified as low risk of bias [31,34–37], and five as high risk of bias [29,30,32,33,38].

In the critical evaluation of the studies regarding allocation secrecy, two studies were allocated to Category A [34,38], because they adequately described the allocation process; eight studies were allocated to Category B [29–31,33–37] because they did not report the allocation process of the groups effectively; however, they described the randomization process of the participants; one study allocated participants inadequately, being designated to Category C [32]. Therefore, no study was allocated to Category D. However, the majority of studies did not declare the way the randomization of individuals was carried out for the intervention and control groups in the right manner.

The performance bias of undetermined levels was observed among the analyzed studies, as they presented differences within the groups. Additionally, regardless of the intervention, there was not adequate reporting about the blinding of the participants.

The clinical trials that composed the review did not present significant losses of individuals among the groups during the proposed time (segment bias). Therefore, the analysis of the outcomes evaluated was not compromised.

The fact that most studies received funding is considered as a conflict of interest and therefore reporting bias, thus attributing to them a high risk of bias. All the studies presented significant clinical heterogeneity, making it impossible to perform a meta-analysis.

4. Discussion

The systematization of the treatment for wounds encompasses an approach with proper protocols for clinical evaluation and treatment management because of the barriers that interfere with healing, including the presence of necrotic tissue, senescent cells, altered extracellular matrix, hypoxia, excess of bacteria, biofilm, and inflammatory enzymes [39].

Different studies have investigated the effectiveness of healing derived from platelet concentrates. One approach is related to molecular healing strategies, which include: migration of chemotactic cytokines that facilitate the process of cell infiltration via the neutrophil-activating peptide (CXCL7), platelet factor 4 (PF4), SDF-1α (factor 1—derived from stromal cells). On the other hand, related to growth factors, it includes the platelet-derived growth factor (PDGF), epidermal growth factor (EGF), transforming growth factor (TGF-b1), vascular endothelial growth factor (VEGF) and hepatocyte growth factor (HGF). Therefore, both approaches are responsible for inducing cell proliferation and angiogenesis within chronic and acute wounds. These molecules can provide activation in different cell phenotypes. After activating the platelet concentrates, in the molecular scenario, it triggers the healing activities of different cell phenotypes. Additionally, it is also antibacterial, mainly for containing thrombocidines [40].

Autologous platelet concentrates have been an innovative approach for treating wounds of various etiologies. The PRF is widely used because it is easy to prepare and is devoid of any synthetic additives in its structure [41].

The fibrin is a protein resulting from the clotting cascade, which forms a tridimensional network, where the platelets and the immunological cells stay attached, forming a blood clotting. The platelets produce growth factors stimulating the migration of fibroblasts and the proliferation of important elements, such as collagen type I and fibronectin. Recently, the products based on fibrin have become popular for wound treatments in both hard and soft tissues, being applied in different forms, such as glue, gel, membrane, or dressings containing fibroblasts [42,43].

Although the fibrin membrane acts as a favorable scaffold to the cells, it has low mechanical resistance, and it is easily degradable. Therefore, in order to enable to support the cells for a longer period, the fibrin membrane has to be reinforced with other natural or synthetic polymers [44].

From this perspective, ten clinical trials were evaluated, including 172 acute wounds of various etiologies and segment time, which were treated with PRF from five to ten days, and chronic wounds from four to eight weeks, resulting in healing or area reduction.

Some of the studies investigated the biological effects of the PRF in acute or chronic wounds with impaired healing. Only one clinical study pointed out the benefits of using the autologous fibrin concentrate in the wound in hands and postoperative of McCash technique, presenting an acceleration of healing after a single application of PRF [42].

One of the clinical studies demonstrated no significant statistical differences in the epithelization of acute surgical wounds between the group that was conventionally treated (control group) and the group treated with PRF. However, there was less pronounced pain related by the patients from the PRF groups [29].

In addition to the acute wounds, the chronic ones provoke intense concerns, since they represent a significant and often underestimated global socioeconomic burden [44]. In the USA, about 6.5 million people are affected by chronic wounds, encouraging the development of autologous biomaterials capable of accelerating wound healing. For instance, a pilot study to evaluate the action and safety of the PRF in chronic wounds of lower limbs caused by diabetes or amputation is reported by Londahl et al. [45] as well. The results showed that the PRF accelerates the angiogenesis, creating a great area of granulation tissue in the wound bed with a reduction of the wound area in a period of treatment from eight to twenty-five days [46].

Besides the acute wounds, the chronic ones cause intense concern, since they represent a significant and sometimes underestimated global socioeconomic burden. In the USA alone, about 6.5 million people suffer from chronic wounds [45].

In this context, the PRF can be considered one of the most versatile delivery systems to the wound bed because it is an excellent carrier of growth factors and leukocytes [47].

Clinical studies reinforce the effectiveness of the PRF in the treatment of chronic wounds, especially those of venous impairment origin. In 2016, 2017 and 2018, accelerated scar processes and cures were observed in public, with the use of PRF. It is reported that the healing process is seven days faster than conventional therapy. Restoration of the tissue is described in 73.3% of the PRF group (against 53.3% of conventional treatment) [32], in 85.51% of wounds after four weeks of PRF application (against 42.74% of conventional treatment) [33], and in 100% of wounds after seven weeks of PRF application (against 42.6% of conventional treatment) [34].

Data from the clinical studies emphasize the use of PRF containing leukocytes as a simple, low-cost, fast, and easy-to-handle alternative that does not require the hospitalization of the patient. It stands out for its potential for healing and protection of soft tissues, tendons, ligaments, and bones, as well as the healing of complex chronic wounds in lower limbs [48].

The PRF can modulate the healing of soft and hard tissues through gradual and prolonged release of growth factors, guaranteeing the homeostasis and stimulating the angiogenesis and cellular proliferation [49,50]. The acceleration of the wound healing based on PRF is promoted by increasing the wound site of the growth factors, then transforming β (TGF-β), insulin-like (IGF), platelet-derived (PDGF), vascular endothelial (VEGF), fibroblastic (FGF), and epithelial (EGF) [51].

Since the autologous PRF is produced from blood, and occasionally the presence of the wound provokes blood loss, laboratory analysis was conducted by [34] to evaluate disorders in the physiology of the blood compounds compared with the blood samples from the PRF receptors. The laboratory standard exams of the hemoglobin, platelets, and albumin in the PRF group, as well as in the control group, were analyzed and compared between the groups. No statistical differences were found, reinforcing the safety of PRF production.

The use of PRF is adequate in different healing mechanisms, such as in the tympanic membrane, which presents a reversal healing cascade in the last two stages, i.e., migration and proliferation [52]. A retrospective and randomized analysis comparing the use of PRF and the conventional treatment (paper patch moistened with polyvinylpyrrolidone) of traumatic perforation of tympanic membrane demonstrated that the PRF accelerated the healing process, promoting better audiological results and removing the need for a second surgical procedure [31].

The tympanic membrane, as well as the injuries in the palatopharyngeal muscle due to the pharyngoplasty for correction of sleep apnea, do not follow the healing cascade of a sharp wound, since they are healed by the second trial, with great deposition of collagen, contraction, and granulation, followed by greater epithelization time compared with the normal healing. This kind of healing is most likely to be opened again, as there are alternative therapies to accelerate healing in this region [53]. A study contemplating the evaluation of the efficacy of the PRF to decrease the incidence of rupture of the wound in pharyngoplasty showed that the autologous PRF reduced the possibility of the wound to be opened again, and it reduced the post-surgery pain. Additionally, the patients returned to their normal diet faster than those who were conventionally treated [37].

Updated psychopathological concepts support the PRF with a dynamic multifunctional hydrogel, working as an active dressing in wound healing, responsible for releasing a large set of healing molecules, in order to be layered on the wound-bed. The degradation of PRF is highly regulated by the plasma serine protease system plasminogen activator inhibitor type 1 and 2 (PAI-1, PAI-2), TATA box-binding protein associated factor (TAF1), and plasmin, and it can be synchronized with the healing process. In contact with the wound-bed, this natural biopolymer changes its configuration and mechanical properties in conjunction with specific secondary dressings (such as hydrogels, polyurethane foams, and hydrocolloids). Furthermore, it has a well-controlled microenvironment mechanism, which is responsible for optimizing its healing activities [54–56].

Numerous therapies have been used to treat wounds. However, the chronicity of the injuries is a challenge for the health and biotechnology professionals because it requires the recovery of homeostasis and the recruitment of molecules to achieve healing. Although the natural hydrogels have shown favorable results in chronic wound treatments, the intrinsic properties of the fibrin matrix are superior in the acceleration of healing because of the angiogenesis stimulation [51].

The search for strategies to control the delivery of therapeutic molecules, such as growth factors, led to the emergence of sophisticated fibrin-based therapeutic delivery systems [16]. Therefore, it is necessary to develop new clinical studies and systematic reviews that may favor precise indexing protocols for targeting the peculiarities of acute and chronic wounds.

5. Conclusions

The outcomes provided some evidence to support the routine using platelet-rich fibrin membrane dressings as the first line of treatment to induce the acceleration of wound healing. However, it seems that the PRF is not so efficient in the treatment of acute post-surgery wounds as in chronic wounds.

The findings do not provide sufficient evidence to support the routine use of the PRF as the first-line treatment for healing acute or chronic wounds of different etiologies. The methodological design of the majority of the clinical trials evaluated herein presented report failures, consequently affecting their results and discussion.

Based on the studies evaluated in this work, we strongly recommend the adoption of the template for intervention description and replication (TIDieR); in order to properly describe the intervention protocol, minimizing the risk of bias. We also emphasize the importance of following the directions proposed by the Consolidated Standards of Reporting Trials (CONSORT) to guarantee the reproducibility of the methodology and the adequate report. These procedures are important to accurately evaluate the benefits of the PRF applied in chronic wounds.

In this review, we found great variability in the application protocols of PRF in wounds, as well as different ways to prepare it, resulting in clinical heterogeneity. As a result, it is impossible to summarize and to classify the evidence of different types of studies to find strong conclusions.

We suggest the Intervention Description and Replication Model (TIDieR) to describe the treatment protocol, allowing for better reproducibility of the clinical trial methods, which uses PRF as an intervention for wound healing.

However, the use of PRF for the treatment of wounds of different etiologies is promising, since there is evidence in the acceleration of healing, reduction of the allergic episode, and spending on ineffective dressings. Therefore, prospective, multicenter, and large-scale clinical studies focused on short and long term therapeutic and economic impacts are necessary for the detailed implementation of this integrative practice as an alternative or adjuvant therapy for acute or chronic wounds.

Author Contributions: Conceptualization: B.L.F.; formal analysis: C.K.L.d.C.; investigation: B.L.F, C.K.L.d.C., and M.A.d.S.; methodology: C.K.L.d.C.; supervision: B.L.F.; validation and visualization: B.L.F., C.K.L.d.C., and M.A.d.S.; writing—original draft preparation: B.L.F., and C.K.L.d.C.; Writing—review & editing: B.L.F. and M.A.d.S. All authors have read and agreed to the published version of the manuscript.

Funding: This research received no external funding.

Conflicts of Interest: The authors declare no conflict of interest.

References

1. Borges, E.L.; Caliri, M.H.L.; Haas, V.J. Revisão sistemática do tratamento tópico da úlcera venosa. *Rev. Latino-Am. Enferm.* **2007**, *15*, 1163–1170. [CrossRef]
2. Garcia, L.K.R.; Rodríguez, M.E.R.; Cabrera, C.G.; Rondón, E.R.; Arboleda, J.C.G. Bioestimulación cutánea periocular con plasma rico em plaquetas. *Rev. Cub. Oftal.* **2015**, *28*, 97–109.
3. Morton, L.M.; Phillips, T.J. Wound healing and treating wounds: Differential diagnosis and evaluation of chronic wounds. *J. Am. Acad. Dermatol.* **2016**, *74*, 589–605. [CrossRef] [PubMed]
4. Sen, C.K.; Gordillo, G.M.; Roy, S.; Kirsner, R.; Lambert, L.; Hunt, T.K.; Gottrup, F.; Gurtner, G.C.; Longaker, M.T. Human skin wounds: A major and snowballing threat to public health and the economy. *Wound Repair Regen.* **2009**, *17*, 763–771. [CrossRef] [PubMed]
5. Gantwerker, E.A.; Hom, D.B. Skin: Histology and Physiology of Wound Healing. *Clin. Plastic Surg.* **2012**, *39*, 85–97. [CrossRef] [PubMed]
6. Sorg, H.; Tilkorn, D.J.; Hager, S.; Hauser, J.; Mirastschijski, U. Skin wound healing: An update on the current knowledge and concepts. *Eur. Surg. Res.* **2017**, *58*, 81–94. [CrossRef]
7. Wang, P.-H.; Huang, B.-S.; Horng, H.-C.; Yeh, C.-C.; Chen, Y.-J. Wound healing. *J. Chin. Med. Assoc.* **2018**, *81*, 94–101. [CrossRef]
8. Boyce, S.T.; Lalley, A.L. Tissue engineering of skin and regenerative medicine for wound care. *Burn Trauma* **2018**, *6*, 4. [CrossRef]
9. Harrison, P. Subcommittee on Platelet Physiology The use of platelets in regenerative medicine and proposal for a new classification system: Guidance from the SSC of the ISTH. *J. Thromb. Haemost.* **2018**, *16*, 1895–1900. [CrossRef]
10. Arora, S.; Kotwal, U.; Dogra, M.; Doda, V. Growth factor variation in two types of autologous platelet biomaterials: PRP Versus PRF. *Indian J. Hematol. Blood Transfus.* **2017**, *33*, 288–292. [CrossRef]
11. Choukroun, J.; Adda, F.; Schoeffler, C.; Vervelle, A. An opportunity in perio-implantology: The PRF. *Implantodontie* **2001**, *42*, 55–62.
12. Yung, Y.L.; Fu, S.C.; Cheuk, Y.C.; Qin, L.; Ong, M.T.; Chan, K.M.; Yung, P.S. Otimização da terapia de concentrado de plaquetas: Composição, localização e duração da ação. *Asia Pac. J. Sports Med. Arthrosc. Reabilitação. Technol.* **2017**, *7*, 27–36. [CrossRef]
13. Chicharro-Alcântara, D.; Rubio-Zaragoza, M.; Damia-Gimenez, E.; Carrillo-Poveda, J.M.; Cuervo-Serrato, B.; Pelaez-Gorrea, P.; Sopena-Juncosa, J.J. Plasma rico em plaquetas: Novos insights para feridas cutâneas Gestão de Cura. *J. Funct. Biomater.* **2018**, *9*, 10. [CrossRef]

14. Varela, H.A.; Souza, J.C.M.; Nascimento, R.M.; Araujo, R.F., Jr.; Vasconcelos, R.C.; Cavalcante, R.S.; Guedes, P.M.; Araujo, A.A. Fibina injetável rica em plaquetas: Conteúdo celular, caracterização morfológica e proteica. *Clin. Oral Investig.* **2018**. [CrossRef]
15. Borie, E.; Oliví, D.G.; Orsi, I.A.; Garlet, K.; Weber, B.; Beltrán, V.; Fuentes, R. Platelet-rich fibrin application in dentistry: A literature review. *Int. J. Clin. Exp. Med.* **2015**, *8*, 7922–7929.
16. Naik, B.; Karunakar, P.; Jayadev, M.; Marshal, V.R. Role of Platelet rich fibrin in wound healing: A critical review. *J. Conserv. Dent.* **2013**, *16*, 284–293. [CrossRef]
17. Miron, R.J.; Zucchelli, G.; Pikos, M.A.; Salama, M.; Lee, S.; Guillemette, V.; Fujioka-Kobayashi, M.; Bishara, M.; Zhang, Y.; Wang, H.L.; et al. Use of platelet-rich fibrin in regenerative dentistry: A systematic review. *Clin. Oral Investig.* **2017**, *21*, 1913–1927. [CrossRef]
18. Barbon, S.; Stocco, E.; Macchi, V.; Contran, M.; Grandi, F.; Borean, A.; Parnigotto, P.P.; Porzionato, A.; De Caro, R. Platelet-rich fibrin scaffolds for cartilage and tendon regenerative medicine: From bench to bedside. *Int. J. Mol. Sci.* **2019**, *20*, 1701. [CrossRef]
19. Polimeni, G.; Xiropaidis, A.V.; Wikesjö, U.M. Biology and principles. of periodontal wound healing/regeneration. *Periodontology 2000* **2006**, *41*, 30–47. [CrossRef]
20. Lourenco, P.A.; Mourão, C.F.A.B.; Leite, P.E.C.; Granjeiro, J.M.; Calasans-Maia, M.D.; Alves, G.G. The in vitro release of cytokines and growth factors from fibrin membrane. produced through horizontal centrifugation. *J. Biomed. Mater. Res.* **2018**, *106*, 1373–1380. [CrossRef]
21. Del Amo, C.; Perez-Valle, A.; Perez-Zabala, E.; Perez-del-Pecho, K.; Larrazabal, A.; Basterretxea, A.; Bully, P.; Andia, I. Wound dressing selection is critical to enhance platelet-rich fibrin activities in wound care settings. *Int. J. Mol. Sci.* **2020**, *21*, 624. [CrossRef] [PubMed]
22. Ding, Y.; Cui, L.; Zhao, Q.; Zhang, W.; Sun, H.; Zheng, L. Platelet- rich fibrin accelerates skin wound healing in diabetic mice. *Ann. Plast. Surg.* **2017**, *79*, e15–e19. [CrossRef] [PubMed]
23. Tsai, H.C.; Chang, G.R.; Fan, H.C.; Ou-Yang, H.; Huang, L.C.; Wu, S.C.; Chen, C.M. A mini-pig model for evaluating the efficacy of autologous platelet patches on induced acute full thickness wound healing. *BMC Vet. Res.* **2019**, *15*, 191. [CrossRef] [PubMed]
24. Ehrenfest, E.D.M.; Del Corso, M.; Diss, A.; Mouhyi, J.; Charrier, J.B. Three-dimensional architecture and cell composition of a Choukroun's platelet-rich fibrin clot and membrane. *J. Periodontol.* **2010**, *81*, 546–555. [CrossRef]
25. Bielecki, T.; Ehrenfest, E.D.M. Platelet-Rich Plasma (PRP) and Platelet-Rich Fibrin (PRF): Surgical adjuvants, preparations for in situ regenerative medicine and tools for tissue engineering. *Curr. Pharm. Biotechnol.* **2012**, *13*, 1121–1130. [CrossRef]
26. Draxler, D.F.; Sashindranath, M.; Medcalf, R.L. Plasmin: A Modulator of Immune Function. *Semin. Thromb. Hemost.* **2017**, *43*, 143–153. [CrossRef]
27. Cochrane Handbook for Systematic Reviews of Interventions Version 6.0. Available online: www.training.cochrane.org/handbook (accessed on 10 February 2020).
28. Moher, D.; Shamseer, L.; Clarke, M.; Ghersi, D.; Liberati, A.; Petticrew, M.; Shekelle, P.; Stewart, L.A. Preferred reporting items for systematic review and meta-analysis protocols (PRISMA-P) 2015 statement. *Syst. Rev.* **2015**, *4*, 1. [CrossRef]
29. Danielsen, P.L.; Jørgensen, B.; Karlsmark, T.; Jorgensen, L.N.; Agren, M.S. Effect of topical autologous platelet-rich fibrin versus no intervention on epithelialization of donor sites and meshed split-thickness skin autografts: A randomized clinical trial. *Plast. Reconstr. Surg.* **2008**, *122*, 1431–1440. [CrossRef]
30. Weber, S.C.; Kauffman, J.I.; Parise, C.; Weber, S.J.; Katz, S.D. Platelet-rich fibrin matrix in the management of arthroscopic repair of the rotator cuff: A prospective, randomized, double-blinded study. *Am. J. Sports Med.* **2012**, *41*, 263–270. [CrossRef]
31. Gür, Ö.E.; Ensari, N.; Öztürk, M.T.; Boztepe, O.F.; Gün, T.; Selçuk, Ö.T.; Renda, L. Use of a platelet-rich fibrin membrane to repair traumatic tympanic membrane perforations: A comparative study. *Acta Otolaryngol.* **2016**, *136*, 1017–1023. [CrossRef]
32. Pravin, A.J.S.; Sridhar, V.; Srinivasan, B.N. Autologous platelet rich plasma (PRP) versus leucocyte-platelet rich fibrin (L-PRF) in chronic non-healing leg ulcers–a randomized, open labeled, comparative study. *J. Evol. Med. Dent. Sci.* **2016**, *5*, 7460–7462. [CrossRef]

33. Somani, A.; Rai, R. Comparison of Efficacy of Autologous Platelet-rich Fibrin versus Saline Dressing in Chronic Venous Leg Ulcers: A Randomised Controlled Trial. *J. Cutan. Aesthet. Surg.* **2017**, *10*, 8–12. [CrossRef] [PubMed]
34. Goda, A.A. Autogenous leucocyte-rich and platelet-rich fibrin for the treatment of leg venous ulcer: A randomized control study. *Egypt J. Surg.* **2018**, *37*, 316–321. [CrossRef]
35. Garrido-Castells, X.; Becerro-de-Bengoa-Vallejo, R.; Calvo-Lobo, C.; Losa-Iglesias, M.E.; Palomo-López, P.; Navarro-Flores, E.; López-López, D. Effectiveness of leukocyte and platelet-rich fibrin versus nitrofurazone on nail post-surgery bleeding and wound cicatrization period reductions: A randomized single blinded clinical trial. *J. Clin. Med.* **2019**, *8*, 1552. [CrossRef] [PubMed]
36. Zhang, S.; Cao, D.; Xie, J.; Li, H.; Chen, Z.; Bao, Q. Platelet-rich fibrin as an alternative adjunct to tendon-exposed wound healing: A randomized controlled clinical trial. *Burns* **2019**, *45*, 1152–1157. [CrossRef]
37. Elkahwagi, M.; Elokda, M.; Elghannam, D.; Elsobki, A. Role of autologous platelet-rich fibrin in relocation pharyngoplasty for obstructive sleep apnea. *Int. J. Oral Maxillofac. Surg.* **2020**, *49*, 200–206. [CrossRef]
38. Vaheb, M.; Karrabi, M.; Khajeh, M.; Asadi, A.; Shahrestanaki, E.; Sahebkar, M. Evaluation of the Effect of Platelet-Rich Fibrin on Wound Healing at Split-Thickness Skin Graft Donor Sites: A Randomized, Placebo-Controlled, Triple-Blind Study. *Int. J. Low. Extrem. Wounds* **2020**, *30*, 1534734619900432. [CrossRef]
39. Barrett, S. Wound-bed preparation: A vital step in the healing process. *Br. J. Nurs.* **2017**, *26*, S24–S31. [CrossRef]
40. Çetinkaya, R.A.; Yenilmez, E.; Petrone, P.; Yılmaz, S.; Bektöre, B.; Şimsek, B.; Kula Atik, T.; Özyurt, M.; Ünlü, A. Platelet-rich plasma as an additional therapeutic option for infected wounds with multi-drug resistant bacteria: In vitro antibacterial activity study. *Eur. J. Trauma Emerg. Surg.* **2019**, *45*, 555–565. [CrossRef]
41. Castro, A.B.; Meschi, N.; Temmerman, A.; Pinto, N.; Lambrechts, P.; Teughels, W.; Quirynen, M. Regenerative potential of Leucocyte- and Platelet Rich Fibrin (L-PRF). Part A: Intrabony defects, furcation defects, and periodontal plastic surgery. A systematic review and meta-analysis. *J. Clin. Periodontol.* **2017**, *44*, 67–82. [CrossRef]
42. Groeber, F.; Holeiter, M.; Hampel, M.; Hinderer, S.; Schenke-Layland, K. Skin tissue engineering—In vivo and in vitro applications. *Adv. Drug Deliv. Rev.* **2011**, *63*, 352–366. [CrossRef] [PubMed]
43. Reinke, J.M.; Sorg, H. Wound repair and regeneration. *Eur. Surg. Res.* **2012**, *49*, 35–43. [CrossRef] [PubMed]
44. Bacakova, M.; Pajorova, J.; Sopuch, T.; Bacakova, L. Fibrin-Modified Cellulose as a Promising Dressing for Accelerated Wound Healing. *Materials* **2018**, *11*, 2314. [CrossRef] [PubMed]
45. Järbrink, K.; Ni, G.; Sönnergren, H.; Schmidtchen, A.; Pang, C.; Bajpai, R.; Car, J. Prevalence and incidence of chronic wounds and related complications: A protocol for a systematic review. *Syst. Rev.* **2016**, *5*, 152. [CrossRef] [PubMed]
46. Londahl, M.; Tarnow, L.; Karlsmark, T.; Lundquist, R.; Nielsen, A.M.; Michelsen, M.; Nilsson, A.; Zakrzewski, M.; Jörgensen, B. Use of an autologous leucocyte and platelet-rich fibrin patch on hard-to-heal DFUs: A pilot study. *J. Wound Care* **2015**, *24*, 172–178. [CrossRef]
47. Heher, P.; Mühleder, S.; Mittermayr, R.; Redl, H.; Slezak, P. Fibrin-based delivery strategies for acute and chronic wound healing. *Adv. Drug Deliv. Rev.* **2018**, *129*, 134–147. [CrossRef]
48. Ozer, K.; Colak, O. Leucocyte- and platelet-rich fibrin as a rescue therapy for small-to-medium-sized complex wounds of the lower extremities. *Burns Trauma* **2019**, *7*, 11. [CrossRef]
49. Maniyar, N.; Sarode, G.S.; Sarode, S.C.; Shah, J. Platelet-Rich fibrin: A "wonder material" in advanced surgical dentistry. *Med. J. DY Patil Vidyapeeth* **2018**, *11*, 287–290. [CrossRef]
50. Viana, M.G.V. Considerações clínicas sobre o uso do L-PRF na terapêutica de osteonecrose medicamentosa dos maxilares: Relato de caso. *Braz. J. Health Rev.* **2019**, *2*, 3318–3327. [CrossRef]
51. Miranda, R.C.; Ferreira Neto, M.D. Plasma rico em fibrina para implante imediato: Revisão de literatura. *Id Line Rev. Mult. Psicol.* **2019**, *13*, 889–899. [CrossRef]
52. Teh, B.M.; Shen, Y.; Friedland, P.L.; Atlas, M.D.; Marano, R.J. A review on the use of hyaluronic acid in tympanic membrane wound healing. *Expert Opin. Biol. Ther.* **2011**, *12*, 23–36. [CrossRef] [PubMed]
53. Sennes, L.U. Músculo palatofaríngeo: O foco das faringoplastias no tratamento da apneia do sono. *Braz. J. Otorhinolaryngol.* **2019**, *85*, 397–398. [CrossRef] [PubMed]
54. Burgos-Alonso, N.; Lobato, I.; Hernández, I.; Sebastian, K.S.; Rodríguez, B.; Grandes, G.; Andia, I. Adjuvant Biological Therapies in Chronic Leg Ulcers. *Int. J. Mol. Sci.* **2017**, *18*, 2561. [CrossRef]

55. Del Pino-Sedeño, T.; Trujillo-Martín, M.M.; Andia, I.; Aragón-Sánchez, J.; Herrera-Ramos, E.; Iruzubieta Barragán, F.J.; Serrano-Aguilar, P. Platelet-rich plasma for the treatment of diabetic foot ulcers: A meta-analysis. *Wound Regen.* **2019**, *27*, 170–182. [CrossRef] [PubMed]
56. Xia, Y.J.; Zhao, J.; Xie, J.; Lv, Y.; Cao, D.S. The effectiveness of the platelet-rich plasma dressing for chronic non-healing ulcers: A meta-analysis of 15 randomized clinical trials. *Plast. Rebuild. Surg.* **2019**, *144*, 1463–1474. [CrossRef] [PubMed]

© 2020 by the authors. Licensee MDPI, Basel, Switzerland. This article is an open access article distributed under the terms and conditions of the Creative Commons Attribution (CC BY) license (http://creativecommons.org/licenses/by/4.0/).

Review

Biodegradable Polymeric Micro/Nano-Structures with Intrinsic Antifouling/Antimicrobial Properties: Relevance in Damaged Skin and Other Biomedical Applications

Mario Milazzo [1,*], Giuseppe Gallone [2], Elena Marcello [3], Maria Donatella Mariniello [4], Luca Bruschini [5], Ipsita Roy [6] and Serena Danti [1,2,4,*]

1. Department of Civil and Environmental Engineering, Massachusetts Institute of Technology, Cambridge, MA 02142, USA
2. Department of Civil and Industrial Engineering, University of Pisa, 56126 Pisa, Italy; giuseppe.gallone@unipi.it
3. School of Life Sciences, University of Westminster, London W1W 6UW, UK; elenamarcello@outlook.com
4. Doctoral School in Clinical and Translational Sciences, Department of Translational Research and New Technologies in Medicine and Surgery, University of Pisa, via Savi 10, 56126 Pisa, Italy; m.d.mariniello@gmail.com
5. Department of Surgical, Medical, Molecular Pathology and Emergency Medicine, University of Pisa, via Savi 10, 56126 Pisa, Italy; luca.bruschini@unipi.it
6. Department of Materials Science and Engineering, Faculty of Engineering, University of Sheffield, Sheffield S1 3JD, UK; i.roy@sheffield.ac.uk
* Correspondence: milazzo@mit.edu (M.M.); serena.danti@unipi.it (S.D.)

Received: 14 June 2020; Accepted: 12 August 2020; Published: 19 August 2020

Abstract: Bacterial colonization of implanted biomedical devices is the main cause of healthcare-associated infections, estimated to be 8.8 million per year in Europe. Many infections originate from damaged skin, which lets microorganisms exploit injuries and surgical accesses as passageways to reach the implant site and inner organs. Therefore, an effective treatment of skin damage is highly desirable for the success of many biomaterial-related surgical procedures. Due to gained resistance to antibiotics, new antibacterial treatments are becoming vital to control nosocomial infections arising as surgical and post-surgical complications. Surface coatings can avoid biofouling and bacterial colonization thanks to biomaterial inherent properties (e.g., super hydrophobicity), specifically without using drugs, which may cause bacterial resistance. The focus of this review is to highlight the emerging role of degradable polymeric micro- and nano-structures that show intrinsic antifouling and antimicrobial properties, with a special outlook towards biomedical applications dealing with skin and skin damage. The intrinsic properties owned by the biomaterials encompass three main categories: (1) physical–mechanical, (2) chemical, and (3) electrostatic. Clinical relevance in ear prostheses and breast implants is reported. Collecting and discussing the updated outcomes in this field would help the development of better performing biomaterial-based antimicrobial strategies, which are useful to prevent infections.

Keywords: skin; antifouling; antimicrobial; antiviral; electrospinning; breast implant; ear prosthesis; biomedical device; chronic wound

---

1. Introduction

Bacterial colonization of implanted biomedical devices and prostheses (e.g., orthopedic prostheses, heart valves, breast implants, catheters, stents) is the main cause of healthcare-associated and nosocomial post-surgical infections, estimated to be 8.8 million per year in Europe. Some superbugs, like those encoding New Delhi metallo-beta-lactamase 1 (NDM-1), have sparked panic in several hospital areas, as they are resistant even to carbapenems, the antibiotics reserved as a last defense line [1]. Therefore, new antibacterial treatments are becoming vital to control nosocomial infections arising as surgical and post-surgical complications [2]. Many infections originate from damaged skin, which lets bacteria exploit injuries and surgical accesses as passageways to get to inner organs. Thanks to percutaneous implants and catheters used for draining, skin-resident microorganisms can migrate into the body and generate infections close to the implanted devices and/or in spare body tissues. Therefore, an effective treatment of skin damage is of utmost importance for the success rate of many biomaterial-related surgical procedures, and in general for healthcare and wellbeing [3].

Skin is the largest organ in the human body, and is the principal interface between the body and the surrounding environment. As such, it represents the first defense line against armful entities. Skin is considered as a bi-layered structure composed of epithelial (i.e., epidermal) and connective (i.e., dermal) tissues (Figure 1) [4]. The thin superficial layer of about 0.1 mm called the epidermis is made up of epithelial cells forming a multilayered squamous structure led by keratinocytes located above the basement membrane and corneocytes at the air interface, as well as including melanocytes and mechanoreceptors (i.e., Merkel cells) [5]. Beneath the epidermis is the dermis, made up by fibroblasts with their secretome (e.g., collagens, elastin), a dense network of innervation and vasculature, the latter providing the nutrients to the cells and appendages. Underneath, the subcutis (or hypodermis) is mainly composed of fat that acts as a buffer for both insulation and energy storage [6].

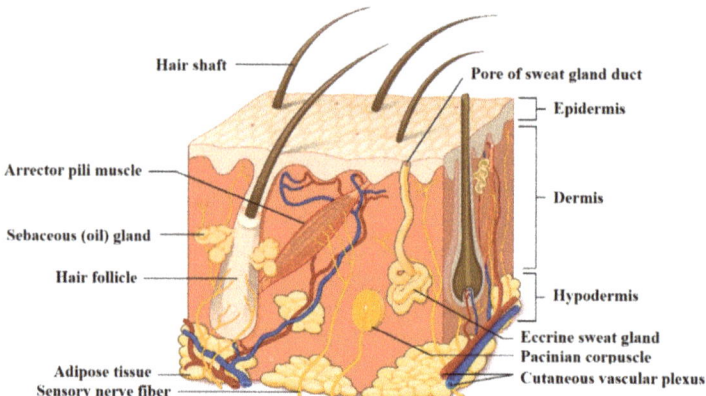

Figure 1. Structure of the skin: a superficial thin layer called the epidermis is the first barrier towards the external environment, provided with cells apt for protection and sensation. Beneath is the dermis layer, which provides both mechanical consistency and nourishing functions; it also hosts sensory receptors and vasculature. The underneath subcutaneous layer is mainly composed of fat for thermic insulation and energy storage. Reprinted with permission from [7] © 16 January 2020 OpenStax. Textbook content produced by OpenStax is licensed under a Creative Commons Attribution License 4.0 license.

Skin, the epidermis layer primarily, is permanently exposed to different chemical and physical factors that can potentially affect its structure or the tissues beneath it, causing the deterioration of the tissue and, potentially, the breakthrough of infections that can spread into the whole body [8]. Per se, human skin has a variegated microbiota, whose specific populations depend on the particular individual. The most common bacterial residents include *Staphylococcus*, *Corynebacterium*, *Cutibacterium*, *Micrococcus*, *Streptococcus*, *Brevibacterium*, *Acinetobacter*, and *Pseudomonas* [9]. In healthy conditions, the microbiota exerts a primary beneficial protective function; however, imbalances are associated with a number of skin diseases and infections. The most common and important class of skin diseases is caused by bacteria, either belonging to one's microbiota (including small intestinal barrier overgrowth) or coming from an external source. Bacterial-related pathologies can either be localized on the skin (e.g., cellulitis, impetigo) or affect other inner organs using the damaged skin as a passageway, and they emerge in a large number of diagnoses, being among the most frequent causes of infections in hospitalized patients [3]. Such clinical issues may even become enhanced when skin is damaged from burns, abrasions, or wounds [10]. Tissue engineering has developed different strategies to heal and replace the skin [11]. If the damage is limited to just the epidermis, no surgery is needed since self-regeneration processes take place thorough the keratinocytes still present in the site. In case the number of cells is not sufficient, it is possible to empower the regeneration process by the use of epithelial stem cells, e.g., from follicles. In contrast, if the defect extends deeper in the skin, homologous/synthetic grafts are employed on injured areas, possibly with surgical intervention to place them and close the wound [12].

The healing process is a delicate procedure, during which there is a high risk of infection since the damaged skin is not able to oppose a proper barrier to chemicals or bacteria. This is the motivation for using antibiotics to cure infections that have greatly reduced the mortality from bacterial pathologies worldwide. Unfortunately, the mutation of bacteria, leading them to become antibiotic-resistant, has significantly changed the curing trend [13,14]. Alternative approaches to antibiotics are nowadays represented by antimicrobial and antifouling solutions (Figure 2).

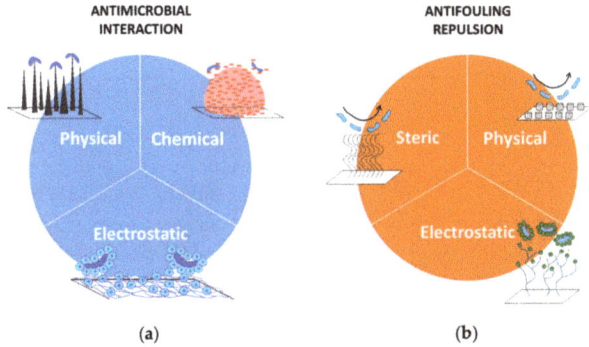

Figure 2. Antimicrobial and antifouling strategies. (**a**) Antimicrobial solutions aim at killing living microorganisms through physical (mechanical), chemical, or electrostatic interactions. (**b**) Antifouling systems contrast living organisms or inert bodies (e.g., dirt) through steric, physical (morphological), or electrostatic repulsive actions. Parts of the figure are reprinted from [15]—Open Access under Creative Commons CC BY 4.0 license.

The first class includes systems aimed at killing undesired living organisms by means of either chemical interactions (e.g., drug release [16]), physical contact through a modification of the nanostructure, or electrostatic neutralization of the activity of the living microbes. Once microbes have been killed, they lose their harmful potential and are promptly removed by the immunity system that, thanks to phagocytes, prevent the permanence of such biological structures. Antifouling solutions have a broader impact because they prevent the adhesion on a surface of microorganisms, but also inert bodies (e.g., dust). Their working principles are slightly different from those of the antimicrobials', and are based on the triggering a repulsion activity of some kind. A first case is the steric repulsion, in which adsorption is prevented by the compression of polymeric chains that cover a surface and behave as mechanical springs. A second class includes the modification of the surface through patterns that establish morphological conditions unfavorable to the attachment of foreign bodies. This is the case, for instance, of the so-called superhydrophobic structures [17]. The last two approaches consist in exploiting physical properties of surfaces that are able to promote repulsion forces, such as, for example, promoting a low surface energy or inducing an electrostatic polarization.

Antimicrobial and antifouling solutions possess a great appeal since they do not require the use of drugs, and thus they are likely to be healthier for the patient, and also they are to a large extent insensitive to mutation of the targeted bacteria. Starting from such considerations, in our review we focus on the design and fabrication of biodegradable structures that exhibit antimicrobial and antifouling properties without releasing additional substances.

Usually, most of the microorganisms that make up the skin flora establish a mutually beneficial relationship with the body. In fact, they hinder the colonization of pathogens by subtracting their nourishment by producing antimicrobial substances and by lowering the skin pH via the degradation of the sebum they feed on. Others, such as *Staphylococcus aureus* or *Candida albicans*, although potentially pathogenic, do not form numerically sufficient colonies to cause problems for a healthy organism.

Nevertheless, under some circumstances, protein absorption and bacterial adhesion can boost the formation of cell clots and biofilms, which can lead to implant rejection or failure, up to death if they migrate to vital organs or become resistant. Bacterial colonization phenomenon at the implant site is enhanced by the formation of a non-vascularized fibrotic scar around the device and is known as biofouling. Once a site becomes infected, an antibiotic therapy is applied. However, since only a low dose can reach the infected area across the avascular fibrotic scar, poor efficacy is observed after a first healing, which in time is followed by the development of chemically resistant species and consequent microbial reinforcement. Infections with *S. aureus* are in fact the prevalent device-related infections. Recently, infections associated with heart valves contaminated by nontuberculous mycobacteria (NTM) highlighted the hydrophobic ability of these cells, which preferentially attach to surfaces to form thick biofilms containing high numbers of cells, and thus the layers of cells and extracellular materials in biofilms substantially develop disinfectant-resistance.

Surface coatings can avoid biofouling by using appropriate physical and morphological properties of materials, thus inhibiting bacterial colonization by means of non-chemical mechanisms, which therefore do not induce bacterial resistance. By providing unfavorable surfaces for bacterial adhesion and/or presenting physical barriers at relevant orders of magnitude to isolate single cells or cell groups, reduced numbers of bacteria are finally present with a difficulty to grow and generate biofilms, thus becoming more vulnerable to the immune system. In this respect, super-hydrophobicity is a highly desired property to limit the biofouling phenomenon. A super-hydrophobic surface prevents wettability (therefore, protein absorption), forming single water drops with a static contact angle >150° with the surface. The physico-chemical properties of superhydrophobic materials have been a subject of study in the last decade due to the relevant implications in industry and healthcare sectors. Nanostructured surfaces, like pillar or sharklet-like patterns, offer valuable anti-biofouling properties [17]. However, some issues have to be solved to obtain industrial-scale applications, including costs and durability of micro/nanostructured surfaces [18]. Recently, electrospinning has been used to fabricate superhydrophobic materials, as the small fiber diameter has shown to contribute to super-hydrophobicity by the surface roughness/texture resulting from the superposition of fibrous layers [19]. Electrospinning is a cheaper and up-scalable technique that, when compared to other nanopatterning systems (e.g., etching, photolithography), results in being more effectively applicable to industrial-scale manufacturing of devices. However, many parameters (voltage, feed rate, viscosity, electrical properties of the solution, set-up geometry, and environmental conditions) affect the final result and must be optimized in order to obtain the desired fiber size, interspace/porosity, alignment, surface roughness, and other specific properties, which greatly enhance the difficulties in setting up the process. Among the different biomaterials reported with antimicrobial properties, biodegradable polymers offer interesting features in biomedical applications. A scientifically recognized definition for "biodegradable polymer" is "a material for which the degradation is mediated, at least, partly from a biological system" [20]. The disappearance of a biomaterial over time after implantation has been related to three distinct phenomena: biodegradation, bioresorption, and bioabsorption. Biodegradation is intended when the polymer chain breaks into natural byproducts processed by biological agents present in the microenvironment, such as enzymes, leading to material disintegration, erosion, or dissolution. Bioresorption occurs when the degradation products of polymers are resorbed in the body by a metabolic process. Instead, bioabsorption involves polymers that dissolve in biofluids and are eliminated without chain scission, such as poly(vinyl alcohol) (PVA) and poly(ethylene glycol) (PEG).

The focus of this review is to highlight the potential role of degradable polymeric micro- and nano-structures (most of all biodegradable, but including other degradable polymers) that show intrinsic antifouling and antimicrobial properties, with a special outlook towards biomedical applications dealing with skin and skin damage. Collecting and discussing the updated outcomes in this field would pave the development of better performing biomaterial-based antimicrobial strategies, useful to prevent and control surgical and post-surgical infections. As clinically relevant examples, this review shows usefulness in the complications of breast implants and otorhinolaryngology prostheses, such as tympanic membrane and nasal septum repair devices. The most widely used breast implants are gel-filled silicone shells, categorized into different surface types according to surface roughness and dimensionality ratio. The most common complication with a 10.6% overall incidence is the capsular contraction, which is foreign body reaction. As the scar is thick and avascular, microbes migrating from catheters to the implant surface are very difficult to eradicate with systemic antibiotic therapy and exacerbate the inflammatory process. Ear and nose are easily contaminated and stressed by bacteria, for example in cases of recurrent otitis and rhinitis. Many types of prosthetic devices are available to repair ear and nose damaged tissues; however, the two most common end-fates for synthetic and biological materials in such infected sites are extrusion and resorption, respectively, which are both driven by inflammatory processes [21,22]. Overall, new approaches to reduce bacterial contamination of surfaces would enable better-performing biomedical devices to treat skin damage and prevent post-surgical complications in many medical areas.

2. Antimicrobial Approaches

Antimicrobial approaches to treat damaged skin aim to kill microbes by exploiting either peculiar surface topologies or the physico-chemical properties of cationic polymers. Although the chemical approach represented by antibiotic-loaded hydrogels have been a reliable approach to dress wounds due to their antimicrobial properties [16,23], we have decided to exclude this topic since our focus concerns materials/topologies with intrinsic antimicrobial features. Before going into detail of each class, we have displayed Table 1 below, which summarizes the main materials/features later discussed.

2.1. Surface Topology

The antibacterial effect based on mechanical interactions was originally observed in nature [59]. For instance, wings of insects such as cicada or dragonflies possess a nanopatterned surface with high-aspect-ratio cone-like nanopillars that is lethal for bacteria such as *Pseudomonas aeruginosa*. It has been speculated that this is the effect of an evolutionary adaptation to the environment, preventing the formation of biofilms that affect the aerodynamicity of such insects [24,26–28].

The antimicrobial mechanisms of nano-structured/patterned surfaces have been deeply investigated by studying the influence of specific parameters, such as the distribution density and the shape of nanopillars. Scientists have shared different opinions—from one side, Li et al. stated that an increased number of nanopillars per surface unit is beneficial to promote a larger adhesion area and, thus, the lethal effect [29]. This result was also confirmed by Linklater et al., as well as by Kelleher et al., who assessed the performance of both synthetic nanostructures made of silicon and of cicada wings [24,30]. In contrast, Xue et al. [33] theorized that a lower density associated with the use of sharp nanopillar could be more effective, and later Fisher et al. [31] experimentally demonstrated this using *P. aeruginosa*.

Table 1. Summary of the antimicrobial materials/features reported in this work.

Classification	Material/Feature	Advantages	Disadvantages	Reference
Surface topology	Made-up nanopillars Diamond Gold Polymethyl methacrylate (PMMA) Titanium	Independent of the material properties	Made-up structures are potentially costly in terms of manufacturing	[24–33] [31] [34] [35] [36,37]
Cationic polymers	Polymers with ammonium ions Polymers with sulfonium ions Polymers with phosphonium ions α–helical structures Chitin Chitin nanofibrils (CN) Polymer/CN composites Polyhydroxyalkanoates (PHAs) with thioether/thioester moieties (PHACOS)	No additional fabrication costs to implement antimicrobial properties (intrinsic features)	Dependent on the physico-chemical properties of the materials	[38–42] [43] [44,45] [46–50] [51–53] [54,55] [56] [57,58]

However, scientists agree on two aspects: antimicrobial efficacy is improved if the nanopillars have a high aspect ratio (e.g., diameter of 20–80 nm and height of 500 nm [32]) and if the bacteria have a high motility, since it promotes a larger contact [24,25]. Bacteria are, in fact, not killed by a sudden interaction, but by the prolonged contact that leads to the elongation of the cellular structures and, ultimately, to unbearably high shear stresses (Figure 3) [26].

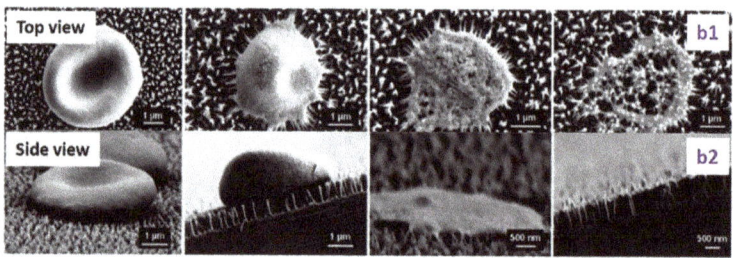

Figure 3. SEM images of the interaction of an erythrocytes with nano-pillared surfaces. Top (**b1**) and side (**b2**) views of the morphological changes of the erythrocyte. Reprinted with permission [60] © 2014 Royal Society of Chemistry.

A large number of antimicrobial surfaces have been developed by taking inspiration from natural systems and employing different materials (e.g., diamond [31], gold [34], polymethyl methacrylate (PMMA) [35]), but only silicon-based structures have demonstrated high in vitro and in vivo biocompatibility [61–63]. Studies in mice have demonstrated how the same material, but with a different patterning, is able to reduce inflammations after surgery [61]. Another interesting material is titanium, which has been used in many prosthetic applications as a substitute of the native cortical bone (e.g., [36,37]). However, the antimicrobial properties of titanium-based structures have been proven only in vitro, and an in vivo assessment is still missing [64–66].

2.2. Antimicrobial Cationic Polymers

Antimicrobial cationic polymers are usually employed in devices for clinical applications, ranging from medical devices to wound treatments [67–69]. This class of polymers base their antimicrobial action through two main functional components: cationic and hydrophobic groups. The first class allows adsorption onto the membranes of the microbes while the hydrophobic components cause cytoplasm leakage and the ultimate death of the pathogenic cells [70,71]. Given this peculiar mechanism, the structural design of such polymers assumes a key role in determining and tuning the antimicrobial properties [72]. The primary structure of antimicrobial cationic polymers concerns cationic groups that are usually ammonium [38–42], sulfonium [43], and phosphonium ions [44,45]. As for the hydrophobic groups, the most used are alkyl chains whose length may or not trigger the lethal effects on microbes [73]. In fact, non-optimized long chains can weaken the biocidal activity and, at the same time, increase the hemolytic activity [39,74]. The other important aspect is the spatial arrangement of the polymer molecules, whose chains are mainly organized in α-helices [75]. A correct control of such a secondary structure can, indeed, boost the antimicrobial properties of the polymer by properly positioning the cationic and hydrophobic functional groups along the molecular chain [76]. In this respect, very effective α-helical structures have been found, which, due to the organization of their hydrophilic and hydrophobic groups, are able to promote a cationic antimicrobial effect [46–50]—hydrophilic regions with positive charges are absorbed by the microbe while the hydrophobic ones cause the death of the pathogens [77,78]. Many studies have been carried out to mimic the behavior observed in nature [79–82]; for instance, Gellman et al. synthetized an artificial α-helical polymer that was effective against *Escherichia coli*, *Bacillus subtilis*, *Staphylococcus aureus*, and *Enterococcus faecium* [83].

Chitin is a natural polysaccharide found particularly in the shells of crustaceans, cuticles of insects, and cell walls of fungi. A deacetylated form of chitin with deacetylation greater than 50% is chitosan, a well-known biopolymer that has been reported to have antibacterial properties [51–53]. Chitosan is metabolized by some human enzymes, such as lysozyme, and thus it is considered bioresorbable. Owing to the large quantities of amino groups on its chain, chitosan dissolves at low pH (i.e., acidic solutions), showing a pH-sensitive behavior as a weak polybase due to the protonization of swollen amine groups under low pH conditions. Due to its pro-inflammatory activity, chitosan plays a role in the wound healing process. Since blood cells have a negatively charged surface, contact with chitosan adheres tightly to the wound and stops bleeding.

Differently from chitin, which exhibits immunogenic properties, and chitosan, which is pro-inflammatory, nano-sized (<0.2 µm) chitin shows anti-inflammatory along with antibacterial properties [54,55]. Chitin nanofibrils (CN) represent the purest crystal form of chitin and show positive surface charges due to the amino groups (Figure 4). In skin contact application, CN exhibited very good anti-inflammatory properties; for example, upon in vitro administration to human keratinocytes (HaCaT cells) at 10 µg/mL, CN reduced many pro-inflammatory interleukins (IL)—IL-1α, IL-1β, IL-6, and IL-8—and tumor necrosis factor alpha (TNF-α) in 6–24 h [84]. CN also showed good direct and indirect antimicrobial properties, the latter through stimulating the innate immune response of skin cells [85,86]. One peptide involved in skin-mediated immunity is the human beta defensin 2 (HBD-2), which acts as an endogenous antibiotic against Gram-positive and Gram-negative bacteria, fungi, and the envelope of some viruses [87]. Danti et al. [84] reported that CN–nanolignin microcomplexes administrated in culture at 0.2 µg/mL to human dermal keratinocytes (HaCaT cells) were able to increase HBD-2 expression of about 140% with respect to basal conditions, which can be relevant in skin self-defense. The indirect antimicrobial properties of CN could also be found in biodegradable polymeric nanocomposites proposed for skin contact applications. However, a high quantity of CN is needed on the surface to obtain a direct antimicrobial effect, which is difficult to achieve using polymer/CN composites [56]. Therefore, CN surface coatings seem to be more suitable for inducing antimicrobial activity than CN bulk incorporation. In bionanocomposites, CN can be released upon polymer degradation to sustain anti-inflammatory activity. CNs can be metabolized from the many families of chitotriosidases present into the human body.

Sulfur has been exploited as an antimicrobial agent since ancient times. Nevertheless, limited research has been conducted on antimicrobial polymers incorporating this element. Alongside the aforementioned sulfonium groups, recently a class of natural polymers containing sulfur, i.e., polyhydroxyalkanoates (PHAs) with thioester moieties (i.e., PHACOS), have been proven to exhibit antibacterial properties [57]. PHAs are bacteria-derived polyesters, an attractive class of biocompatible materials for a range of medical applications, including skin [88,89]. They are bioresorbable and biobased polymers that can be obtained from renewable resources, making them a valuable alternative to synthetic petroleum-derived plastics [90]. PHACOS are PHAs with thioester linkages in their side chains, possessing antimicrobial activity selectively against *S. aureus*. The presence of the thioester group has been associated with the biocidal activity of these polymers [57,58]. Dinjaski et al. showed that a direct contact between the bacteria and the material was essential for the antimicrobial activity of PHACOS, suggesting a possible interaction between the thioester groups and the bacterial membrane [57,58]. The alkyl chains present in the structure of PHACOS (Figure 5) could also participate in the biocidal activity of the material, as described in the previous section.

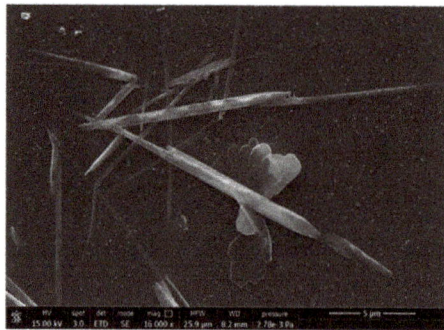

Figure 4. Scanning electron microscopy (SEM) image showing CN. Reprinted from [84]—Open Access under Creative Commons CC BY 4.0 license.

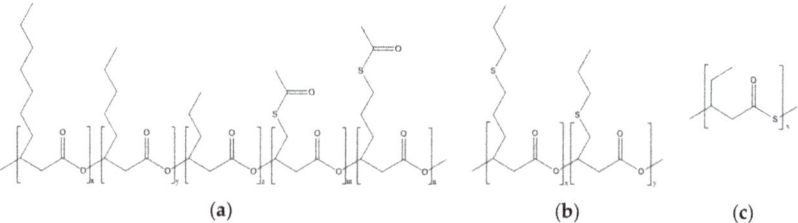

Figure 5. Chemical structure of polyhydroxyalkanoates (PHAs) containing sulfur. (**a**) PHAs with thioester linkages in their side chains (PHACOS). (**b**) PHAs with thioether linkages in their side chains (poly(3HPTB-co-3HPTHx)). (**c**) PHAs with thioester groups in their main chains (poly(3MV)).

Nevertheless, the exact mode of action of PHACOS and thioester groups has not yet been elucidated. To date, PHACOS are the first and only PHAs to show inherent antibacterial properties. Other sulfur-containing PHAs have been produced, containing thioester groups in their main chains [91] or thioether linkages in their side chains [92], but their possible antimicrobial activity has not yet been analyzed.

Another interesting class of polymers containing sulfate groups is polysaccharides from seaweeds. Different subclasses of sulfated polysaccharides have been identified according to the algal origin: ulvan from green seaweeds, fucoidan from brown seaweeds, and galactan and carragenan from red seaweeds [93]. The sulfate groups, and in particular their position in the polymer chain, were found to be responsible for a set of bioactive properties, encompassing antioxidant, anti-inflammatory, and antiviral activities [94,95].

3. Antifouling Approaches

In contrast to antimicrobial approaches, antifouling materials/features (summarized in Table 2) prevent the adhesion of microbes/external bodies. The following subsections aim to describe the main applications for treating damaged skin.

Table 2. Summary of the antifouling materials/features reported in this work.

Classification	Material/Feature	Advantages	Disadvantages	Reference
Steric	Polymer zwitterions	No additional fabrication costs to implement antifouling properties (intrinsic features)	Dependent on the physico-chemical properties of the materials	[96–102]
	Ulvan			[103–107]
Surface topology	Superhydrophobic surfaces	Independent of the material properties	Potentially costly in terms of manufacturing	[108–122]

3.1. Steric Repulsion

The mechanism behind steric repulsion involves long-chain polymers stabilized on a surface, which prevent bacteria adsorption. Among the most common polymers used for this purpose, PEG and poly(ethylene oxide) (PEO) are two polyethers that possess the same repeating unit but with different molecular weight. PEG/PEO are important materials for producing antifouling surfaces since they have a low surface energy (below 5 mJ·m^{-2}) and make weak bonds with proteins [123]. These polymers are bioabsorbable materials, and thus they are eliminated by the excretory organs without chain scission. As stand-alone, they are not advisable for human body-related applications, since they can trigger anaphylactic reactions and hypersensitivity reactions [124,125]. PEG and PEO are thus used to obtain polymer-based composites as dispensing agents for the filler, such as CN/Polylactic acid (PLA) composites [56,126].

Polymer zwitterions are a second class of materials that are exploitable for antifouling purposes thanks to their low energy and structure [99,100]. They are chemically stable and are able to pull water and foreign organisms away, creating a barrier, and thus being more suitable for tissue engineering applications [101,102]. Polymer zwitterions are usually electrospun from precursor solutions with low concentration and viscosity [97,98]. Remarkable applications include the fabrication of vascular grafts using polyurethanic matrices [96].

The antimicrobial activity of ulvan has been recently reported [105,107]. Ulvan shows good biocompatibility and antifouling properties, thus appearing as a green material with prospects in functional coating biomedical implants and devices. Sulfate groups specifically impart ulvan with antiviral activity against herpes simplex virus 1 (HSV-1) and paramyxoviridae [103,104]. By means of its antiadhesive character led by its negative charges, and independently of surface roughness, ulvan was able to inhibit biofilm formation and bacterial adhesion by *P. aeruginosa* on titanium surfaces [106].

3.2. Surface Topology

As stated and described in Section 2.1., the antifouling effect based on surface topology also has its roots in the observation of nature. Scientists have observed and studied a large number of natural systems that use this feature as a mean to adapt to the surrounding environment and to enable specific features such as protection against external organisms or improvement of dynamic or self-cleaning performances. Scientists have identified different mechanisms under the umbrella of micro-topology modification. The most important, and common, is the superhydrophobicity—a surface is defined as superhydrophobic if the contact angle generated by the deposition of a water droplet is between 90° and 150°. Without a perfect attachment, water can remove contaminants by rolling off the surface [110,115]. Similar effects have been investigated at different scales to prevent the attachment of cellular tissues or macro-organisms while dynamically exposed to drag forces [112–114]. In contrast, a completely different mechanism is represented by animal secretions that dynamically influences the antifouling properties of the surfaces [111,122].

Natural antifouling surfaces exploiting topological features have been found in about 900 marine animals of the Elasmobranchii family [121]. This group includes sharks, skates, and rays, whose placoid scales are composed of a vascular core of dentine beneath an acellular layer of enamel [119,120]. These structures have a undulated topology that enables a passive cleaning induced by drag forces and, at the same time, improves the hydrodynamicity of the animal. On the basis of the species, the specific topology of such structures, and thus the associated features, slightly change for the best adaptation to the surrounding environment (Figure 6a,b).

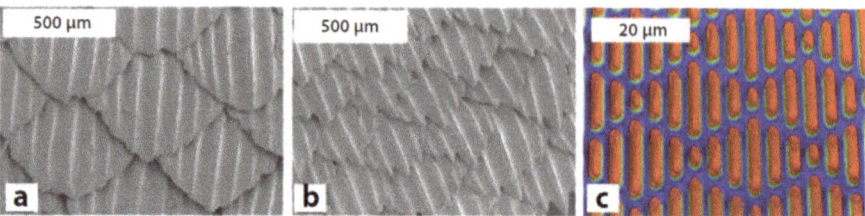

Figure 6. SEM micrographs of (**a**) spinner shark skin and (**b**) Galapagos shark skin surfaces. (**c**) Optical profilometry of sharklet surface. Reprinted with permission [127] © 2010 Elsevier.

The self-cleaning mechanisms activated by superhydrophobicity have been observed on lotus and rice leaves, above which water drops are able to move freely, collecting foreign living and inert bodies and transporting them passively thanks to inertial forces. The so-called lotus effect is due to the microstructure of the leaf surface that presents a pattern of shaped cones with nano-scale hair-like structures, which induces the superhydrophobic effect (Figure 7a) [116–118].

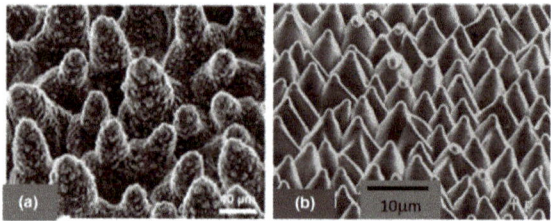

Figure 7. (**a**) SEM images of the micro-surface of a leaf of a lotus plant with the characteristic conical cell pattern. Reprinted from [118]—Open Access under Creative Commons CC BY 4.0 license. (**b**) Bioinspired cone-shaped patterned surface resembling the microstructure of the lotus leaf. Reprinted with permission [109] © 2006 American Institute of Physics.

Artificial attempts to shape surfaces with bioinspired antifouling properties have been carried out using lithography and self-assembly techniques [108,109,115]. Following this approach, a number of synthetic products have been developed and commercialized, exploiting the peculiar antifouling characteristics. Remarkable examples are the devices made in either polydimethylsiloxane or thermoplastic polyurethane polymers by Sharklet Technologies, whose micropatterned surfaces aimed at mimicking shark skin (Figure 6c) or lotus micro-surface (Figure 7b) were effective in preventing bacterial adhesion with direct implications on skin-related applications [128] or laser-ablated films.

4. Clinical Relevance and Future Perspectives

We present a review on emerging approaches to prevent skin and skin-derived infections using biodegradable polymeric micro/nano-structures with intrinsic antifouling/antimicrobial properties as a valuable alternative to traditional pharmacological antibiotic treatments or drug release materials [16] that can generate resistant bacteria. Degradable biomaterials, such as properly biodegradable as well as bioresorbable and bioabsorbable, play a key role in a number of biomedical applications, including, but not limited to, skin repair [23]. Moreover, the cosmetic and sanitary industries offer a set of skin care products and procedures to clean, restore, reinforce, protect, and maintain skin wellbeing, which can all benefit from the possibility of antifouling and antimicrobial surfaces. The functionalities expected by sanitary and cosmetic products are less strict than those required by biomedical devices, however, they may require materials/additives to treat potentially contaminated skin and bacterial proliferation (e.g., diapers); allow skin cleaning from dirt, sebum, microorganisms, and dead cells (e.g., beauty masks); reduce inflammation; and allow healing.

Depending on the extent of damage and presence of systemic pathologies (e.g., diabetes), skin wounds caused by burns, soars, or cuts are exposed to microbial contamination for several weeks and need to be properly treated. The physiological process that leads to the repair of a wound relies on the activation of a series of specific events involving a large number of cells and molecules. The process can be divided into three phases that overlap each other (Figure 8): (a) inflammation, (b) tissue formation, and (c) tissue remodeling. Some types of wounds, such as diabetic foot ulcers, venous ulcers, and some surgical wounds fail to complete this chain of events, thus becoming chronic wounds. These types of wounds often remain stationary at the first stage [129] and show a high presence of proinflammatory cytokines, high protease levels, many neutrophils, and senescent cells that do not respond to stimuli [129,130]. The wound bed can remain open for months or years, leading to repeated bacterial infections [131]. It is a fact that damaged skin offers a passageway for skin-resident and nosocomial bacteria to migrate to inner organs due to continuity between sterile and non-sterile compartments of the body. In skin dressings used to treat chronic wounds, a weak antimicrobial activity combined with defensin (e.g., HBD-2) expression by keratinocytes is desirable to maintain a safe microenvironment and allow the skin self-repair process.

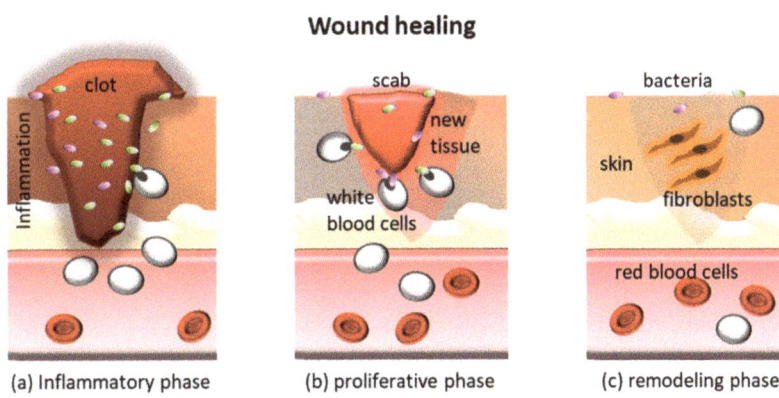

Figure 8. Schematic of the three phases of the wound healing process.

Chronic wounds are indeed very difficult to treat, representing sites at risk of infections since they are usually unhealed for more than 3 months [129]. In these cases, often secondary to other pathologies such as diabetes that decrease the skin regenerative capacity, locally aggressive antimicrobial treatments, including those containing silver, can further hamper the wound repair, whereas long-term antibiotic treatments are not recommended [4]. Therefore, biodegradable polymers with intrinsic antifouling/antimicrobial activity are highly suitable to produce wound dressings. One expected action of these materials is to combine three key properties, which all concur to the optimal wound healing cascade: (a) immunomodulation (including indirect antibacterial effects), thus controlling the inflammatory phase; (b) biocompatibility combined with antimicrobial properties, thus allowing the optimal timeline of tissue formation phase; and (c) degradability to allow space for tissue growth and remodeling. Since the cells involved in wound healing process, i.e., keratinocytes, fibroblasts, Langerhans cells, and recruited immune cells, concur to the eradication of bacteria, properly designed bio (nano)-materials can drive the healing process by regulating the immune cell trafficking and cross-talk at the wound site [132,133].

Many body parts have been reported as targets of infections and, in particular, of bacterial biofilms, as a consequence of device implantation [134] (Figure 9). In fact, permanent implants and endoprostheses (e.g., breast implants, orthopedic prostheses, implantable ear devices, and many others) pose the problem of surgical and post-surgical contaminations. In these cases, the stronger the immune reaction, the thicker the avascular fibrotic layer and the less efficient the effect of systemic antimicrobial treatments [135]. Therefore, bacteria that have reached the implant surface can proliferate undisturbed. Under these circumstances, intrinsically antimicrobial degradable materials could be used as surface coatings with a primary role in controlling immune reactions and bacterial infiltration in the initial stage post-implantation, namely, when fibrotic encapsulation takes place. Emerging robotic implants also raise the question of controlling undesired reactive phenomena such as contaminations and inflammatory reactions, which ultimately affect biocompatibility [136].

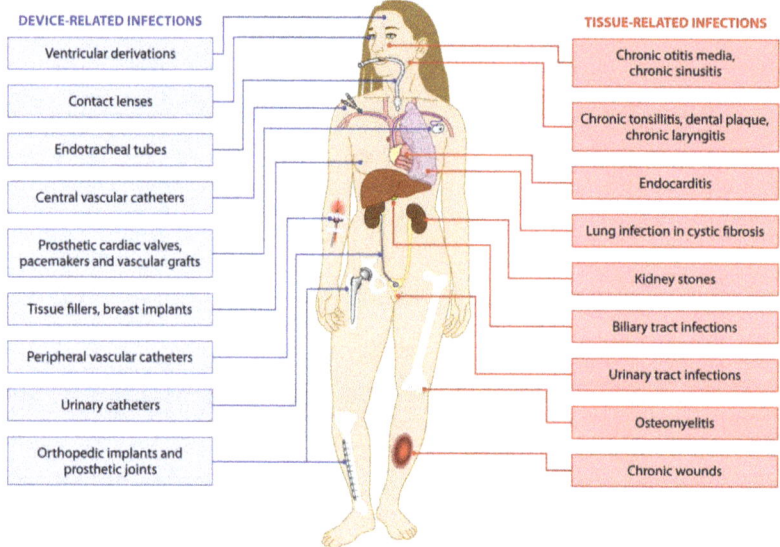

Figure 9. Reported sites of device-related and tissue infections. Reprinted with permission [134] © 2014 American Society for Microbiology.

Depending on the specific clinical issue, the antifouling/antimicrobial features of such structures can be exploited either as surface coatings or as actual structural medical devices. Usually, different biomedical devices and implants have been reported to show relevant incidence of infection [137]: 16–36% of left ventricular assist devices, 5–20% of cardiac implantable electronic devices, catheters (20% of peritoneal, 4.2–5.3% of dialysis, 2.5–4% of central venous, 1–3.4% of arterial, and 3.1–7.5% of urinary), 1–10% of meshes for ventral hernial repair, and 5% of elbow joint prostheses. Catheters, in particular, play a main role in infections due to the permanence of a percutaneous access to skin and environmental bacteria. In this review, we report here specifically on other settings dealing with complications for bacterial infections, namely, breast implants and otorhinolaryngologic implantable devices.

Breast implant infections are the leading cause of morbidity after breast surgery, with an incidence rate of 1% to 2.5% [138]. The incidence is higher after surgery for breast cancer (especially mastectomy followed by radiotherapy) than after breast augmentation and surgical technique, and patient's conditions are considered the most important determinants. In most cases, this complication develops within the acute post-operative period, but it is reported also years after surgery. In the majority of cases, it is caused by Gram-positive pathogens, such as *Staphylococcus aureus*, and empiric initial treatment is based on the prevalence of methicillin-resistant staphylococci [138]. Determining the origin of implant infections is still a challenge but potential sources of early post-operative infections include a contaminated prosthesis, the surgical environment, the patient's mammary ducts or skin, and the prosthesis contamination from remote infection sites. On the contrary, late infections usually are consequent to secondary bacteremia or an invasive procedure in other body districts [139]. Depending on severity of the infection, patients are treated with empiric intravenous or oral antibiotics and closely monitored [138]. All attempts should be made to salvage the implant. In cases of failure of the medical therapy, it is necessary to combine implant removal and antibiotic treatment followed by delayed reconstruction with a new operation weeks/months after the onset of the complication [138]. This approach requires multiple surgical procedures with increased risks and costs. Furthermore, delayed implant re-insertion may be technically more difficult due to tissue fibrosis. Breast implant infections are overall responsible for 2.0–2.5% of re-interventions performed in breast surgery [139] and low-grade or subclinical infections probably have a role also in the origin of capsular contracture. Therefore, new materials to coat breast prostheses able to reduce contaminations are expected to improve the success rate of these prostheses.

Antifouling/antimicrobial coating seems to be useful for a number of biomedical devices including the auricular implants. The aim is avoiding the infections and the consecutive extrusion of the implanted devices, which is a frequent event due to frequent otitis affecting these patients. Preliminary testing of a thin antifouling coating for a new type of cochlear implant have been evaluated [140]. Similar materials could be utilised for middle ear implants such as bone-anchored hearing aids. These devices often cause a skin reaction, and an antifouling/antimicrobial coating could solve this problem.

Moreover, these materials could find an application as surgical devices in ear, nose, and throat (ENT) procedures. At the end of the ear and nose surgery, de-epithelialized areas of the external or middle ear or the nasal septum or fossa have to be protected with silicone sheets or other materials to avoid scar formation. A layer of antifouling/antimicrobial material could improve the protection and facilitate the skin restoration. Equally, thin sheets coated with antifouling/antimicrobial biomaterials show potential to seal eardrum or nasal septum perforations. The chronic perforation of the tympanic membrane or of the nasal septum needs a difficult surgical reparation and the new antifouling/antimicrobial materials could easily close the hole. In particular, for nasal septum perforation, a hole greater than 2 cm significantly increases the surgical failure rate [141]. In case of surgical failure, a nasal button available on the market can be a solution. However, as a foreign body, this material can promote the formation of mucous scabs, creating an obstruction to the air flow. On the contrary, a nasal button covered with an antifouling/antimicrobial coating could seal the septal perforation, avoiding the surgical procedure and the formation of mucous scabs, thus restoring tissue function.

Proper material selection, design, and manufacturing of the surfaces of biomedical devices and implants, in particular considering their physico-mechanical, chemical, and electrostatic characters, are of paramount importance to reduce infection-related complications, thus allowing a better performance of the implanted devices. In light of this, additive manufacturing technologies may represent the key to synergistically exploit topologies and biomaterials as a powerful alternative to the traditional methodologies (e.g., etching), creating optimized surface topologies made of tuned materials able to either kill or repulse microbes. Examples of such an approach are the development of optimized antimicrobial/antifouling fibers fabricated via electrospinning [15,142], tuned hydrogels fabricated with 3D printing technologies [16,23], or the employment of stereolithography to modify hydroxyapatite-based composites for manufacturing antimicrobial dental bites [143].

The process by which bacteria adhere to, colonize, and infect a device is not fully understood, as it occurs on a multi-length scale and involves several physical, chemical, and biological aspects [137]. Therefore, biomaterial science integrated with nanotechnology and microfabrication could empower the intrinsic ability of some natural and biosynthetic polymers towards safer and better performing biomedical devices.

Author Contributions: Conceptualization, M.M. and S.D.; methodology, M.M. G.G., and S.D.; investigation, M.M. and S.D.; writing—original draft preparation, M.M., E.M., M.D.M., L.B., and S.D.; writing—review and editing, M.M., G.G., and S.D.; visualization, M.M.; supervision, G.G. and S.D.; project administration, I.R. and S.D.; funding acquisition, I.R. and S.D. All authors have read and agreed to the published version of the manuscript.

Funding: The project received funding from the Bio-Based Industries Joint Undertaking (BBI JU) under the European Union's Horizon 2020 research and innovation program by POLYBIOSKIN (grant agreement no. 7N5839). S.D. and L.B. acknowledge the 4NanoEARDRM project (EuroNanoMed III co-funded action by the Italian Ministry of University and Research—MIUR). M.M. was supported by the European Union's Horizon 2020 research and innovation program under the Marie Skłodowska-Curie grant agreement COLLHEAR no. 794614. E.M. was supported by the European Union's Horizon 2020 research and innovation program under the Marie Skłodowska-Curie grant agreement HyMedPoly no. 643050.

Conflicts of Interest: The authors declare no conflict of interest.

References

1. Ventola, C.L. The antibiotic resistance crisis: Part 1: Causes and threats. *Pharm. Ther.* **2015**, *40*, 277.
2. Stewart, P.S.; Bjarnsholt, T. Risk factors for chronic biofilm-related infection associated with implanted medical devices. *Clin. Microbiol. Infect.* **2020**, *26*, 1034–1038. (in press). [CrossRef] [PubMed]
3. Saginur, R.; Suh, K.N. Staphylococcus aureus bacteraemia of unknown primary source: Where do we stand? *Int. J. Antimicrob. Agents* **2008**, *32*, S21–S25. [CrossRef] [PubMed]
4. Danti, S.; D'Alessandro, D.; Mota, C.; Bruschini, L.; Berrettini, S. Applications of bioresorbable polymers in skin and eardrum. In *Bioresorbable Polymers for Biomedical Applications*; Elsevier: Amsterdam, The Netherlands, 2017; pp. 423–444.
5. Fuchs, E. Scratching the surface of skin development. *Nature* **2007**, *445*, 834–842. [CrossRef] [PubMed]
6. Shier, D.; Butler, J.; Lewis, R. *Hole's Essentials of Human Anatomy & Physiology*; McGraw-Hill Education New York: New York, NY, USA, 2015.
7. Betts, J.G.; Young, K.A.; Wise, J.A.; Johnson, E.; Poe, B.; Dean, H.K.; Korol, O.; Jody, E.J.; Womble, M.; DeSaix, P. *Anatomy and Physiology*; OpenStax: Houston, TX, USA, 2013.
8. Bickers, D.R.; Athar, M. Oxidative stress in the pathogenesis of skin disease. *J. Investig. Dermatol.* **2006**, *126*, 2565–2575. [CrossRef]
9. O'Sullivan, J.N.; Rea, M.C.; Hill, C.; Ross, R.P. Protecting the outside: Biological tools to manipulate the skin microbiota. *FEMS Microbiol. Ecol.* **2020**, *96*, 1–14. [CrossRef]
10. Elixhauser, A.; Steiner, C. *Most Common Diagnoses and Procedures in US Community Hospitals, 1996. Healthcare Cost and Utilization Project*; HCUP Research Note; Agency for Health Care Policy and Research; AHCPR Pub: Rockville, MD, USA, 1999; pp. 46–99.
11. Böttcher-Haberzeth, S.; Biedermann, T.; Reichmann, E. Tissue engineering of skin. *Burns* **2010**, *36*, 450–460. [CrossRef]
12. Enoch, S.; Roshan, A.; Shah, M. Emergency and early management of burns and scalds. *BMJ* **2009**, *338*, b1037. [CrossRef]
13. O'neill, J.I.M. Antimicrobial resistance: Tackling a crisis for the health and wealth of nations. *Rev. Antimicrob. Resist.* **2014**, *20*, 1–16.
14. Solomon, S.L.; Oliver, K.B. Antibiotic resistance threats in the United States: Stepping back from the brink. *Am. Fam. Physician* **2014**, *89*, 938–941.
15. Kurtz, I.S.; Schiffman, J.D. Current and emerging approaches to engineer antibacterial and antifouling electrospun nanofibers. *Materials* **2018**, *11*, 1059. [CrossRef]
16. Li, S.; Dong, S.; Xu, W.; Tu, S.; Yan, L.; Zhao, C.; Ding, J.; Chen, X. Antibacterial hydrogels. *Adv. Sci.* **2018**, *5*, 1700527. [CrossRef]
17. Lima, A.C.; Mano, J.F. Micro/nano-structured superhydrophobic surfaces in the biomedical field: Part II: Applications overview. *Nanomedicine* **2015**, *10*, 271–297. [CrossRef]
18. Simpson, J.T.; Hunter, S.R.; Aytug, T. Superhydrophobic materials and coatings: A review. *Rep. Prog. Phys.* **2015**, *78*, 86501. [CrossRef] [PubMed]
19. Nuraje, N.; Khan, W.S.; Lei, Y.; Ceylan, M.; Asmatulu, R. Superhydrophobic electrospun nanofibers. *J. Mater. Chem. A* **2013**, *1*, 1929–1946. [CrossRef]

20. Guillet, J.E.; Huber, H.X.; Scott, J. Biodegradable polymers and plastics. In Proceedings of the Second International Scientific Workshop on Biodegradable Polymers and Plastics, Montpellier, France, 25–27 November 1991; pp. 55–70.
21. Danti, S.; D'Alessandro, D.; Pietrabissa, A.; Petrini, M.; Berrettini, S. Development of tissue-engineered substitutes of the ear ossicles: PORP-shaped poly(propylene fumarate)-based scaffolds cultured with human mesenchymal stromal cells. *J. Biomed. Mater. Res. Part A* **2010**, *92*, 1343–1356. [CrossRef]
22. Mota, C.; Danti, S. Ear Tissue Engineering. In *Comprehensive Biotechnology*, 3rd ed.; Moo-Young, M., Ed.; Elsevier: Amsterdam, The Netherlands, 2019; pp. 270–285.
23. Zhang, X.; Shu, W.; Yu, Q.; Qu, W.; Wang, Y.; Li, R. Functional Biomaterials for Treatment of Chronic Wound. *Front. Bioeng. Biotechnol.* **2020**, *8*, 516. [CrossRef] [PubMed]
24. Kelleher, S.M.; Habimana, O.; Lawler, J.; O'reilly, B.; Daniels, S.; Casey, E.; Cowley, A. Cicada wing surface topography: An investigation into the bactericidal properties of nanostructural features. *ACS Appl. Mater. Interfaces* **2016**, *8*, 14966–14974. [CrossRef]
25. Diu, T.; Faruqui, N.; Sjöström, T.; Lamarre, B.; Jenkinson, H.F.; Su, B.; Ryadnov, M.G. Cicada-inspired cell-instructive nanopatterned arrays. *Sci. Rep.* **2014**, *4*, 7122. [CrossRef]
26. Bandara, C.D.; Singh, S.; Afara, I.O.; Wolff, A.; Tesfamichael, T.; Ostrikov, K.; Oloyede, A. Bactericidal effects of natural nanotopography of dragonfly wing on Escherichia coli. *ACS Appl. Mater. Interfaces* **2017**, *9*, 6746–6760. [CrossRef]
27. Ivanova, E.P.; Hasan, J.; Webb, H.K.; Truong, V.K.; Watson, G.S.; Watson, J.A.; Baulin, V.A.; Pogodin, S.; Wang, J.Y.; Tobin, M.J.; et al. Natural bactericidal surfaces: Mechanical rupture of Pseudomonas aeruginosa cells by cicada wings. *Small* **2012**, *8*, 2489–2494. [CrossRef] [PubMed]
28. Mainwaring, D.E.; Nguyen, S.H.; Webb, H.; Jakubov, T.; Tobin, M.; Lamb, R.N.; Wu, A.H.-F.; Marchant, R.; Crawford, R.J.; Ivanova, E.P. The nature of inherent bactericidal activity: Insights from the nanotopology of three species of dragonfly. *Nanoscale* **2016**, *8*, 6527–6534. [CrossRef] [PubMed]
29. Li, X. Bactericidal mechanism of nanopatterned surfaces. *Phys. Chem. Chem. Phys.* **2016**, *18*, 1311–1316. [CrossRef]
30. Linklater, D.P.; Nguyen, H.K.D.; Bhadra, C.M.; Juodkazis, S.; Ivanova, E.P. Influence of nanoscale topology on bactericidal efficiency of black silicon surfaces. *Nanotechnology* **2017**, *28*, 245301. [CrossRef] [PubMed]
31. Fisher, L.E.; Yang, Y.; Yuen, M.-F.; Zhang, W.; Nobbs, A.H.; Su, B. Bactericidal activity of biomimetic diamond nanocone surfaces. *Biointerphases* **2016**, *11*, 11014. [CrossRef]
32. Ivanova, E.P.; Hasan, J.; Webb, H.K.; Gervinskas, G.; Juodkazis, S.; Truong, V.K.; Wu, A.H.F.; Lamb, R.N.; Baulin, V.A.; Watson, G.S.; et al. Bactericidal activity of black silicon. *Nat. Commun.* **2013**, *4*, 1–7. [CrossRef]
33. Xue, F.; Liu, J.; Guo, L.; Zhang, L.; Li, Q. Theoretical study on the bactericidal nature of nanopatterned surfaces. *J. Theor. Biol.* **2015**, *385*, 1–7. [CrossRef]
34. Wu, S.; Zuber, F.; Brugger, J.; Maniura-Weber, K.; Ren, Q. Antibacterial Au nanostructured surfaces. *Nanoscale* **2016**, *8*, 2620–2625. [CrossRef]
35. Dickson, M.N.; Liang, E.I.; Rodriguez, L.A.; Vollereaux, N.; Yee, A.F. Nanopatterned polymer surfaces with bactericidal properties. *Biointerphases* **2015**, *10*, 21010. [CrossRef] [PubMed]
36. Milazzo, M.; Danti, S.; Inglese, F.; Jansen van Vuuren, G.; Gramigna, V.; Bonsignori, G.; De Vito, A.; Bruschini, L.; Stefanini, C.; Berrettini, S. Ossicular replacement prostheses from banked bone with ergonomic and functional geometry. *J. Biomed. Mater. Res. Part B Appl. Biomater.* **2016**, *105*, 2495–2506. [CrossRef] [PubMed]
37. Milazzo, M.; Muyshondt, P.G.G.; Carstensen, J.; Dirckx, J.J.J.; Danti, S.; Buehler, M.J. De novo topology optimization of Total Ossicular Replacement Prostheses. *J. Mech. Behav.Biomed. Mater.* **2019**, *103*, 103541. [CrossRef] [PubMed]
38. Gottenbos, B.; Grijpma, D.W.; van der Mei, H.C.; Feijen, J.; Busscher, H.J. Antimicrobial effects of positively charged surfaces on adhering Gram-positive and Gram-negative bacteria. *J. Antimicrob. Chemother.* **2001**, *48*, 7–13. [CrossRef]
39. Lu, G.; Wu, D.; Fu, R. Studies on the synthesis and antibacterial activities of polymeric quaternary ammonium salts from dimethylaminoethyl methacrylate. *React. Funct. Polym.* **2007**, *67*, 355–366. [CrossRef]
40. Jacobs, W.A. The bactericidal properties of the quaternary salts of hexamethylenetetramine: I. the problem of the chemotherapy of experimental bacterial infections. *J. Exp. Med.* **1916**, *23*, 563. [CrossRef]

41. Rahn, O.; Van Eseltine, W.P. Quaternary ammonium compounds. *Annu. Rev. Microbiol.* **1947**, *1*, 173–192. [CrossRef]
42. Torres, L., Jr; Bienek, D.R. Use of Protein Repellents to Enhance the Antimicrobial Functionality of Quaternary Ammonium Containing Dental Materials. *J. Funct. Biomater.* **2020**, *11*, 54. [CrossRef] [PubMed]
43. Anderson, R.A.; Feathergill, K.; Diao, X.; Cooper, M.; Kirkpatrick, R.; Spear, P.; Waller, D.P.; Chany, C.; Doncel, G.F.; Herold, B.; et al. Evaluation of Poly (Styrene-4-Sulfonate) as a Preventive Agent for Conception and Sexually Transmitted Diseases. *J. Androl.* **2000**, *21*, 862–875. [PubMed]
44. Popa, A.; Davidescu, C.M.; Trif, R.; Ilia, G.; Iliescu, S.; Dehelean, G. Study of quaternary 'onium'salts grafted on polymers: Antibacterial activity of quaternary phosphonium salts grafted on 'gel-type'styrene-divinylbenzene copolymers. *React. Funct. Polym.* **2003**, *55*, 151–158. [CrossRef]
45. Chang, H.-I.; Yang, M.-S.; Liang, M. The synthesis, characterization and antibacterial activity of quaternized poly (2, 6-dimethyl-1, 4-phenylene oxide) s modified with ammonium and phosphonium salts. *React. Funct. Polym.* **2010**, *70*, 944–950. [CrossRef]
46. Hancock, R.E.W.; Sahl, H.-G. Antimicrobial and host-defense peptides as new anti-infective therapeutic strategies. *Nat. Biotechnol.* **2006**, *24*, 1551–1557. [CrossRef]
47. Peschel, A.; Sahl, H.-G. The co-evolution of host cationic antimicrobial peptides and microbial resistance. *Nat. Rev. Microbiol.* **2006**, *4*, 529–536. [CrossRef] [PubMed]
48. Fjell, C.D.; Hiss, J.A.; Hancock, R.E.W.; Schneider, G. Designing antimicrobial peptides: Form follows function. *Nat. Rev. Drug Discov.* **2012**, *11*, 37–51. [CrossRef] [PubMed]
49. Melo, M.N.; Ferre, R.; Castanho, M.A.R.B. Antimicrobial peptides: Linking partition, activity and high membrane-bound concentrations. *Nat. Rev. Microbiol.* **2009**, *7*, 245–250. [CrossRef]
50. Brogden, K.A. Antimicrobial peptides: Pore formers or metabolic inhibitors in bacteria? *Nat. Rev. Microbiol.* **2005**, *3*, 238–250. [CrossRef] [PubMed]
51. El-Mekawy, A.; Hudson, S.; El-Baz, A.; Hamza, H.; El-Halafawy, K. Preparation of chitosan films mixed with superabsorbent polymer and evaluation of its haemostatic and antibacterial activities. *J. Appl. Polym. Sci.* **2010**, *116*, 3489–3496. [CrossRef]
52. Milazzo, M.; Contessi Negrini, N.; Scialla, S.; Marelli, B.; Farè, S.; Danti, S.; Buehler, M.J. Additive Manufacturing Approaches for Hydroxyapatite-Reinforced Composites. *Adv. Funct. Mater.* **2019**, *29*, 1903055. [CrossRef]
53. Scialla, S.; Barca, A.; Palazzo, B.; D'Amora, U.; Russo, T.; Gloria, A.; De Santis, R.; Verri, T.; Sannino, A.; Ambrosio, L.; et al. Bioactive Chitosan-Based Scaffolds with Improved Properties Induced by Dextran-Grafted Nano-Maghemite and L-Arginine Amino Acid. *J. Biomed. Mater. Res. Part A* **2019**, *107*, 1244–1252. [CrossRef]
54. Morganti, P.; Del Ciotto, P.; Stoller, M.; Chianese, A. Antibacterial and anti-inflammatory green nanocomposites. *Chem. Eng. Trans.* **2016**, *47*, 61–66.
55. Alvarez, F.J. The effect of chitin size, shape, source and purification method on immune recognition. *Molecules* **2014**, *19*, 4433–4451. [CrossRef]
56. Coltelli, M.-B.; Aliotta, L.; Vannozzi, A.; Morganti, P.; Panariello, L.; Danti, S.; Neri, S.; Fernandez-Avila, C.; Fusco, A.; Donnarumma, G.; et al. Properties and skin compatibility of films based on poly (lactic acid)(PLA) bionanocomposites incorporating chitin nanofibrils (CN). *J. Funct. Biomater.* **2020**, *11*, 21. [CrossRef]
57. Dinjaski, N.; Fernández-Gutiérrez, M.; Selvam, S.; Parra-Ruiz, F.J.; Lehman, S.M.; San Román, J.; Garcìia, E.; Garcìa, J.L.; Garcìa, A.J.; Prieto, M.A. PHACOS, a functionalized bacterial polyester with bactericidal activity against methicillin-resistant Staphylococcus aureus. *Biomaterials* **2014**, *35*, 14–24. [CrossRef] [PubMed]
58. Escapa, I.F.; Morales, V.; Martino, V.P.; Pollet, E.; Avérous, L.; Garc\'\ia, J.L.; Prieto, M.A. Disruption of β-oxidation pathway in Pseudomonas putida KT2442 to produce new functionalized PHAs with thioester groups. *Appl. Microbiol. Biotechnol.* **2011**, *89*, 1583–1598. [CrossRef] [PubMed]
59. Lin, N.; Berton, P.; Moraes, C.; Rogers, R.D.; Tufenkji, N. Nanodarts, nanoblades, and nanospikes: Mechano-bactericidal nanostructures and where to find them. *Adv. Colloid Interface Sci.* **2018**, *252*, 55–68. [CrossRef] [PubMed]
60. Pham, V.T.H.; Truong, V.K.; Mainwaring, D.E.; Guo, Y.; Baulin, V.A.; Al Kobaisi, M.; Gervinskas, G.; Juodkazis, S.; Zeng, W.R.; Doran, P.P.; et al. Nanotopography as a trigger for the microscale, autogenous and passive lysis of erythrocytes. *J. Mater. Chem. B* **2014**, *2*, 2819–2826. [CrossRef]

61. Pham, V.T.H.; Truong, V.K.; Orlowska, A.; Ghanaati, S.; Barbeck, M.; Booms, P.; Fulcher, A.J.; Bhadra, C.M.; Buividas, R.; Baulin, V.; et al. "Race for the surface": Eukaryotic cells can win. *ACS Appl. Mater. Interfaces* **2016**, *8*, 22025–22031. [CrossRef]
62. Wang, X.; Bhadra, C.M.; Dang, T.H.Y.; Buividas, R.; Wang, J.; Crawford, R.J.; Ivanova, E.P.; Juodkazis, S. A bactericidal microfluidic device constructed using nano-textured black silicon. *RSC Adv.* **2016**, *6*, 26300–26306. [CrossRef]
63. Vassallo, E.; Pedroni, M.; Silvetti, T.; Morandi, S.; Toffolatti, S.; Angella, G.; Brasca, M. Bactericidal performance of nanostructured surfaces by fluorocarbon plasma. *Mater. Sci. Eng. C* **2017**, *80*, 117–121. [CrossRef]
64. Sengstock, C.; Lopian, M.; Motemani, Y.; Borgmann, A.; Khare, C.; Buenconsejo, P.J.S.; Schildhauer, T.A.; Ludwig, A.; Köller, M. Structure-related antibacterial activity of a titanium nanostructured surface fabricated by glancing angle sputter deposition. *Nanotechnology* **2014**, *25*, 195101. [CrossRef]
65. Sjöström, T.; Nobbs, A.H.; Su, B. Bactericidal nanospike surfaces via thermal oxidation of Ti alloy substrates. *Mater. Lett.* **2016**, *167*, 22–26. [CrossRef]
66. Bhadra, C.M.; Truong, V.K.; Pham, V.T.H.; Al Kobaisi, M.; Seniutinas, G.; Wang, J.Y.; Juodkazis, S.; Crawford, R.J.; Ivanova, E.P. Antibacterial titanium nano-patterned arrays inspired by dragonfly wings. *Sci. Rep.* **2015**, *5*, 1–12. [CrossRef]
67. Hall-Stoodley, L.; Costerton, J.W.; Stoodley, P. Bacterial biofilms: From the natural environment to infectious diseases. *Nat. Rev. Microbiol.* **2004**, *2*, 95–108. [CrossRef] [PubMed]
68. Pavithra, D.; Doble, M. Biofilm formation, bacterial adhesion and host response on polymeric implants—Issues and prevention. *Biomed. Mater.* **2008**, *3*, 34003. [CrossRef] [PubMed]
69. Harding, J.L.; Reynolds, M.M. Combating medical device fouling. *Trends Biotechnol.* **2014**, *32*, 140–146. [CrossRef] [PubMed]
70. Ganewatta, M.S.; Tang, C. Controlling macromolecular structures towards effective antimicrobial polymers. *Polymer* **2015**, *63*, A1–A29. [CrossRef]
71. Carmona-Ribeiro, A.M.; de Melo Carrasco, L.D. Cationic antimicrobial polymers and their assemblies. *Int. J. Mol. Sci.* **2013**, *14*, 9906–9946. [CrossRef]
72. Yang, Y.; Cai, Z.; Huang, Z.; Tang, X.; Zhang, X. Antimicrobial cationic polymers: From structural design to functional control. *Polym. J.* **2018**, *50*, 33–44. [CrossRef]
73. Engler, A.C.; Tan, J.P.K.; Ong, Z.Y.; Coady, D.J.; Ng, V.W.L.; Yang, Y.Y.; Hedrick, J.L. Antimicrobial polycarbonates: Investigating the impact of balancing charge and hydrophobicity using a same-centered polymer approach. *Biomacromolecules* **2013**, *14*, 4331–4339. [CrossRef]
74. Dizman, B.; Elasri, M.O.; Mathias, L.J. Synthesis and antimicrobial activities of new water-soluble bis-quaternary ammonium methacrylate polymers. *J. Appl. Polym. Sci.* **2004**, *94*, 635–642. [CrossRef]
75. Takahashi, H.; Palermo, E.F.; Yasuhara, K.; Caputo, G.A.; Kuroda, K. Molecular Design, Structures, and Activity of Antimicrobial Peptide-Mimetic Polymers. *Macromol. Biosci.* **2013**, *13*, 1285–1299. [CrossRef]
76. Horne, W.S.; Gellman, S.H. Foldamers with heterogeneous backbones. *Acc. Chem. Res.* **2008**, *41*, 1399–1408. [CrossRef]
77. Chen, H.; Li, M.; Liu, Z.; Hu, R.; Li, S.; Guo, Y.; Lv, F.; Liu, L.; Wang, Y.; Yi, Y.; et al. Design of antibacterial peptide-like conjugated molecule with broad spectrum antimicrobial ability. *Sci. China Chem.* **2018**, *61*, 113–117. [CrossRef]
78. Oren, Z.; Shai, Y. Mode of action of linear amphipathic α-helical antimicrobial peptides. *Pept. Sci.* **1998**, *47*, 451–463. [CrossRef]
79. Chongsiriwatana, N.P.; Patch, J.A.; Czyzewski, A.M.; Dohm, M.T.; Ivankin, A.; Gidalevitz, D.; Zuckermann, R.N.; Barron, A.E. Peptoids that mimic the structure, function, and mechanism of helical antimicrobial peptides. *Proc. Natl. Acad. Sci. USA* **2008**, *105*, 2794–2799. [CrossRef] [PubMed]
80. Epand, R.F.; Raguse, L.; Gellman, S.H.; Epand, R.M. Antimicrobial 14-helical β-peptides: Potent bilayer disrupting agents. *Biochemistry* **2004**, *43*, 9527–9535. [CrossRef] [PubMed]
81. Liu, D.; DeGrado, W.F. De novo design, synthesis, and characterization of antimicrobial β-peptides. *J. Am. Chem. Soc.* **2001**, *123*, 7553–7559. [CrossRef]
82. Hamuro, Y.; Schneider, J.P.; DeGrado, W.F. De novo design of antibacterial β-peptides. *J. Am. Chem. Soc.* **1999**, *121*, 12200–12201. [CrossRef]
83. Porter, E.A.; Wang, X.; Lee, H.-S.; Weisblum, B.; Gellman, S.H. Non-haemolytic β-amino-acid oligomers. *Nature* **2000**, *404*, 565. [CrossRef]

84. Danti, S.; Trombi, L.; Fusco, A.; Azimi, B.; Lazzeri, A.; Morganti, P.; Coltelli, M.-B.; Donnarumma, G. Chitin nanofibrils and nanolignin as functional agents in skin regeneration. *Int. J. Mol. Sci.* **2019**, *20*, 2669. [CrossRef]
85. Panariello, L.; Coltelli, M.-B.; Buchignani, M.; Lazzeri, A. Chitosan and nano-structured chitin for biobased anti-microbial treatments onto cellulose based materials. *Eur. Polym. J.* **2019**, *113*, 328–339. [CrossRef]
86. Donnarumma, G.; Fusco, A.; Morganti, P.; Palombo, M.; Anniboletti, T.; Del Ciotto, P.; Baroni, A.; Chianese, A. Advanced medications made by green nanocomposites. *Int. J. Res. Pharm. Nano Sci.* **2016**, *5*, 261–270.
87. Donnarumma, G.; Paoletti, I.; Fusco, A.; Perfetto, B.; Buommino, E.; de Gregorio, V.; Baroni, A. β-defensins: Work in progress. In *Advances in Microbiology, Infectious Diseases and Public Health*; Springer: Berlin/Heidelberg, Germany, 2015; pp. 59–76.
88. Elmowafy, E.; Abdal-Hay, A.; Skouras, A.; Tiboni, M.; Casettari, L.; Guarino, V. Polyhydroxyalkanoate (PHA): Applications in drug delivery and tissue engineering. *Expert Rev. Med Devices* **2019**, *16*, 467–482. [CrossRef] [PubMed]
89. Rai, R.; Roether, J.A.; Knowles, J.C.; Mordan, N.; Salih, V.; Locke, I.C.; Gordge, M.P.; McCormick, A.; Mohn, D.; Stark, W.J.; et al. Highly elastomeric poly (3-hydroxyoctanoate) based natural polymer composite for enhanced keratinocyte regeneration. *Int. J. Polym. Mater. Polym. Biomater.* **2017**, *66*, 326–335. [CrossRef]
90. Keshavarz, T.; Roy, I. Polyhydroxyalkanoates: Bioplastics with a green agenda. *Curr. Opin. Microbiol.* **2010**, *13*, 321–326. [CrossRef] [PubMed]
91. Lütke-Eversloh, T.; Bergander, K.; Luftmann, H.; Steinbüchel, A. Identification of a new class of biopolymer: Bacterial synthesis of a sulfur-containing polymer with thioester linkages. *Microbiology* **2001**, *147*, 11–19. [CrossRef] [PubMed]
92. Ewering, C.; Lütke-Eversloh, T.; Luftmann, H.; Steinbüchel, A. Identification of novel sulfur-containing bacterial polyesters: Biosynthesis of poly (3-hydroxy-S-propyl-ω-thioalkanoates) containing thioether linkages in the side chains. *Microbiology* **2002**, *148*, 1397–1406. [CrossRef] [PubMed]
93. Morelli, A.; Puppi, D.; Chiellini, F. Perspectives on biomedical applications of ulvan. In *Seaweed Polysaccharides*; Elsevier: Amsterdam, The Netherlands, 2017; pp. 305–330.
94. Liang, W.; Mao, X.; Peng, X.; Tang, S. Effects of sulfate group in red seaweed polysaccharides on anticoagulant activity and cytotoxicity. *Carbohydr. Polym.* **2014**, *101*, 776–785. [CrossRef]
95. Patel, S. Therapeutic importance of sulfated polysaccharides from seaweeds: Updating the recent findings. *3 Biotech* **2012**, *2*, 171–185. [CrossRef]
96. Ye, S.-H.; Hong, Y.; Sakaguchi, H.; Shankarraman, V.; Luketich, S.K.; D'Amore, A.; Wagner, W.R. Nonthrombogenic, biodegradable elastomeric polyurethanes with variable sulfobetaine content. *ACS Appl. Mater. Interfaces* **2014**, *6*, 22796–22806. [CrossRef]
97. Brown, R.H.; Hunley, M.T.; Allen, M.H., Jr; Long, T.E. Electrospinning zwitterion-containing nanoscale acrylic fibers. *Polymer* **2009**, *50*, 4781–4787. [CrossRef]
98. Lalani, R.; Liu, L. Synthesis, characterization, and electrospinning of zwitterionic poly (sulfobetaine methacrylate). *Polymer* **2011**, *52*, 5344–5354. [CrossRef]
99. Evers, L.H.; Bhavsar, D.; Mailänder, P. The biology of burn injury. *Exp. Dermatol.* **2010**, *19*, 777–783. [CrossRef] [PubMed]
100. Schlenoff, J.B. Zwitteration: Coating surfaces with zwitterionic functionality to reduce nonspecific adsorption. *Langmuir* **2014**, *30*, 9625–9636. [CrossRef] [PubMed]
101. Hasan, J.; Crawford, R.J.; Ivanova, E.P. Antibacterial surfaces: The quest for a new generation of biomaterials. *Trends Biotechnol.* **2013**, *31*, 295–304. [CrossRef] [PubMed]
102. Hower, J.C.; Bernards, M.T.; Chen, S.; Tsao, H.-K.; Sheng, Y.-J.; Jiang, S. Hydration of "nonfouling" functional groups. *J. Phys. Chem. B* **2009**, *113*, 197–201. [CrossRef]
103. El-Baky, H.H.A.; Baz, F.K.E.; Baroty, G.S.E. Potential biological properties of sulphated polysaccharides extracted from the macroalgae Ulva lactuca L. *Acad. J. Cancer Res.* **2009**, *2*, 1–11.
104. Aguilar-Briseño, J.A.; Cruz-Suarez, L.E.; Sassi, J.-F.; Ricque-Marie, D.; Zapata-Benavides, P.; Mendoza-Gamboa, E.; Rodr\'\iguez-Padilla, C.; Trejo-Avila, L.M. Sulphated polysaccharides from Ulva clathrata and Cladosiphon okamuranus seaweeds both inhibit viral attachment/entry and cell-cell fusion, in NDV infection. *Mar. Drugs* **2015**, *13*, 697–712. [CrossRef]
105. Gadenne, V.; Lebrun, L.; Jouenne, T.; Thebault, P. Role of molecular properties of ulvans on their ability to elaborate antiadhesive surfaces. *J. Biomed. Mater. Res. Part A* **2015**, *103*, 1021–1028. [CrossRef] [PubMed]

106. Gadenne, V.; Lebrun, L.; Jouenne, T.; Thebault, P. Antiadhesive activity of ulvan polysaccharides covalently immobilized onto titanium surface. *Colloids Surf. B Biointerfaces* **2013**, *112*, 229–236. [CrossRef] [PubMed]
107. Junter, G.-A.; Thebault, P.; Lebrun, L. Polysaccharide-based antibiofilm surfaces. *Acta Biomater.* **2016**, *30*, 13–25. [CrossRef]
108. Bixler, G.D.; Theiss, A.; Bhushan, B.; Lee, S.C. Anti-fouling properties of microstructured surfaces bio-inspired by rice leaves and butterfly wings. *J. Colloid Interface Sci.* **2014**, *419*, 114–133. [CrossRef]
109. Murthy, N.S.; Prabhu, R.D.; Martin, J.J.; Zhou, L.; Headrick, R.L. Self-assembled and etched cones on laser ablated polymer surfaces. *J. Appl. Phys.* **2006**, *100*, 23538. [CrossRef]
110. Bixler, G.D.; Bhushan, B. Biofouling: Lessons from nature. *Philos. Trans. R. Soc. A Math. Phys. Eng. Sci.* **2012**, *370*, 2381–2417. [CrossRef] [PubMed]
111. Bixler, G.D.; Bhushan, B. Bioinspired rice leaf and butterfly wing surface structures combining shark skin and lotus effects. *Soft Matter* **2012**, *8*, 11271–11284. [CrossRef]
112. Salta, M.; Wharton, J.A.; Stoodley, P.; Dennington, S.P.; Goodes, L.R.; Werwinski, S.; Mart, U.; Wood, R.J.K.; Stokes, K.R. Designing biomimetic antifouling surfaces. *Philos. Trans. R. Soc. A Math. Phys. Eng. Sci.* **2010**, *368*, 4729–4754. [CrossRef]
113. Hoipkemeier-Wilson, L.; Schumacher, J.F.; Carman, M.L.; Gibson, A.L.; Feinberg, A.W.; Callow, M.E.; Finlay, J.A.; Callow, J.A.; Brennan, A.B. Antifouling potential of lubricious, micro-engineered, PDMS elastomers against zoospores of the green fouling alga Ulva (Enteromorpha). *Biofouling* **2004**, *20*, 53–63. [CrossRef]
114. Chandra, P.; Lai, K.; Sung, H.-J.; Murthy, N.S.; Kohn, J. UV laser-ablated surface textures as potential regulator of cellular response. *Biointerphases* **2010**, *5*, 53–59. [CrossRef]
115. Nishimoto, S.; Bhushan, B. Bioinspired self-cleaning surfaces with superhydrophobicity, superoleophobicity, and superhydrophilicity. *Rsc Adv.* **2013**, *3*, 671–690. [CrossRef]
116. Reddy, S.T.; Chung, K.K.; McDaniel, C.J.; Darouiche, R.O.; Landman, J.; Brennan, A.B. Micropatterned surfaces for reducing the risk of catheter-associated urinary tract infection: An in vitro study on the effect of sharklet micropatterned surfaces to inhibit bacterial colonization and migration of uropathogenic Escherichia coli. *J. Endourol.* **2011**, *25*, 1547–1552. [CrossRef]
117. Koch, K.; Barthlott, W. Superhydrophobic and superhydrophilic plant surfaces: An inspiration for biomimetic materials. *Philos. Trans. R. Soc. A Math. Phys. Eng. Sci.* **2009**, *367*, 1487–1509. [CrossRef]
118. Ensikat, H.J.; Ditsche-Kuru, P.; Neinhuis, C.; Barthlott, W. Superhydrophobicity in perfection: The outstanding properties of the lotus leaf. *Beilstein J. Nanotechnol.* **2011**, *2*, 152–161. [CrossRef]
119. Raschi, W.; Tabit, C. Functional aspects of placoid scales: A review and update. *Mar. Freshw. Res.* **1992**, *43*, 123–147. [CrossRef]
120. Lang, A.W.; Motta, P.; Hidalgo, P.; Westcott, M. Bristled shark skin: A microgeometry for boundary layer control? *Bioinspiration Biomim.* **2008**, *3*, 46005. [CrossRef] [PubMed]
121. Bone, Q.; Moore, R. *Biology of Fishes*; Taylor & Francis Group: New York, NY, USA, 2008.
122. Kirschner, C.M.; Brennan, A.B. Bio-inspired antifouling strategies. *Annu. Rev. Mater. Res.* **2012**, *42*, 211–229. [CrossRef]
123. Krishnan, S.; Weinman, C.J.; Ober, C.K. Advances in polymers for anti-biofouling surfaces. *J. Mater. Chem.* **2008**, *18*, 3405–3413. [CrossRef]
124. Baum, C.L.; Arpey, C.J. Normal cutaneous wound healing: Clinical correlation with cellular and molecular events. *Dermatol. Surg.* **2005**, *31*, 674–686. [CrossRef] [PubMed]
125. Thomas, J. Wound healing and ulcers of the skin. Diagnosis and therapy—The practical approach. By A Shai and HI Maibach.© Springer-Verlag Berlin Heidelberg, 2005. ISBN: 3 540 2127520470 848987. Hardcover, 270 pages. *Pract. Diabetes Int.* **2005**, *22*, 283. [CrossRef]
126. Coltelli, M.-B.; Cinelli, P.; Gigante, V.; Aliotta, L.; Morganti, P.; Panariello, L.; Lazzeri, A. Chitin nanofibrils in poly (lactic acid)(PLA) nanocomposites: Dispersion and thermo-mechanical properties. *Int. J. Mol. Sci.* **2019**, *20*, 504. [CrossRef]
127. Magin, C.M.; Cooper, S.P.; Brennan, A.B. Non-toxic antifouling strategies. *Mater. Today* **2010**, *13*, 36–44. [CrossRef]
128. Mann, E.E.; Manna, D.; Mettetal, M.R.; May, R.M.; Dannemiller, E.M.; Chung, K.K.; Brennan, A.B.; Reddy, S.T. Surface micropattern limits bacterial contamination. *Antimicrob. Resist. Infect. Control.* **2014**, *3*, 28. [CrossRef]

129. Frykberg, R.G.; Banks, J. Challenges in the treatment of chronic wounds. *Adv. Wound Care* **2015**, *4*, 560–582. [CrossRef]
130. Wolcott, R.D.; Rhoads, D.D.; Dowd, S.E. Biofilms and chronic wound inflammation. *J. Wound Care* **2008**, *17*, 333–341. [CrossRef] [PubMed]
131. Costerton, J.W.; Stewart, P.S.; Greenberg, E.P. Bacterial biofilms: A common cause of persistent infections. *Science* **1999**, *284*, 1318–1322. [CrossRef] [PubMed]
132. Li, B.; Xie, J.; Yuan, Z.; Jain, P.; Lin, X.; Wu, K.; Jiang, S. Mitigation of inflammatory immune responses with hydrophilic nanoparticles. *Angew. Chem. Int. Ed.* **2018**, *57*, 4527–4531. [CrossRef] [PubMed]
133. Shi, G.; Chen, W.; Zhang, Y.; Dai, X.; Zhang, X.; Wu, Z. An antifouling hydrogel containing silver nanoparticles for modulating the therapeutic immune response in chronic wound healing. *Langmuir* **2018**, *35*, 1837–1845. [CrossRef]
134. Lebeaux, D.; Ghigo, J.-M.; Beloin, C. Biofilm-related infections: Bridging the gap between clinical management and fundamental aspects of recalcitrance toward antibiotics. *Microbiol. Mol. Biol. Rev.* **2014**, *78*, 510–543. [CrossRef]
135. Witherel, C.E.; Abebayehu, D.; Barker, T.H.; Spiller, K.L. Macrophage and Fibroblast Interactions in Biomaterial-Mediated Fibrosis. *Adv. Healthc. Mater.* **2019**, *8*, 1801451. [CrossRef]
136. Cristallini, C.; Danti, S.; Azimi, B.; Tempesti, V.; Ricci, C.; Ventrelli, L.; Cinelli, P.; Barbani, N.; Lazzeri, A. Multifunctional Coatings for Robotic Implanted Device. *Int. J. Mol. Sci.* **2019**, *20*, 5126. [CrossRef]
137. VanEpps, J.S.; Younger, J.G. Implantable Device Related Infection. *Shock (Augusta Ga.)* **2016**, *46*, 597. [CrossRef]
138. Lalani, T. Breast implant infections: An update. *Infect. Dis. Clin.* **2018**, *32*, 877–884. [CrossRef]
139. Pittet, B.; Montandon, D.; Pittet, D. Infection in breast implants. *Lancet Infect. Dis.* **2005**, *5*, 94–106. [CrossRef]
140. Griffo, A.; Liu, Y.; Mahlberg, R.; Alakomi, H.-L.; Johansson, L.-S.; Milani, R. Design and Testing of a Bending-Resistant Transparent Nanocoating for Optoacoustic Cochlear Implants. *Chem. Open* **2019**, *8*, 1100–1108. [CrossRef] [PubMed]
141. Kim, S.-W.; Rhee, C.-S. Nasal septal perforation repair: Predictive factors and systematic review of the literature. *Curr. Opin. Otolaryngol. Head Neck Surg.* **2012**, *20*, 58–65. [CrossRef] [PubMed]
142. Azimi, B.; Milazzo, M.; Lazzeri, A.; Berrettini, S.; Uddin, M.J.; Qin, Z.; Buehler, M.J.; Danti, S. Electrospinning Piezoelectric Fibers for Biocompatible Devices. *Adv. Healthc. Mater.* **2019**, *9*, 1901287. [CrossRef] [PubMed]
143. Makvandi, P.; Esposito Corcione, C.; Paladini, F.; Gallo, A.L.; Montagna, F.; Jamaledin, R.; Pollini, M.; Maffezzoli, A. Antimicrobial modified hydroxyapatite composite dental bite by stereolithography. *Polym. Adv. Technol.* **2018**, *29*, 364–371. [CrossRef]

© 2020 by the authors. Licensee MDPI, Basel, Switzerland. This article is an open access article distributed under the terms and conditions of the Creative Commons Attribution (CC BY) license (http://creativecommons.org/licenses/by/4.0/).

Review

Bio-Based Electrospun Fibers for Wound Healing

Bahareh Azimi [1,2], Homa Maleki [3,*], Lorenzo Zavagna [1], Jose Gustavo De la Ossa [4], Stefano Linari [5], Andrea Lazzeri [1,2] and Serena Danti [1,2,*]

[1] Interuniversity National Consortium of Materials Science and Technology (INSTM), 50121 Florence, Italy; b.azimi@ing.unipi.it (B.A.); lorenzo@zavagna.it (L.Z.); andrea.lazzeri@unipi.it (A.L.)
[2] Department of Civil and Industrial Engineering, University of Pisa, 56126 Pisa, Italy
[3] Department of Carpet, University of Birjand, Birjand 9717434765, Iran
[4] Doctoral School in Life Sciences, University of Siena, 53100 Siena, Italy; josegustavo.delao@student.unisi.it
[5] Linari Engineering s.r.l., 56121 Pisa, Italy; stefano.linari@linarisrl.com
* Correspondence: hmaleki@birjand.ac.ir (H.M.); serena.danti@unipi.it (S.D.)

Received: 22 August 2020; Accepted: 14 September 2020; Published: 22 September 2020

Abstract: Being designated to protect other tissues, skin is the first and largest human body organ to be injured and for this reason, it is accredited with a high capacity for self-repairing. However, in the case of profound lesions or large surface loss, the natural wound healing process may be ineffective or insufficient, leading to detrimental and painful conditions that require repair adjuvants and tissue substitutes. In addition to the conventional wound care options, biodegradable polymers, both synthetic and biologic origin, are gaining increased importance for their high biocompatibility, biodegradation, and bioactive properties, such as antimicrobial, immunomodulatory, cell proliferative, and angiogenic. To create a microenvironment suitable for the healing process, a key property is the ability of a polymer to be spun into submicrometric fibers (e.g., via electrospinning), since they mimic the fibrous extracellular matrix and can support neo- tissue growth. A number of biodegradable polymers used in the biomedical sector comply with the definition of bio-based polymers (known also as biopolymers), which are recently being used in other industrial sectors for reducing the material and energy impact on the environment, as they are derived from renewable biological resources. In this review, after a description of the fundamental concepts of wound healing, with emphasis on advanced wound dressings, the recent developments of bio-based natural and synthetic electrospun structures for efficient wound healing applications are highlighted and discussed. This review aims to improve awareness on the use of bio-based polymers in medical devices.

Keywords: skin; tissue engineering; biopolymers; biodegradable; nanofiber; wound dressing

---

1. Introduction

Skin is an important multifunctional organ of the human body, which acts as a natural barrier against environmental factors and protects the interior tissues from physical, chemical, and biological influences. As the skin performs vital functions, any structural damage, such as large and deep wounds, can be troublesome and requires prompt and effective treatment [1–3]. Over the past decades, wound care has progressively become a major worldwide public health concern. Because, inefficient and defective treatment of skin damages can even be fatal. Hence, intensive research has been performed in this area focusing on the development of efficient therapeutic approaches and design of new dressing materials that can improve the wound healing procedure [4–6]. For the restoration of the injured tissue, the wound healing process consists of a cascade of events, including hemostasis, inflammation, and proliferation as well as remodeling of the tissue [5,7].

Restoring a protective barrier is the main part of the wound therapy. Depending on the severity and acuteness of the wound, an appropriate dressing with effective protection of the wound site and efficient skin regeneration is clinically essential for accelerating the wound healing [4,8].

Since ancient times, the wound care process has included covering of the injured site using a dressing material that prevents dehydration and infections [9]. In the past decade, wound healing treatments have progressed from traditional treatments using ointments and gauze coverage to advanced multifunctional wound dressings and tissue engineered substitutes. Inexpensive and readily available conventional wound dressings such as gauze and bandages can basically protect the wound from external agents, but they create a dry environment locally, which can lead to complications, such as subsequent infections [2,10]. In recent years, considerable advances have been achieved in designing modern dressings to protect the wound from dehydration and infection, and facilitate the healing process instead of just covering the wound [3,11,12]. Wound dressings in the form of hydrogels [13–15], hydrocolloids [16–18], sponges [19–22], alginates [23–25], and transparent films [11] have been developed and some of them are commercially available. These materials are different in their inherent features, such as hydrophobicity, permeability, and adsorption capacity.

An ideal wound dressing should isolate the wound from adverse external factors, absorb the exudates from the wound surface, protect the injured site from bacterial infection, have anti-inflammatory function, and induce cell proliferation to facilitate tissue regeneration and boost the healing process. It should provide a suitable moist environment at the wound site, and preserve the tissue from further damage. In the case of non-compressible type of wounds, elasticity is another important requirement for a dressing to avoid wound compression. In addition, the wound dressing material should be soft, biocompatible, non-toxic, and non-allergenic [19,26–28]. The effective healing process will improve by applying multifunctional wound dressings. Wound type and its features, healing time, and chemical, physical, and mechanical characteristics of the dressing material should be considered in designing a functional wound dressing [5,7].

Recently, great attention has been placed on fabricating biopolymer-based nanofibrous structures through the electrospinning method. Electrospinning technology is a commonly, low-cost, and tunable procedure for generating ultra-fine fibers with some unique properties. Owing to flexibility in choosing the raw materials and the possibility to tune the ultimate properties, the electrospinning technique has been extensively employed for biomedical materials like tissue engineering scaffolds, wound dressings, and drug delivery systems [7,8]. Electrospun structures composed of fibers with nano-scale diameter are proposed as ideal wound dressings and tissue substitutes due to their similarity to the extracellular matrix (ECM) fibrillar part [11,26,29,30]. Owing to the large specific surface area and high porosity with small pore size, these materials present efficient performance for improving the healing process. Nanofibrous scaffolds with appropriate mechanical properties could be proper substrates for cell proliferation. Besides, the incorporation of therapeutic and pharmaceutical agents is used to functionalize electrospun dressings for efficiently targeting different wound types [12,27,31–34].

The arrival of new polymers and fabrication methods offers advanced wound dressings with tunable functionalities along with excellent structural and mechanical properties. Bio-based polymeric materials are much proposed for skin regeneration as functional skin substitutes, wound healing patches, and dressings in healing of different types of wounds. A wide range of natural biopolymers (e.g., cellulose, chitosan, gelatin, hyaluronic acid, and collagen) have been used for electrospinning of nanofibers to simulate the native tissue matrix and healing of the wounds. Polylactides (PLAs) and polyhydroxyalkanoates (PHAs) are synthetic bio-based polymers which are commonly applied for electrospinning of wound dressings. The mechanical, degradation, and/or morphological characteristics of wound dressings can be tuned by coaxial, multi-nozzle, or blend electrospinning of natural and bio-based synthetic polymers [35].

Highlighting and explaining the recent progress in electrospinning of different synthetic and natural biopolymers, with a focus on their applications related to wound healing, are the intention of this review (Figure 1). Firstly, the fundamental concepts of wound healing, with emphasis on wound dressings, are presented. Then, the electrospinning method, the advantages of electrospun nanofiber for wound healing applications, and the relationship between the electrospinning variables and efficiency of the designed wound dressing are reported. Thereafter, an overview of recently developed multifunctional bio-polymeric based electrospun structures including cellulose, chitosan, PHAs, and PLA for wound healing are presented and discussed. In the last section, the review encompasses the future prospects of electrospinning biopolymer fibers for wound healing applications.

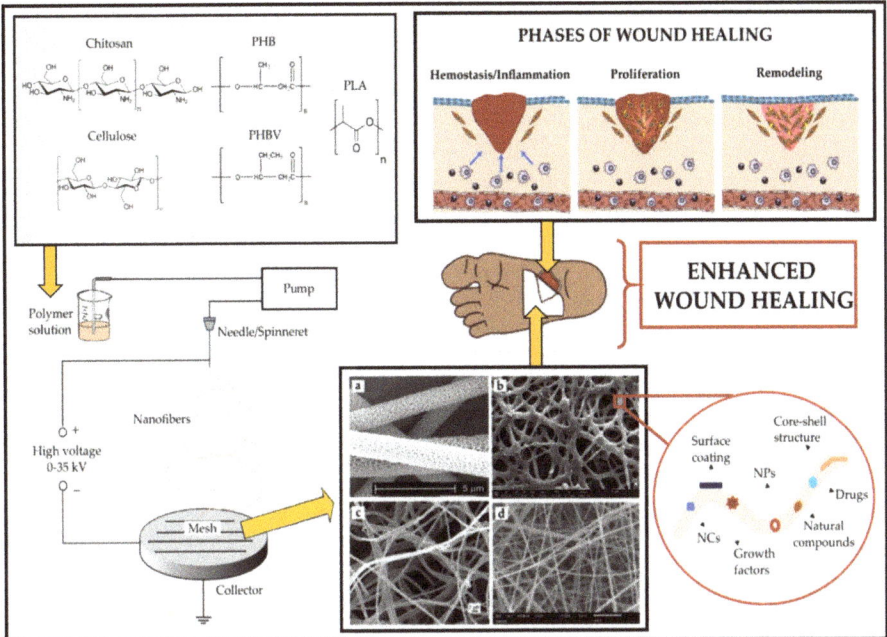

Figure 1. Schematic depicting the topics covered in this review article: biopolymers (natural and synthetics) and their composites processed via electrospinning to produce ultrafine fibers specific for wound healing applications. The schematic of the phases of wound healing reproduced from an open access paper [36] distributed under the terms of the Creative Commons CC BY license, scanning electron microscope (SEM) images (**a**,**b**) are unpublished original pictures by the authors, (**c**) reproduced with permission from [37] (license number: 4873720705310) and (**d**) reproduced with permission from [38] (license number: 4873720243859). The schematic of the foot is adapted with permission from [39] (license numbers: 4875271369979).

2. Wound Healing

Wound healing is a procedure in which the cells can regenerate and treat the injured tissue [2]. This process occurs across four continuous and partially overlapping phases, consisting of hemostasis, inflammation surrounding the injured region, cellular migration and proliferation, and ECM remodeling (Figure 1) [40–42].

The history of wound care dates back to the ancient Sumerian and Egyptian therapies using natural products such as honey, milk, mud, plants, and animal fats. Today, healing approaches have dramatically changed. The wound healing products have evolved over the years from topical ointments to traditional dressings mainly cotton and wool gauzes. Although most of these products provided some benefit to acute wounds they failed to cure chronic and complicated wounds. Therefore, gauze and cotton dressings have been partially replaced with a new generation of dressings. Recently, the main approach relies on preventing, emerging, or eradicating existing infections along with accelerating the healing process with structural and functional regeneration of the skin [20,43].

Wound Dressing

Numerous research works aiming to enhance the healing of wounds have confirmed the significant benefits and effective role of wound dressings in the healing process. Effectual wound care strongly depends on the appropriate selection of dressings for each specific type of wound [2,44].

A proper wound dressing should have good biocompatibility and mechanical flexibility. Besides, it should enable the maintenance of a moist environment, have permeability, and exudate absorption, provide efficient protection against bacterial infections, and external trauma, and have easy removal without adhesion [43,45].

According to types of injury, wound dressings have been classified in to passive, interactive, advanced, and bioactive. Passive wound dressings such as gauze, absorbent pads, and bandages are known to protect the wound site from bacterial penetration and mechanical trauma. However, they are unable to control the amount of moisture in the wound. These types of dressings are useful for minor wounds. Interactive dressings owing to their flexibility are applicable for wounds in joints and other hard to reach areas of the body. Hydrogel dressings, based on polyurethanes and transparent silicones film or foam are some examples of interactive dressings. Advanced dressings fabricated of alginates, hydrofibers, and hydrocolloids are another type of wound dressing, which facilitates the healing process by controlling the wound environment moisture. Recent research attempts have led to improving the functionality of wound dressings to design bioactive dressings (e.g., skin grafts or substitutes, drug-loaded, and antimicrobial dressings), as the new generation of wound healing products. The resultant dressings effectively enhance wound healing and prolong the usage time of the dressing [2,3,46]. Besides, these multifunctional wound dressings are important to accelerate wound healing through the controlled delivery of therapeutic agents [43,45].

Despite the numerous wound dressings available to date, there is still an essential demand to improve the performance and efficiency of wound dressings. In research of advanced wound therapeutics, electrospun fibers have currently gained great attention as multifunctional wound dressings.

3. Electrospinning Process and Its Advantages for Wound Healing Applications

The nanofibrous structures have shown more efficient wound healing compared to traditional dressings, owing to their distinguished features. Among the nanofibers manufacturing methods, electrospinning is a versatile and directly applicable procedure to spin fibrous structures produced from a wide range of materials with controllable features, compositions, shapes, and morphologies [2]. The formation of nanofibers applying the electrospinning technique relies on the uniaxial stretching of the viscoelastic solution or melt under electrostatic forces. Owing to their excellent characteristics, electrospun fibers have the potential to develop textile technology to design advanced biomedical materials in the form of wound dressings, drug carriers, and scaffolds of tissue engineering [47].

The electrospinning process enables the production of interconnected networks from fibers of nano-scale diameter, which are similar to the native structure of the natural ECM, thus promoting the normal functions of the cells, such as supporting attachment and proliferation [48]. Regarding biologically inspired features, large surface area to volume ratio, and the porosity of the electrospun fibrous networks with a small pore size improve the hemostasis at the wound site without the use of a hemostatic agent. Due to the abovementioned physical properties, nanofibers can absorb exudates optimally and provide a moist environment for cell respiration and proliferation. The porous nature of such structures with tiny pores seems to reduce bacterial infection, provides high permeability, and protects the injured tissue from dehydration. The capability and flexibility to incorporate drugs and other bioactive molecules such as growth factors, nanoparticles, antimicrobials, and anti-inflammatory agents, into the nanofibers is another important advantage of the electrospinning process [49]. Electrospun wound dressings can provide compliance and flexibility upon applying and comfortability after placement. In the case of dressings which are fabricated by electrospinning of biodegradable polymers, enhanced patient compliance and comfort may be provided due to less need to change the dressing. The biodegradable electrospun wound dressings also persuade healing and increase cell growth rate due to their high compatibility with blood and tissues. The degradation rate of the scaffolds could be tuned with the rate of tissue regeneration [1,40,46,50,51].

The abovementioned advantages offer electrospun nanofibers as promising materials to improve wound healing and skin regeneration. The electrospinning method promotes the modulation of the physical and mechanical properties of fibers through tuning the corresponding variables. These superior aspects are the main factors to attend to in designing advanced wound dressings that can mimic the native tissue environment and enhance healing of the wound [3,12,28,50].

Effect of Electrospinning Parameters on Biological Applications

Electrospun nanofibers may influence and interact with the injured tissue and its biological environment according to their chemical (e.g., composition, degradation), and physical (e.g., diameter, strength, porosity, etc.) characteristics, as well as via additional incorporated bioactive molecules [52]. Hence, the mentioned features of electrospun nanofibers directly affect the efficiency and performance of dressings produced from them. Moreover, the architecture and structure of dressings also significantly affect the wound healing process [5].

There are different parameters and variables that affect the electrospinning procedure and therewith the ultimate characteristics of micro/nanofibers. These factors are generally classified into three categories of inherent solution properties (e.g., polymer concentration, viscosity, solvent system, and conductivity), process conditions (e.g., voltage, distance, flow rate, and nozzle characteristics), and ambient conditions (e.g., temperature, and humidity). All these parameters concur to tune the diameter, morphology, hydrophobicity, thermal and mechanical properties of the electrospun fiber mesh and thereby its possible end-use application as wound dressings and tissue engineering scaffolds [3,33,50]. Hence, proper control of the working parameters and optimization of different electrospinning variables are necessary to obtain structures with the desired physical and biological performance.

Different morphological structures (e.g., cylindrical or a ribbon shape, porous, beaded, hollow, and core-shell) (Figure 2) and arrangements (e.g., random, oriented, aligned bundle, yarn) (Figure 3) of fiber deposits can be obtained by altering the electrospinning parameters.

Generally, research attempts are focused on finding the optimized electrospinning conditions to produce uniform structures composed of cylindrical shape fibers with smooth surfaces and without any bead formation (Figure 2a). However, it has been reported that the release of a drug from electrospun fibers with flat ribbon morphology (Figure 2b) might occur through diffusion along the shorter route, due to the high surface area/volume ratio. Compared to flat ribbon shape, cylindrical fibers provided more sustained release behavior [53,54]. Hydrophobicity of fibrous structures, mainly determined by surface topology, is important for their performance in the biological interface, such as drug release, cell adhesion, and proliferation. It has been shown that the flat ribbon-like morphology, restricts the volume of trapped air at the fiber-water interface, and thus affects the surface hydrophobicity [54,55]. Furthermore, when compared to flat ribbon-like shapes, cylindrical morphologies have facilitated cell growth on the membrane surface [54,55].

Since the efficiency of the wound dressings mainly depends on their surface topography, when the fiber morphology changes from a smooth solid structure to a porous one (Figure 1c), several features such as specific surface area, network porosity, and functional versatility may improve [56]. The increased network porosity may provide sufficient space for cell penetration and thus enhance their potential as engineering scaffolds. The porous morphology of the electrospun fibers with small pore size can enhance hemostasis, and effectively protect the wound from bacterial permeation [56]. Those with tiny pores and increased surface to volume ratio can also influence the release behavior of the loaded biomolecules in the fibrous structures [57–59]. Although beadless fibers are usually preferred to improve the uniformity of the electrospun nanofibers, bead-on-string nanofibers (Figure 2d), with suitable control of the bead diameter, shape, and surface morphology, have provided efficient encapsulation of biomolecules and controlled release suitable for tissue engineering and wound dressing applications. In vitro release studies have demonstrated that bead- on-string morphology has resulted in a more sustainable release profile with less initial burst release compared to uniform fibers [60–62]. Coaxial electrospinning, as an interesting and efficient technology to produce core-shell, has led to several types of researches in the biomedical field (Figure 2e). A common use of this technology is to encapsulate biomolecules in the core part. In this case, the sheath protects the unstable biological agent from an aggressive environment and delivers it in a sustained way by minimizing the burst release. Superior mechanical properties, and the possibility to functionalize the surface without affecting the core material are other advantages [6,59,63,64].

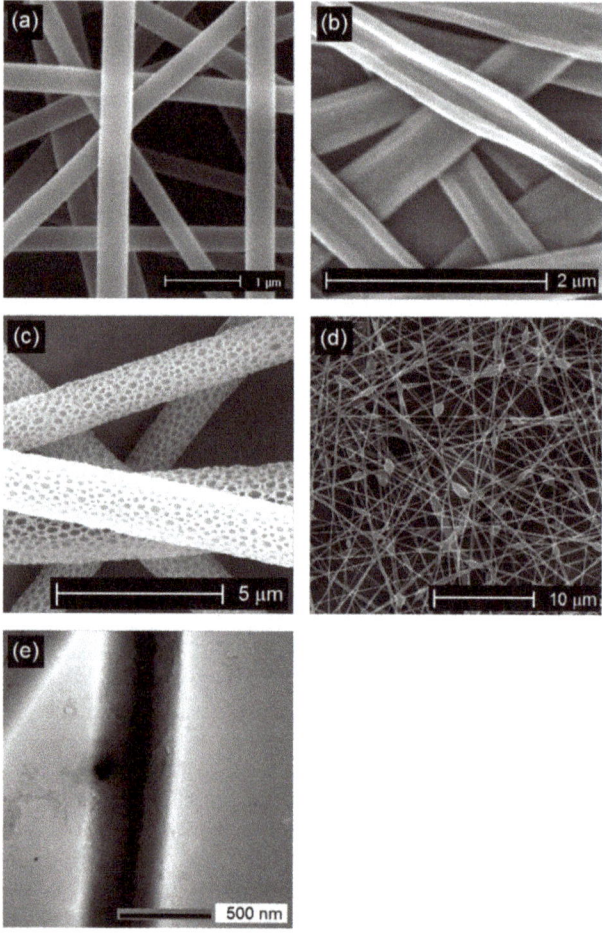

Figure 2. Different morphological structures of fibers can be obtained by altering the electrospinning parameters: (**a**) cylindrical shape with smooth surface (unpublished original picture by the authors), (**b**) flat ribbon like morphology (unpublished original picture by the authors), (**c**) porous (unpublished original picture by the authors), (**d**) bead-on-string morphology (unpublished original picture by the authors), and (**e**) core-shell structure (reproduced from open access article [63] under the terms of the Creative Commons Attribution-Non Commercial License.) All these morphologies are obtained through the electrospinning of polylactide.

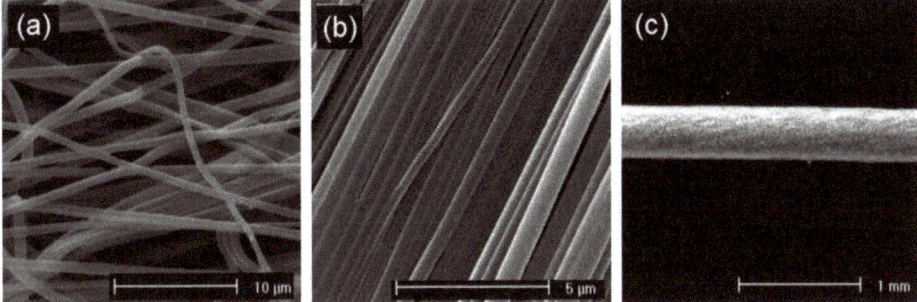

Figure 3. Different arrangements of fiber deposits obtained through electrospinning: (**a**) random, (**b**) oriented, and (**c**) yarn. Unpublished original pictures by the authors. All these structures are obtained through the electrospinning of PLA (original images from the authors).

Besides the morphological features, the architecture of the wound dressings and arrangement of the fibers in their structure also greatly affect the adhesion, proliferation, and penetration of the cells, and release behavior of the bioactive molecules from the structure. By applying different collection devices or by manipulating the electric field during the electrospinning process, fiber deposition can be obtained randomly (Figure 3a), up to uniaxially oriented (Figure 3b). Electrospun scaffolds formed from randomly oriented fibers allow adhesion and proliferation of cells in a random way all over the scaffolds. Despite that, on the aligned structures with fibers deposited in a specific direction, cells adhere and grow in the path of the fiber orientation. Aligned fiber deposition can also control the release of entrapped biomolecules by altering the network porosity [3,11,33].

In addition to the abovementioned features, the mechanical properties of the wound dressings and scaffolds are important and can significantly regulate the arrangement of the fibers in the structure. The tensile strength of the dressing should be sufficient to withstand the dressing handling and exchange during the treatment. In biomedical engineering, primary research has focused on the collection of randomly oriented fibrous webs during the electrospinning. Aligned nanofiber bundles preserve more lateral interaction and friction between fibers with highly improved mechanical properties in compression with nanofiber webs. These interactions can be further enhanced by twisting the nanofibers to form yarns (Figure 3c). Aligned fibrous bundles or twisted yarns have developed feasibilities to design a new generation of medical textiles in various wound healing applications, including woven wound dressings, scaffolds for tissue regeneration, and surgical sutures [4,58,65,66].

4. Multifunctional Wound Dressings

In recent years, researches have been conducted to develop multifunctional dressings which can supply all the requirements for effective wound healing [67]. The multifunctional composite scaffolds can be produced by blending various natural or synthetic polymers and incorporating drugs, nanoparticles, and bioactive agents through the electrospinning process. More recently, the next generation of multifunctional scaffolds has been fabricated via electrospinning of smart materials [2,5]. Smart materials enable their physicochemical properties to be changed managed by external stimuli such as pH, heat, light, and electric field. These types of materials are interesting for wound healing applications because they have demonstrated multiple advantages compared to ordinary materials [2,11].

4.1. Drug Loaded Electrospun Wound Dressings

Multifunctional wound dressings usually need controlled, on-demand release of therapeutic agents, in order to boost restoration within the wound site and prevent undesirable effects such as infection. In this regard, electrospinning offers a great opportunity to prepare drug delivery systems, as it proposes different strategies for incorporating drugs and other biomolecules. Electrospun fibers can deliver agents to the target sites, while reducing the toxic side effects of drugs. Electrospun drug carriers improve the effectuality of drug therapy by controlling the rate, and mechanism of release [58,68–70]. Especially, the stimuli-responsive drug delivery systems are promising materials for generating smart wound dressings. These materials, with the ability to tune drug release in response to stimuli of pH [71–74], temperature [71,75,76], light irradiation [11,75], electric field [77], and oxygen [11,78] hold remarkable promise for wound healing.

Electrospun fibers demonstrate many advantages for the delivery of biomolecules for healthcare. Owing to the small diameter and large surface area to volume ratio, electrospun fibers grant good drug encapsulation efficiency and result in controllable drug delivery to the on-demand sites [58,68,79]. Furthermore, electrospinning affords various choices by altering the morphology, diameter, porosity, and alignment of fibers by controlling the operational variables and type of materials. Tuning these features, provides modulation of drug release behavior and its kinetics [58,68,80]. For example, it has been reported that the drug-loaded dressings composed of fibers with a smaller diameter and higher surface area showed faster release. Higher porosity of the fibers and structure also causes an increase in the release rate. Fiber alignment in the matrix is another parameter influencing the release behavior through regulating the structure porosity. By electrospinning of different materials with diverse properties, varying polymer blend composition, or altering the percentage of amorphous and crystalline segments in copolymers, degradation properties and release behavior can also be controlled [58,68]. Different strategies, including physical adsorption, blend, or coaxial electrospinning, and surface immobilization have been employed to load the bioactive molecules into the electrospun structures (Figure 4).

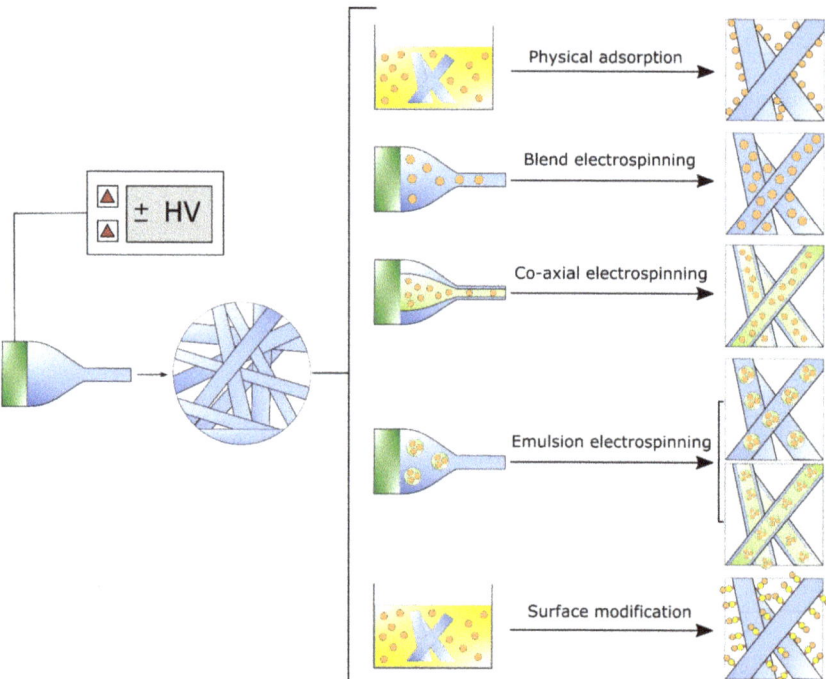

Figure 4. Fabrication techniques of biomolecule-loaded electrospun fibers.

Physical adsorption occurs when the electrospun fibers are immersed in a drug-containing solution bath. This strategy is the simplest process to load the bioactive molecules, since the agents inside the bath are inclined to adsorb at the fiber surface. However, its application is limited owing to an uncontrollable release rate [32,43,68,81]. Blend electrospinning is also a simple process to produce drug-loaded fibers by blending the biomolecules with the polymer solution used for electrospinning. Compared to physical adsorption, this method enables the incorporation of the active agents inside the fibers, and thus offers a more sustained drug release profile with an initial burst release of the agents near the surface of the fibers [46,58]. The biomolecules can be completely entrapped in the fibers through the coaxial electrospinning. The release of the agents from core-shell fibers is sustained and burst release is significantly lower than that obtained using blend electrospinning. In fact, by applying this method, the biomolecules loaded in the core layer are well protected by a sheath [63,82,83]. Blend or coaxial emulsion electrospinning is another approach to encapsulate water-soluble agents [9,32,43]. Coaxial and emulsion electrospinning give rise to well controlled release rate and are able to avoid the initial burst release. Surface immobilization (usually, covalent immobilization) enables the attachment of the biomolecules on the surface of the fibers by a chemical bond. Through this method, the surface features of fibrous membranes may be modified [68,84,85].

4.2. Electrospun Wound Dressings with Antibacterial Activity

Wound infections are a major global concern and designing antibacterial products for wound healing applications is a prominent field of research [2,63]. In order to prevent deleterious effects caused by infections in the injury area, it is necessary to use a wound dressing capable of both barricading bacterial penetration and microbial colonization into the wound site and supporting skin regeneration [9,50]. The developed electrospun scaffolds with antibacterial activity can prevent wound infection. The antibacterial nanofibers are commonly fabricated by incorporating antibacterial agents during electrospinning. Diverse antimicrobial agents (e.g., antibiotics, metallic nanoparticles, and natural extracts derived products) have been embedded into electrospun nanofibers to improve their antibacterial properties. Metallic nanomaterials such as silver nanoparticles (AgNPs) are known as efficient agents for the treatment of wound infections. Nanoscale particles with high surface to volume ratio, afford a significant improvement in antibacterial activity of electrospun wound dressings [3,63,86,87]. Recent strategies rely on using polymers with intrinsic antibacterial activity, due to physical, chemical or morphological cues which cause an obstacle for bacterial colonization and biofilm formation [88]. Bio-based and biopolymers can offer great opportunities for this purpose.

4.3. Electrospun Wound Dressings Loaded with Bioactive Molecules

The incorporation of biological molecules such as growth factors, vitamins, and anti-inflammatory molecules into electrospun fibers is another promising approach to design multifunctional dressings for improving wound healing and skin regeneration [32,89].

Growth factors are biologically active polypeptides that are beneficial to control cell growth, proliferation, and migration during the wound healing process. All the stages of the wound healing process can be regulated using a wide variety of growth factors [90,91]. The role of vitamins in the healing procedure is also considerable. The incorporation of vitamins, particularly vitamins A, C, and, E into electrospun wound dressings aimed at improving healing has been reported in several research works [89,92]. Vitamin A with its proven inflammatory effects can incite angiogenesis and collagenization to enhance wound closure. Vitamin C acts as an antioxidant and accelerates healing through promoting collagen synthesis, acting as an antioxidant, and regulating immune function. Vitamin E relieves free radical detectives in the inflammatory phase of the wound healing process [9,46].

5. Application of Bio-Based Electrospun Fibers in Wound Healing

Bio-based polymers (also referred to as biopolymers) are organic macromolecules synthesized by living organisms. Biopolymers from different sources such as plant (cellulose, lignin), animal (collagen, chitin, chitosan), micro-organisms (bacterial cellulose, PHA) and biotechnological process (polylactides) (Figure 5), have shown promise in biomedical applications, including drug delivery, tissue engineering and wound healing because of their specific properties. Many of them possess antibacterial, antifungal, antiviral, non/low-immunogenic, renewable, biodegradable, and biocompatible characters [88]. These polymers are expected to soon replace plastic goods in the biomedical sector, not only as proper bioactive devices, but also as sterile packaging and consumables [93,94]. The latter, after inactivation treatments, can indeed provide a better end-of-life option to incineration of biohazard disposables in hospital settings. In addition, wound care products can be developed from biopolymers using different nanotechnology strategies [95]. Biopolymers can be made into fibrous scaffolds in the pure form or blended with other polymers which makes them good candidates for skin substitutes.

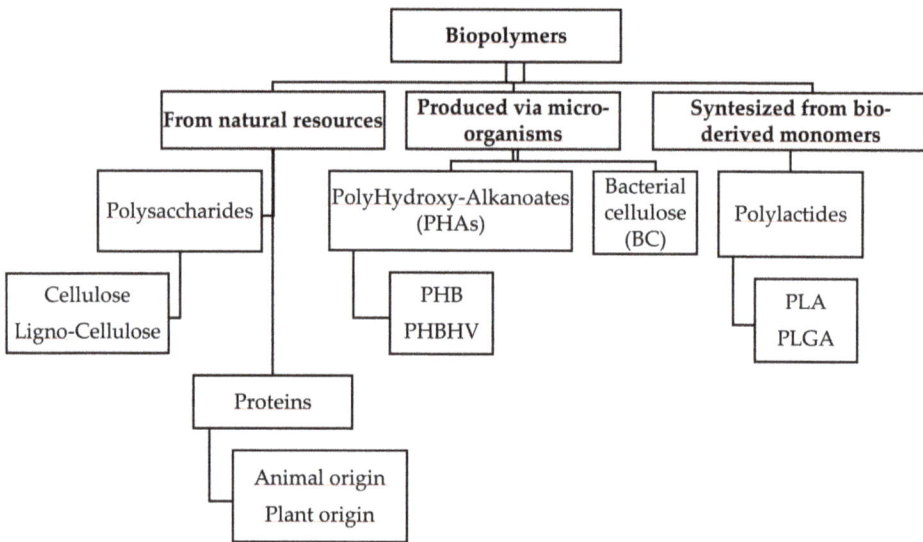

Figure 5. Flow chart showing biopolymers according to diverse origins and types.

5.1. Application of Cellulose-Electrospun Nanofibers in Wound Healing (Including Its Composite)

Cellulose is a natural, biocompatible, biodegradable, and environmentally friendly, biopolymer which plays an important role in various biomedical applications, including scaffolds in tissue repair and reconstruction, wound dressing, artificial tissue/skin, controlled drug delivery, blood purification as well as cell culture materials [96–98]. The moisture-retaining characteristic of cellulose is one of the main reasons for its application in wound care since moist wounds can be treated faster as a result of sufficient growth factors that can be supplied to the healing tissues. The porous structure of cellulose also helps in tissue regeneration via mimicking the skin ECM [99]. Good mechanical properties, high permeability, low toxicity, and adequate conformability are other advantages of cellulose in the application for biological dressings. Using different technologies, such as nanotechnology, biotechnology, three dimensional (3D)- and bio-printing, cellulose properties can be easily tuned to meet the bioengineering demands. For example, different types of nanosized celluloses, especially bacterial cellulose (BC), which is obtained from microorganisms, represent promising functional materials [100,101]. Bacterial cellulose-based scaffolds were investigated in pre- clinical and clinical trials, like wound dressings for skin lesions [102]. To speed up the skin healing of a patient who was suffering with second-degree burns on his face, a bacterial cellulose- based scaffold named Nanocell® was applied and successfully adhered to the wound sites without using other bandages (Figure 6a) [103]. The wounds on the face demonstrated complete re- epithelialization after two weeks

(Figure 6b) and during the treatment, no irritation or allergic reaction was observed, indicating the suitability of bacterial cellulose dressings for healing the burned skin. Cellulose nanofibril (CNF) derived from wood has also shown excellent results when used as a biological dressing of skin wounds—it connected well to the wound bed and after skin recovery, separated easily from the surface of the wound itself [100].

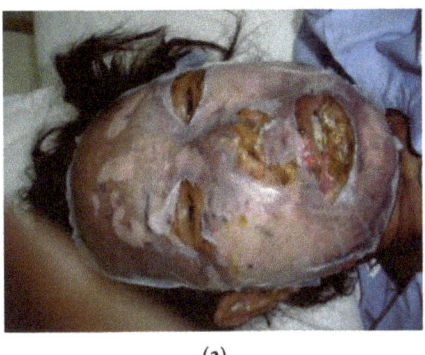
(a)

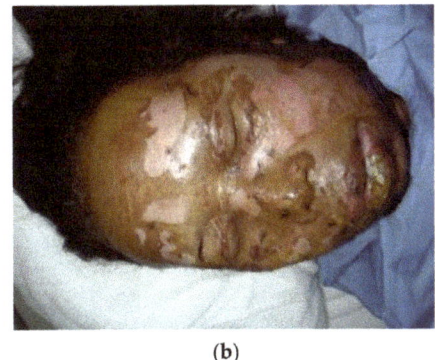
(b)

Figure 6. (a)The application of a bacterial cellulose-based scaffold as a biological dressing on burned facial skin and (b) complete epithelialization obtained after two weeks by using bacterial cellulose (BC) as a temporary skin substitute. Reproduced with permission from [103] (License number: 4875380107950).

Electrospinning is an emerging technique, which can also be used to produce cellulose nanofibers or different polymer/cellulose blends or blends of cellulose with nanoparticles with improved functional properties, most importantly, antimicrobial properties in order to avoid wound site infection [104]. High surface areas and extremely interconnected porous structures of nanofibrous nonwovens are naturally appropriate for wound healing application, since they possess a high capacity for exudate absorption and adequate gas exchange [105,106]. Furthermore, cellulose scaffolds are able to carry different bioactive components such as anti-inflammatory and antimicrobial agents [107]. However, only a few studies have focused on electrospinning of cellulose as a control of the solution and electrospinning parameters are required to produce cellulose nanofibers with particular characteristics. The low solubility of cellulose in several solvents, due to strong intramolecular hydrogen bonds is one of the most critical features to be considered [108]. Some solvent systems which can directly dissolve cellulose have low volatility, which is not desirable in the electrospinning procedure [109]. Since cellulose derivatives have better solubility in common electrospinning solvents, about 70% of researches regarding electrospinning of cellulosic materials use cellulose derivatives, such as cellulose acetate (CA) [109]. Liu et al. used blend-electrospinning to produce a series of membranes using polyester urethane (PEU) and CA for wound dressing applications [110]. Hydrophilicity and air permeability of the membrane was improved due to the presence of CA and a humid environment was created, which accelerated wound recovery. A long- term antimicrobial effect was observed for the membranes with controlled diffusion. Electrospun nanofiber composites have also been produced from cellulose blends or derivatives with other biopolymers. Using 1-ethyl-3-methylimidazoliumacetate [EmIm] [Ac] as an ionic solvent, Park et al. [111], developed non-derivatized electrospun chitosan-cellulose composite mesh with satisfactory antibacterial properties, which can be used for treating skin ulcers as a bandage or via incorporation into other absorbents or gauzes. Miao et al. used electrospinning to produce cellulose, cellulose-poly(methylmethacrylate) (PMMA) and cellulose-chitosan fibers (Figure 7a–c) for anti- infective bandage uses [38]. The fibers were functionalized through covalent immobilization of lysostaphin (Lst) and functionalized fibers showed antibacterial activity against S. aureus and low toxicity toward human keratinocytes (HaCaT cells). Electrospun CA/gelatin scaffolds with different CA/gelatin ratio were fabricated by Vatankhah et al. and the best ratios of CA/gelatin for wound care application and for skin regeneration of injured tissues were determined (CA/gelatin l75:25) and (CA/gelatin 25:75) respectively [105]. Roy et al. proved the potential of paclitaxel incorporated poly (2-hydroxy ethyl methacrylate)/bamboo (pHEMA-bamboo) cellulose electrospun fibers as an anticancer structure for coating skin cancers and wound healing (Figure 7d) [112].

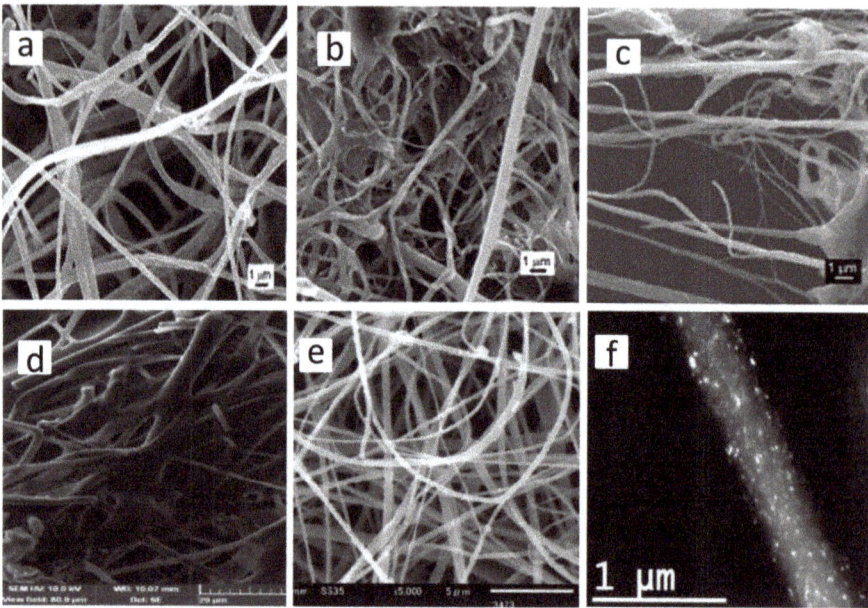

Figure 7. SEM images of (**a**) oxidized cellulose fibers, (**b**) cellulose-chitosan fibers, (**c**) cellulose-poly (methylmethacrylate) (PMMA) reproduced with permission from [38] (License Number: 4873720243859), (**d**) poly (2-hydroxy ethyl methacrylate) (pHEMA)-bamboo cellulose nanocomposite fiber reproduced with permission from [108] (License Number: 4873710894891), (**e**) CA/honey nanofibrous mesh reproduced with permission from [37] (License Number: 4873720705310), (**f**) Transmission electron microscopy (TEM) image of the ZnO/CA composite fiber reproduced with permission from [113] (License Number: 4873720979763).

Very recently, Ullah et al. successfully fabricated nanofibrous CA meshes (Figure 7e) with different quantities of Manuka honey for potential wound care applications [37]. The high porosity of this nonwoven mesh aided wound breathability, and the presence of honey improved the hydrophilicity of the manufact.

Furthermore, in vitro results have generally shown high efficacy of these fiber meshes in avoiding the growth of bacteria at the wound surface and high cytocompatibility to effectively promote wound healing. Fabrication of cellulose nanofibers containing antimicrobial nanoparticles is an interesting alternative for wound management. Anitha et al. demonstrated that the presence of ZnO nanoparticles inside the CA fibrous membrane improved its antimicrobial properties (Figure 7f) [113]. ZnO-loaded CA fiber meshes demonstrated antibacterial activity against, Escherichia coli, Citrobacter freundii, and Staphylococcus aureus. Song et al. fabricated cellulose, carboxymethylated cellulose (CMC) and ribbon-shaped CA electrospun fibers, and functionalized their surface with Ag nanoparticles at different pH [114]. Silver nanoparticles covered the surface of fibers according to the following order; CMC > cellulose > CA at the same pH conditions. The presence of nanoparticles finally improved the antibacterial properties and the capacity of CMC fibers for wound healing application.

5.2. Application of Chitosan Electrospun Fibers in Wound Healing

Chitin and chitosan; the deacetylated form of chitin are polysaccharides which can be used for wound healing applications because of their antimicrobial, biocompatible, and hemostatic properties [67,115]. Due to the presence of a large number of amino groups on its chain, chitosan behaves as a weak polybase and for this reason exerts antibacterial activity [88]. The pro-inflammatory properties of chitosan have been invoked to play a fundamental role in wound healing procedure. Chitosan is able to accelerate the wound healing process via macrophage activation. Moreover, chitosan is able to develop granulation tissue construction by inducing the migration of polymorphonuclear neutrophils (PMNs) at the start of the wound healing process. Jayakumar et al. demonstrated the potential of chitosan on the regeneration and re-epithelialization of the skin granular layer [116]. Moreover, chitosan proficiently interacts with negatively charged blood cells and with such an efficient adhesion to the wound, it stops the bleeding [88]. Min et al. used electrospinning to fabricate chitin and chitosan nanofibrous matrices (Figure 8a,b) for wound dressings application, using 1,1,1,3,3,3-hexafluoro-2-propanol (HFIP) as a spinning solvent.

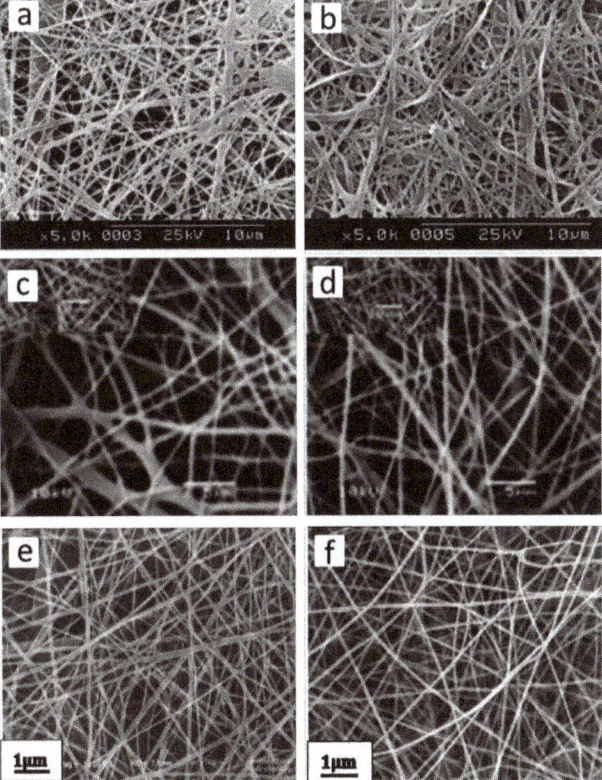

Figure 8. SEM images of (**a**) chitin and (**b**) deacetylated chitin (chitosan) nanofibrous matrix reproduced with permission from [117] (License Number: 4875380936021), (**c**) randomly oriented fibrous mesh of chitosan/polyethylene oxide (PEO) (1:1) (**d**) randomly oriented fibrous mesh of chitosan/PEO/chitin nanocrystals (ChNC) reproduced with permission from [118] (License Number: 4875390863950), (**e**) chitosan/PEO (90/10) nanofibrous matrix, (**f**) chitosan/PEO (90/10) nanofibrous matrix blend with 1 wt% Henna extract reproduced with permission from [119] (License Number: 4875401354853).

However, the applicability of pure electrospun chitosan is limited due to its insufficient mechanical properties [96]. Blend-electrospinning can be used to overcome this limitation [117]. For example, the application of 50 wt% of chitin nanocrystals (ChNC) as a reinforcement in electrospun chitosan fibrous meshes improved tensile strength, modulus, the moisture stability of the as-spun meshes and facilitated water-mediated crosslinking processes. Stable and bead-free random meshes of chitosan/polyethylene oxide (PEO) and chitosan/PEO/ChNC were obtained after spinning for about 30 min (Figure 8c,d). These meshes were cytocompatible toward adipose-derived stem cells after 7 days, and they can be used for wound dressing application [118]. Similarly, electrospun chitosan/sericin [120], and chitosan/silk fibroin [121] composites, showed good antibacterial properties and potential application for wound dressing. In another study, Ardila et al. fabricated electrospun meshes containing chitosan and bacterial nanocellulose using two different approaches including simultaneous spinning of the solutions using two separate syringes and coaxial electrospinning to produce core-shell structures [122]. Coaxial electrospinning led to the formation of nanofibers containing both chitosan and bacterial nanocellulose with a noticeable antimicrobial property which reduced 99.9% of an Escherichia coli population in comparison to the control. As such, it is a promising material for wound healing dressing. Datta et al. developed oleoyl chitosan (OC)/gelatin nanofibrous scaffolds with good mechanical strength, tunable wettability, desirable biocompatibility, and suitable degradation rate for ameliorating full-thickness wound healing in a rat model [123]. Wound contraction and skin tissue regeneration in terms of enhanced collagen deposition and re-epithelialization were significantly improved by the aid of these nanofibers The presence of the long oleoyl tail of OC makes these nanofibers useful for burn wounds, acting as a drug-releasing matrix and at the same time impeding bacterial infection during healing. Different agents including antibiotics, nanoparticles, and/or natural products such as honey and plant extracts have been incorporated within the chitosan electrospun fibers to further improve antibacterial properties, avoid the entry of pathogens into the wound, and kill the invading microorganisms [124]. For example, Yousefi et al., incorporated Henna leaf extract in chitosan nanofibers meshes in order to increase the antibacterial and wound healing efficacy (Figure 8e,f) [119]. Continuous and bead-free nanofibers with a diameter of 64–87 nm were prepared with notable antibacterial activity, and can be considered as biodegradable, biobased, and antibacterial wound healing dressings.

5.3. Application of PHA Electrospun Fibers in Wound Healing

PHAs are a family of linear thermoplastic bio-polyesters. They are synthesized by many microorganisms under unbalanced growth conditions, as an alternative nutrient (carbon) reserve [125,126]. Over the last three decades, many studies have investigated the physicochemical properties of this new class of biopolymers, revealing innumerable advantages in using PHAs as a biomaterial in medical applications thanks to their biocompatibility, mechanical stability, strength, and biodegradability under physiological conditions with non-toxic degradation products [127–129]. Studies also explored pharmaceutical applications of some degradation products of PHA, finding that they can evoke an inhibitory effect on microbial growth [128–131]. To date over 100 units of PHA monomers have been identified and no studies have reported carcinogenesis induced by any PHA or their biodegradation products [125,128,129]. The properties of PHAs vary considerably depending on their monomer content and hence can be tailored by controlling their composition [126,132]. Furthermore, PHAs can be surface modified, blended with other polymers, and composite with inorganic materials such as nanoparticles (NPs), nanocrystals (NCs), drugs, and biomolecules to enhance their biocompatibility, antimicrobial activity, mechanical and thermal properties, and degradation rates, depending on the required application [125,128,132].

Electrospinning of PHAs has also been extensively investigated [133–135]. This technique enabled a simple, scalable production of ultrafine fibrous meshes, eventually incorporated, and/or blended, by direct mixing in the electrospinning solution with different organic and inorganic materials, to reach the targeted properties [133,136].

There are many advantages in using electrospun nonwoven membranes in wound healing applications, mainly associated with their structural porosity, wettability, and similarity to the natural ECM, which has shown to effectively promote cellular migration, attachment, and proliferation [125,137,138].

Towards the last decade, electrospun PHA meshes started gaining high attention for potential applications as a wound dressing in skin regeneration. A wide range of PHAs have been used to produce electrospun fiber meshes with different morphology and alignment for wound healing application [125,137–140], but in recent years, most studies have focused on the application of PHA- based blends and composites, or functionalized electrospun fiber meshes with improved physicochemical and bioactive properties [137,139,141–143]. For example, Shishatskaya et al. used poly(3-hydroxybutyrate-co-4-hydroxybutyrate) [P(3HB-co-4HB)], considered one of the best choices among the PHAs to produce electrospun fibers for wound healing applications, due to its low degree of crystallinity and high elasticity [137]. They demonstrated that the presence of fibroblasts inside the fibers had a significant effect on the amount of hyperemia and purulent exudate, and considered use of composite fibers a better candidate for wound healing application. The wounds under the cell- loaded P(3HB-co-4HB) membrane showed 1.4 times faster healing with respect to the wounds under the cell-free membrane and a 3.5 times faster healing than the wound healing under the eschar (control). Complete healing was achieved after 14 days in the cell-loaded membrane group while, approximately 90% and 70% area reduction were observed in the pure P(3HB-co-4HB) meshes and control groups, respectively. The use of keratin-loaded-poly (3-hydroxybutyrate-co-3-hydroxyvalerate) (PHBV) electrospun fibrous meshes for wound dressing applications has also been investigated [142]. These fiber meshes were tested for cell viability on NIH 3T3 cells and on a wound closure follow-up, conducted on athymic nude mice, consisting of five-time points (0, 2, 4, 7, and 9 days), using gauze as negative control and pure PHBV ultrafine electrospun mats for comparison. On day-3 and day-5, cell viability on the keratin-loaded PHBV mesh was significantly higher than that of the PHBV control. The dimensions of the wound and histology of the healing tissue were observed, and the results showed that on day-9, only the wound under the keratin-loaded PHBV mesh was almost closed (i.e., 94% wound size reduction) and, in the newly formed skin tissue, complete and uniform re-epithelization of the epidermis was formed. At the same time, only a small epidermal layer was observed in the PHBV mesh and the gauze controls. Some studies explored the influence of blending poly (3-hydroxybutyrate) (P3HB) or PHBV with collagen and/or gelatin on the scaffold properties, cell viability, and/or wound closure tests [139,143–145]. In fact, adding a hydrophilic natural protein (e.g., collagen) to pure PHBV (or to PHAs in general) highly increases the scaffold wound exudate absorption capacity and cell-scaffold interactions. In 2007, two different groups demonstrated that the presence of collagen in PHBV electrospun nonwoven mesh significantly improved cell adhesion and proliferation of NIH3T3 and dermal sheath (DS) cells/epithelial outer root sheath (ORS) cells [139,144]. Han et al. further evaluated the effectiveness of these scaffolds in vivo using Athymic nude mice for an open wound-healing model and, interestingly they pointed out a faster wound closure and better healing for pure PHBV mesh than for the blended one. They assessed the result by checking that filaggrin (protein that binds to keratin fibers in epithelial cells) was strongly expressed on the upper epidermal layer of PHBV-grafted skin, while only a little filaggrin was observed on the PHBV/collagen-grafted tissue [139]. According to the better cell viability of PHBV/collagen meshes and better wound closing and healing of pure PHBV meshes, they concluded that the physical property and mechanical stability of the matrices, given by the PHBV component, seemed to be a more important factor for early-stage wound dressings than their cell culture activity.

Salvatore et al. successfully produced and investigated the potential of electrospun PHB/collagen type I meshes for tissue engineering applications at PHB/collagen weight ratios of 100/0, 70/30, and 50/50 *w/w* (Figure 9a) [143]. The meshes were demonstrated to be effective substrates for viability and proliferation of murine fibroblasts for up to six days of culture at all the three different weight ratio compositions. The introduction of collagen improved the wettability and thermal stability of the scaffold and decreased the crystallinity degree, also enhancing the sensitivity to hydrolytic degradation. The morphological, mechanical, and degradation properties of the obtained meshes were suitable for wound dressing applications and all tunable to some extent by adjusting the PHB/collagen weight ratios. Interestingly, the 50/50 *w/w* sample displayed the highest wettability and degradation rate, while maintaining an elastic modulus comparable to that of pure PHB samples. Overall, although preliminary, the study revealed the possibility of tuning different physicochemical properties of electrospun PHB meshes by blending them with a specific amount of collagen directly in the electrospinning solution, keeping good cell viability and proliferation at all the different ratios.

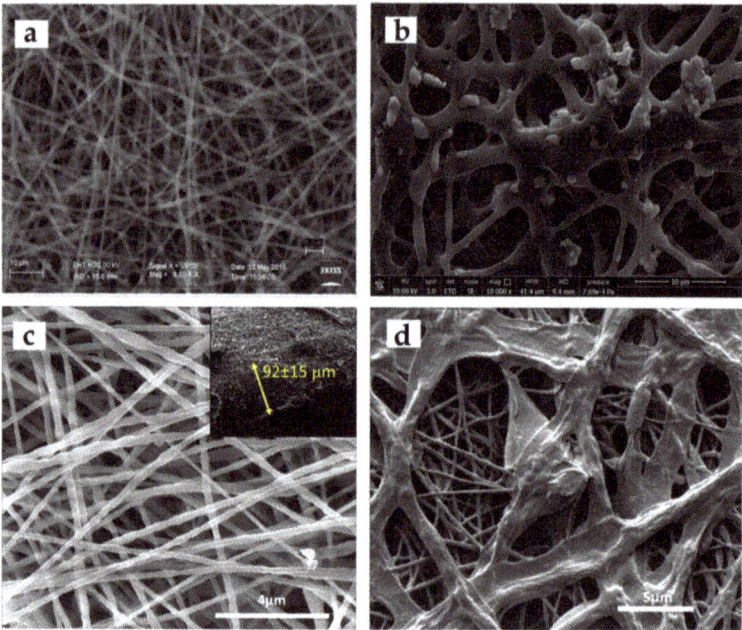

Figure 9. SEM images of (**a**) poly(hydroxybutyrate) (PHB)/Collagen (50/50 wt%) electrospun fiber meshes reproduced from open access article [143] distributed under the Creative Commons Attribution License, (**b**) poly(3-hydroxybutyrate-co-4-hydroxybutyrate) P(3HB)/P(3HO-co-3HD) electrospun fiber meshes functionalized with electrosprayed chitin-lignin/glycyrrhizin acid (CLA) (**c**) poly (3-hydroxybutyrate-co-3-hydroxyvalerate) (PHBV)/Curcumin (0.5) electrospun fiber meshes and cross-sections of drug-loaded nanofibers and (**d**) SEM images of L929 fibroblast cells cultured on PHBV/Curcumin (0.5 w/v%) electrospun fiber meshes after incubation for 14 days reproduced from [39] (License Number: 4876940110497).

Azimi et al. used electrospinning in order to produce blended poly (3-hydroxybutyrate)/poly(3-hydroxyoctanoate-co-3-hydroxydecanoate) [P(3HB)/P(3HO-co-3HD)] fiber meshes surface-decorated via electrospray of chitin-lignin/glycyrrhizin acid (CLA) complexes (Figure 9b) [146]. CLA complexes are bio-based micro-compounds that can be considered useful bioactive agents for functionalizing skin contact substrates, as they showed a proficient interaction in an in vitro skin model [147]. These bio-based and biodegradable functional nonwovens showed strong anti-inflammatory activity which is promising in wound healing applications. Kandhasamy et al. developed a composite scaffold for effective wound healing treatment based on a PHB/gelatin/ostholamide (OSA) electrospun blend coated with collagen. The obtained scaffold showed great mechanical stability, stable enzymatic degradation, and efficient antimicrobial activity against Pseudomonas aeruginosa and Staphylococcus aureus due to the sustained release of OSA [145]. In vitro and in vivo analysis displayed excellent cytocompatibility, determined by NIH 3T3 fibroblast proliferation studies, and good wound healing efficacy confirmed by using an open wound model in Wistar rats. In fact, after 15 days, complete wound closure was achieved using the PHB/gelatin/OSA/collagen scaffold, meanwhile wound size reductions of approximately 75%, 65%, and 45% were respectively observed for PHB/gelatin/OSA scaffold, pure collagen scaffold, and the cotton gauze positive control.

Recent studies have shown that many additives (e.g., drugs, NPs, NCs, essential oils, biomolecules) can be incorporated into nanofibers to enhance their mechanical and antimicrobial properties [148,149], and the nanofiber structure enables tunable sustained drug release, which in turn can promote/regulate the wound healing process. The increasing outcome of antibiotic resistance during the last years has also brought a big interest in developing new antimicrobial agents. The antimicrobial activity of some NCs and metal/metal oxides NPs, together with the capability of dispersing them in electrospun matrices, have been extensively studied and reviewed [148,150–153]. In this regard, recent studies focused on the wound dressing application of these nanocomposites. For example, Abdalkarim et al. reported the potential of cellulose nanocrystal-ZnO nanohybrids (CNC-ZnO) incorporated in PHBV electrospun fibers at different concentrations (3–15 wt%) for antibacterial wound dressings [154]. Results showed that the presence of CNCs-ZnO inside the fibers led to a reduction in fiber diameter and crystallinity and increased the porosity, thermal stability, and mechanical stiffness of the fibrous meshes, especially for 5 wt% CNC-ZnO, with respect to the pure ones reporting an increase of 150% in tensile strength and 112.5% in Young's modulus. Furthermore, enhanced absorbency capacity of the fresh blood model (4 g/g for PHBV meshes and 8.4 g/g for the membrane with 5 wt% CNC-ZnO), better barrier properties, and excellent antimicrobial activity were observed in CNC-ZnO/loaded PHBV fibers [154,155]. The inclusion of natural bioactive substances into PHA nano/micro-fibers have also been explored. For example, Mutlu et al. incorporated curcumin as an antioxidant, anti-inflammatory, and antitumor agent-into the PHBV nanofibers (Figure 9c) [39]. These matrices showed potential for wound dressing application since they were not toxic towards the L929 mouse fibroblast cell line (Figure 9d). The presence of curcumin particles inside the fibers increased swelling capacity, cell attachment, and proliferation while due to its plasticizer effect, decreased the mechanical properties of the matrix but still provided enough mechanical strength during the wound healing process. Diabetic wounds can often become chronic, with nonhealing mainly associated with insufficient cell proliferation and angiogenesis. Furthermore, reactive oxygen species (ROS) have been shown to delay the healing process, exacerbating the chronicity status; therefore, they cannot be used as antimicrobial agents. On the other hand, antioxidant therapies have been shown to improve the healing of chronic diabetic wounds by inhibiting ROS generation [156,157]. Cerium oxide nanoparticles ($nCeO_2$) show effective antioxidant activity, so their application in diabetic wound dressings can be a promising approach to promote angiogenesis and healing. Augustine et al. developed for the first time a novel electrospun PHBV membrane containing $nCeO_2$ for diabetic wound healing [158]. They reported that an optimum load of 1% w/w $nCeO_2$ can improve the tensile strength of the neat PHBV membrane without a significant loss of elongation at break. They assessed excellent in vitro cytocompatibility and cell adhesion properties, especially for the 1% w/w $nCeO_2$ load, and enhanced cell migration and angiogenesis were observed for all the nanocomposites in comparison to pure PHBV meshes. Finally, they demonstrated the efficiency of $nCeO_2$ containing membranes in enhancing the diabetic wound healing process on full-thickness excision wound model performed on Male Sprague–Dawley rats [158].

Electrospun PHBV fibers have shown the possibility of being incorporated in hydrogels, thus generating a fiber-based composite material apt for controlled release of drugs [159]. Cristallini et al. proposed this method for generating a surface coating able to release dexamethasone, thus modulating the fibrotic encapsulation of implanted devices.

Finally, we would like to report (even if not fully biobased) a smart exploitation of the hydrophobic nature and the specific biotic enzymatic degradation mechanism of some PHAs. Specifically, Liu S. et al. used a blend composed of PHBV/polyethylene succinate (PES)/poly(3-hydroxyoctanoate-co-3-hydroxyhexanoate) (PHOHHx), in a ratio of 3:2:1 (*w/w*) to produce core-shell fibers which can effectively bring on a burst release of antibiotic triggered by bacterial presence. The use of broad-spectrum potent biocides can play an active role in the wound healing process by preventing or treating infections but, on the other hand, its use can develop bacterial resistance against antibiotics and attack pathogens and host cells non-selectively, further delaying the wound healing process. For these reasons, core-shell ultrafine fibers were developed using coaxial electrospinning, applying the mentioned blend as shell, and a core consisting of polyvinylpyrrolidone (PVP) as a polymer carrier and dodecyl trimethyl ammonium chloride (DTAC) as a model biocide. In vitro results showed that this composite structure effectively prevented degradation of the PHAs/PES based shell in abiotic solutions, with slow releasing rates of the biocide. An opposite trend was observed in the presence of the model bacteria Pseudomonas aeruginosa. The in vitro release data demonstrated the bacteria-triggered drug delivery property of PHAs/PES based core-shell nanofibrous membranes [160].

5.4. Application of PLA-Based Electrospun Fibers in Wound Healing

PLA is a synthetic biopolymer made from renewable resources. PLA can be derived from natural raw materials such as corn starch, rice, and sugar cane through ring-opening polymerization or condensation polymerization of the lactic acid. It has been approved in diverse applications by the Food and Drug Administration (FDA). Lactic acid (2-hydroxypropionic acid, CH3–CHOHCOOH) is the constituent monomer of PLA. Poly (l-lactide) (PLLA), Poly (d-lactide) (PDLA), racemic poly(dl-lactide) (PDLLA) are three isomeric forms of PLA [51,69,161].

PLA is a biocompatible, biodegradable, and bioabsorbable aliphatic thermoplastic polyester with valuable thermo-mechanical properties. These features, in addition to its non-toxicity to the human body, make it a suitable candidate for use in bioengineering [6,26,31,162]. In some applications, such as medical implants, PLA stereocomplex has attracted growing attention. PLA stereocomplex can be obtained by the strong interaction formed by the side-by-side arrangement of the molecular chains of enantiomeric PLA polymers such as PLLA and PDLA. The formation of stereocomplex also endows PLA-based materials with enhanced mechanical function, thermal stability, and hydrolysis resistance [163,164].

Electrospun nanofibers based on PLA have been widely considered for biomedical applications as tissue engineering scaffolds, wound dressings, and drug carriers [7,165]. The related researches are briefly reviewed in this section, with a focus on different aspects of the electrospinning process of PLA, biomolecule-loaded PLA-based nanofibers, and clinical applications of the electrospun PLA- based structures as wound healing materials.

The electrospinning process of PLA has broadly been discussed; a lot of research efforts have been focused on the systematic investigation and optimizing the parameters that influence the ultimate properties of electrospun PLA fibers and thus their possible end-use. Depending on the characteristics of the PLA solutions, the process parameters and ambient conditions, the diameter and morphology of the PLA fibers produced by electrospinning will differ. As an example, in research works performed by Maleki et al. the influence of the solvent system on the electrospinning of PLLA fibers is investigated [57,66]. The results demonstrated that the dielectric constant and vapor pressure of the solvent are important parameters, which significantly affect the fiber morphology, diameter, and crystallinity and hence influence the mechanical properties of the fibrous structures. Electrospinning of PLLA solutions using solvents with high vapor pressure and lower boiling points, like chloroform and dichloromethane, formed fibers with porous surfaces. Whereas the electrospinning of PLLA solution based on solvents such as hexafluoro isopropanol (HFIP) and 2,2,2- trifluoroethanol (TFE) let to formation of uniform fibers with a smooth surface (Figure 2a) [51,57,66]. Solvents with a low vapor pressure like 2,2,2-trifluoroethanol (TFE) provided enough time for crystals to grow during fiber formation resulting in a high crystallinity. Solutions obtained from chloroform with lower conductivity caused less stretching of the charged jet in the electrostatic field and resulted in larger diameters of fibers [57,66]. In another study by Maleki et al., in order to investigate the effect of solvent type on the mechanical properties of electrospun PLLA structures, a double nozzle electrospinning set-up was used to fabricate continuous twisted yarns from PLLA fibers. The tensile strength and modulus of electrospun yarns produced from a solvent with low vapor pressure like TFE showed higher tensile strength and modulus than yarns produced from fibers electrospun from PLLA-dichloromethane or chloroform solutions [66]. The PLA-based solution properties like concentration and viscosity are the other factors that affect the ultimate characteristics of the electrospun PLA fibers and have been considered by several researchers. It was shown that, by increasing the solution concentration, uniform fibers were produced with growing average diameters [57,65].

The applications of electrospun PLA-based structures for wound healing and skin regeneration have also been investigated in several recently published research works. Most of these efforts were made towards the development of bioactive PLA-based electrospun membranes for the designing of wound healing products with multifunctional properties. In this context, electrospun PLA fibers were employed as drug delivery carriers.

Alves et al. recently confirmed that PLA electrospun membranes are promising drug delivery systems for sustained release for developing wound dressings [69]. They employed both techniques of physical adsorption and blend electrospinning to load anti-inflammatory agents, dexamethasone acetate (DEX), and betamethasone to PLA electrospun fibers and compare their release efficiency. The influence of both drugs on the features of the electrospun PLA fibers, such as morphological and mechanical properties, was studied. The drug-loaded fibers produced from the blend electrospinning of the PLA solution containing drugs presented a better sustained release profile after a burst release during the first five hours. Samples prepared by physical adsorption of the drug on the PLA electrospun membranes showed a significant burst release.

Pankongadisak et al., developed electrospun curcumin incorporated PLLA fibers as carriers to control curcumin release with anti-inflammatory, and antioxidant properties for wound dressing materials. The fabricated membrane through blend electrospinning was non-toxic to human adult dermal fibroblast (HDF) cells and protected cell attachment and proliferation [166].

In a research attempt performed by Moradkhannejhad et al., the blending of PLA and poly(ethylene glycol) (PEG) aiming to increase the hydrophilicity of PLA for wound dressing applications is reported. In their study, the hydrophobicity of curcumin (CUR) loaded PLA electrospun nanofibers was modified through the addition of PEG with different contents and molecular weights [31]. The incubation of PLA/CUR/PEG nanofibers in PBS indicated that the increase in concentration and a decrease in the molecular weight of PEG caused an increase in the weight loss values. The prepared electrospun CUR-loaded PLA/PEG membranes provided desirable conditions for cell growth and to accelerate the drug release through balancing the hydrophilicity– hydrophobicity of the medium.

In order to improve the elasticity and hydrophilicity of the PLA nanofibrous membranes as wound dressings, in a study by Zou et al., PLA/Poly (1, 8-octanediol-co-citric acid) (POC) nanofibers were fabricated using the blend electrospinning method. Aspirin loading and its release behavior were also investigated. In these authors' view, the developed elastic, hydrophilic, and biodegradable PLA-based nanofibrous membrane can modify several disadvantages of PLA dressings, namely, limited tensile deformation, uncontrollable degradation rate, and low hydrophilicity, which hinder its application in wound dressings [167].

Owing to its biological features, PLA is often used as the main component of electrospun composites for biomedical applications. In a research study, by Bi et al., PLA and PLA/PVA/SA (sodium alginate) membranes were produced via electrospinning for wound healing applications. The obtained results suggested these membranes could be used as novel wound dressings. In vitro experiments indicated that PLA and PLA/PVA/SA electrospun membranes were able to provide good support for the growth of mouse fibroblasts (L929 cell line). The fibroblasts displayed relatively better adhesion and proliferation on PLA/PVA/SA than on the PLA r membranes. The in vivo assay also confirmed that the PLA and PLA/PVA/SA electrospun membranes significantly enhanced wound healing compared to commercially available gauzes [26]. Ilomuanya et al., designed PLA and collagen-PLA electrospun fibrous scaffolds with antibacterial activity for wound healing applications [12]. They used silver sulphadiazine

(Ag+S) and Aspalathus linearis (AL) fermented extract to improve antibacterial properties and cellular biocompatibility. They were nontoxic to the cells and provided favorable substrates for cell attachment and proliferation. In another study by Cui et al., doxycycline (DCH), a broad-spectrum antibiotic, selected as a model drug, and DCH-loaded PLA nanofibers were produced by the blend electrospinning process [42]. Results of cytotoxicity and antibacterial tests revealed that DCH-loaded PLA nanofibers showed favorable cytocompatibility to L929 mouse fibroblasts and exhibited good antibacterial activity, suggesting its potential as wound dressings for chronic wound healing. With the purpose of avoiding disadvantages of conventional antibiotic-loaded wound dressings, Zhang et al. in their research work employed the electrospinning process to develop fibrous composite based on silver (I) metal-organic frameworks-PLA antibiotic- free wound dressing with the efficient antibacterial capability to promote tissue regeneration, simultaneously. The in vivo experiments indicated that the electrospun fibrous composite could significantly accelerate the healing rate of infected wounds in rats [10]. Fang et al. employed the coaxial electrospinning method to develop core-shell nanofibers based on PLA and γ- PGA for wound healing as tissue engineering scaffolds. The in vitro cell culture study and in vivo animal experiment on PLA/γ-PGA core-shell nanofiber membrane demonstrated favorable biocompatibility with good ability to wound repair [6]. A study by Yang et al. used poly (glycerol sebacate) (PGS)/PLLA fibrous scaffold with a PGS core and a PLLA shell via coaxial electrospinning. The fabricated fibers with porous morphology of the shell surface showed excellent ability to repair tissues of the skin wound. In comparison to pure PLLA scaffold, the core-shell structure exhibited superior cell proliferation, with a lower inflammatory response [51]. In another work performed by Yang et al., the coaxial electrospinning technique was applied to fabricate the core-shell structured PLLA/chitosan nanofibrous scaffolds for wound dressing applications. In this study, graphite oxide (GO) nanosheets were coated on core PLLA-shell chitosan nanofibers to provide a synergistic microenvironment for wound healing. The coating with GO nanosheets significantly enhanced the hydrophilicity of the electrospun membrane. GO coated chitosan/PLLA nanofibrous scaffolds indicated desirable antibacterial activity. In addition, they promoted the growth of pig iliac endothelial cells (PIECs). GO-coated chitosan/PLLA nanofibrous scaffolds possessed favorable wound healing in rats [168]. In a research work performed by Augustine et al., the wound healing membranes composed of core-shell fibers were produced by coaxial electrospinning. Connective tissue growth factor (CTGF) was encapsulated within the PVA core which was covered by a thin layer of PLA. The developed CTGF loaded core (PVA)-shell (PLA) membranes boosted its sustained release for diabetic wound healing applications [165].

The effect of scaffold architecture on healing efficiency was lately investigated by Majchrowicz et al. [169]. In this work, a nanofibrous composite was made of PLA matrix and calcium phosphate organically modified glass (CaP ormoglass) nanoparticles fabricated by blend electrospinning having two arrangement of the aligned and random orientation of fiber deposits. The addition of CaP ormoglass nanoparticles in electrospun composites promoted bone regeneration by improving the degradation process. Random fiber orientation seems to be superior for CaP compounds released during in vitro degradation. Differently, for their aligned architecture only small amounts of CaP deposits were observed after 21 days of immersion in PBS. Recently, a wet electrospinning technique with a liquid coagulation bath collector was applied in order to produce 3D porous PLA nanofibrous scaffolds to seed rat bone-marrow stem cells (BMSCs). Wet electrospun nanofibers have distinguished advantages such as high porosity and surface area compared with the dry electrospinning technique. The results of in vitro and in vivo assay proved that the 3D electrospun fibrous PLA can be a suitable dressing for wound repair [40].

6. Translational Approaches

6.1. Drug Delivery Electrospun Fibers

Controlled drug delivery systems (DDSs) display one of the advancing areas of biomedical and pharmaceutical sciences [170]. In biodegradable polymers, the drug release system encompasses a faster drug desorption from the surface and a slower drug diffusion from the bulk, which is accelerated during polymer degradation. The velocity of these phenomena also depends on the surface to volume ratio and chemistry of drug versus polymer. Some unique advantages of electrospun fibers including impressive drug loading and encapsulation, easy modulation of the release rate, and its simple processability aid DDSs to achieve controlled drug release. Electrospun nanofibers can be used for transdermal drug delivery systems (TDDS), in which target agents and sensitive drugs, which cannot be transferred orally are delivered through the skin into the body. Indeed, different herbal pharmaceutical compounds (e.g., aloesin, curcumin, thymol, etc.) can be incorporated in electrospun nanofibers in order to expand TDDS and/or active wound dressings. The release mechanism understanding is essential for developing the DDS systems. Mainly release mechanisms are diffusion (through water-filled pores of the polymer bulk), degradation/erosion, and osmotic pumping.

Single-nozzle electrospinning produces fiber meshes which usually act as degradation-based DDSs since the drugs are dispersed/dissolved within the dissolvable/degradable polymeric matrix in which dissolution or degradation of the polymeric matrix leads to drug release. Degradation occurs in the bulk since the average molecular weight of polymer decreases while erosion occurs at the surface when the polymer undergoes a decrease in total mass [171]. On the other hand, a reservoir DDSs can usually be produced by coaxial electrospinning in which the drug molecules are encapsulated in the inner solution core, surrounded with the outer shell barrier, and their release profiles are usually governed by diffusion through the shell membrane which helps to delay fast drug diffusion. The shell thickness and core diameter play an important role in the extent of the drug release delay [172].

While prior approaches are based on the encapsulation of drug molecules within the nanofiber bulk phase, surface-modified electrospun fibers have also opened up a new possibility of constructing more complicated drug delivery platforms. Indeed, premier adhesiveness toward biological surfaces lead nanofibers to be a suitable nominate for topical drug delivery devices [173]. Physical immobilization of drug molecules on the surface of nanofibrous meshes with high surface area to volume ratio leads to higher drug loading amount per unit mass in comparison to any other device. For some particular applications in which prevention of bacterial infection is required, the instant release of drugs from the surface of the nanofiber is advantageous since it can provide facile dosage control of some specific pharmaceutical agents [174]. Furthermore, some specific structures such as drug-loaded nanoparticles located on the surface of nanofibers in addition to the high drug loading capacity are able to provide unique drug-releasing profiles, which the nanofiber itself cannot obtain [175].

Table 1 summarized the use of different biopolymers, incorporating or not active ingredients, to have a specific function and target specific wound types, which include the types chronic/acute, localized/extended, infected/inflamed, due to burning, diabetes, ulcers and trauma. It can be observed that the most searched features in wound dressing function rely on antimicrobial and anti-inflammatory properties to allow efficient wound healing in complicated wound types. Using conventional wound dressings, the most frequent side effect reported is dermatitis, especially in patients with leg ulcers [176]. Allergic contact dermatitis is an individual host response to a material, hardly predictable. However, a careful choice of biomaterials, in vitro and preclinical investigation on immune response may help in developing better biocompatible wound dressings.

Table 1. Specific function and wound type targets for different electrospun biopolymer dressings.

Electrospun Mesh	Incorporated Therapeutics	Function & Wound Type	References
Cellulose acetate (CA)/Manuka honey (MH)	-	Antibacterial activity for infection in the burn wounds	[37]
CA/polyester urethane	Polyhexamethylene biguanide (PHMB)	Antimicrobial activity	[110]
Chitosan/bacterial nano cellulose	-	Antimicrobial properties	[122]
Chitosan/silk fibroin	-	Antibacterial properties Acute wounds	[121]
Chitosan/sericin	-	Biocompatibility and antibacterial properties	[120]
Chitosan/Poly (l-lactide) (PLLA)	Graphene oxide	Antimicrobial activities for chronic infected wounds	[50]
Chitosan/keratin/polycaprolactone (PCL)	Aloevera extract	Anti-inflammatory, antibacterial, antiviral, and antioxidant properties for acute and burn wounds	[35]
Chitosan/Polyvinyl alcohol (PVA)	Nanobioglass (nBG)	Biocompatibility, antibacterial activity and regeneration promotion effect for chronic wound	[5]
Gelatin/Oleoyl chitosan	-	Large skin defects or chronic wounds	[123]
Polyhydroxyalkanoate (PHA)	Dodecyltrimethylammonium chloride (DTAC) biocide	Antimicrobial effects for chronic wounds	[160]
PHA	Graphene/decorated silver nanoparticles (GAg)	Antimicrobial activity for chronic wounds	[141]
Poly (3-hydroxybutyrate-co-3-hydroxyvalerate) (PHBV)	Cerium Oxide Nanoparticle	Antioxidant and angiogenic properties Diabetic wounds	[8]
PHBV	Cerium oxide nanoparticles (nCeO2)	Antioxidant and angiogenic properties for diabetic wounds	[8]
PHBV	Curcumin	Antioxidant, anti-inflammatory, and antitumor properties chronic wounds including burns, diabetic foot ulcers, venous leg ulcers, and pressure ulcers	[39]
PHBV/cellulose	ZnO nanocrystals	Antibacterial activity for acute and infected wounds	[154]
Polylactides (PLA)	AgNPs	Antimicrobial activity for burn wounds and diabetic ulcers	[87]
PLA	Curcumin, Enrofloxacin	Antioxidant activity, antimicrobial activity, and biocompatibility	[70]
PLA	Doxycycline (DCH)	Antibacterial activity, Chronic wounds, diabetic wounds	[42]
PLA	Silver (I) metal–organic framework Ag2[HBTC][im]	Antibacterial feature	[10]
PLLA	Curcumin	Anti-inflammatory antioxidant effects	[166]
PLA/hyperbranched polyglycerol (HPG)	Curcumin	Antioxidant, anti-inflammatory and anti-infective properties for acute and chronic wound	[31]
PLA/PVA	Connective tissue growth factor (CTGF)	Diabetic wounds	[165]
poly(lactic-co-glycolic acid) (PLGA)/gelatin	Recombinant human epidermal growth factor (rhEGF), gentamicin sulfate	- Antibacterial activity and rhEGF supply - Diabetic wound	[90]
PLGA/polydopamine	Basic fibroblast growth factor (bFGF), ponericin G1	Antibacterial and cell proliferation-promoting properties for skin tissue regeneration	[89]

6.2. Clinical Trials of Electrospun Wound Dressings

Despite the great number of produced and ongoing researches highlighting the potential of electrospun meshes as wound dressings and, more in general, for pharmaceutical applications, an actual impact in the clinics with any therapeutic product for use in humans has not yet been achieved [177]. Among the many proposed products, a few have reached an actual preclinical level [178] and very few an in-human clinical trial [177,179,180]. Of three of these products, regarding wound dressing applications, namely SurgiCLOT® [181], Pathon [182,183], and TPP-fibers (Tecophilic™) [184], to our knowledge, only SurgiCLOT® is currently available on the market. Table 2 reports the in-human tested and commercially available products related to the electrospinning technology for wound dressing. In this table, we also reported SpinCare™, which is a portable device for electrospin directly onto the wound. It has also to be noted that actually, only SurgiCLOT® made from dextran, a polysaccharide, is purely bio-based. Dextran is the collective name given to a large class of α1→6- linked glucose polymers. It is synthesized from sucrose by certain lactic acid bacteria, such as *Leuconostoc mesenteroides*, *Lactobacillus* spp., and *Streptococcus mutans*. Hydrogels of this polysaccharide are particularly suitable as scaffolds for soft tissue engineering applications because dextran is resistant to both protein adsorption and cell adhesion. Incidentally, this biopolymer is used in the clinics for its antithrombotic (antiplatelet), properties. Polyurethanes are a class of synthetic polymers. However, one of the special features of polyurethanes is that their monomers can be derived from natural sources. The principal natural source of polyols are vegetable oils, like castor oil, olive oil, and canola oil. Thus, it may be expected that green polyurethanes will enter into biomedical products in the future. The biodegradation of polyurethanes in the physiological environment is complex, but involves three main mechanisms: hydrolytic, enzymatic, and oxidative. The degradation products depend strongly on the chemistry of the polymer but also on the degradation mechanism.

Table 2. Electrospinning-related products for wound dressing.

Product	Polymer	Device	Current Status	Case Studie or Clinical Trial Performed/Ongoing	References
Pathon	Polyurethane	Composite mesh, NO drug delivery	Clinical trial	Two double blind, randomized controlled clinical trials	[182,183]
Tecophilic™	Polyurethane-PEG	Photosensitizing-loaded mesh	Clinical trial	Comparative, 3-group based study over a total of 162 patients	[184]
SurgiCLOT®	Dextran	Fibrin Sealant Patch	On the market	Clinical safety and performance study in UK and Norway, Clinical safety and performance study in India, Pre-clinical comparative study of bone bleeding treated with SurgiCLOT® and standard gauze, Pre-clinical Performance study of SurgiCLOT® compared to Dextran-only dressing	[181,185–188]
SpinCare™	Various electrospinnable polymers	Portable electrospinning wound dressing device	On the market	Donor site wound single case Partial thickness burns single case	[189–191]

Mulholland, E.J., in his recent work presented great perspectives in using electrospun meshes for scar treatment, highlighting the clinical absence of any effective electrospun anti-scarring device [179]. He also reported that there are currently five clinical trials exploring the use of electrospun nanofibers scaffolds, but no current recruiting trials [192].

The problems that have slowed down the translation of clinical electrospun nanofiber products into the market have been associated with two main different aspects: the first is related to the manufacturing and technical part. In fact, although highly scalable, the industrial throughput is (still) not competitive and there is a need for highly skilled workers to produce and develop new products [193]. In biomedical applications, due to the reduced area of the devices, this problem seems more easily solvable. The second aspect is the difficulty to achieve the legal standards posed by the regulatory bodies for these novel technologies which exploit state-of-the-art research, such as the sterilization part and the application of devices containing living cells and/or bioactive molecules [177,179,194].

7. Future Perspective

Wounds have been affecting people of all genders, race, and age throughout the history of humankind. The integration of textile technology and medical science presents a new field known as "medical textile" which is a substantial and growing section of the textile industry. The noticeable advancement of biomaterials science, especially in the field of bio-nanotechnology, will generate a great number of opportunities for medical textiles in the coming decade. Wound dressings are an important component of the "on the patient" medical textiles [195], as effective wound dressings can convert painful days to comfortable days for patients. Since the nineties, several biomaterials and technologies have been developed for producing wound dressings. As such, the numerous research efforts in this field performed by research institutes and industries around the world, are expected to have a considerable effect on the quality and expectation of life.

The market for bioactive wound care products is growing fast, as various biomaterials, superabsorbent and multifunctional dressings are continuously being proposed instead of conventional gauzes. Multiple and specific functionalities, such as antimicrobial, hemostatic, bio- responsive and biomimetic properties are required for future wound care products to provide a suitable micro-environment. Unveiling chemical, physical, and mechanical properties of existing biopolymers and even designing new ones, would push forward the availability of the desired efficacy in wound dressings. To this purpose, the recent advancements in regenerative medicine, nanotechnology, and bioengineering technologies, allow the development of specific key structures for wound healing applications. Electrospinning and bioprinting enable the development of 3D biopolymer-based scaffolds and 3D artificial skin mimicking with the desired pore size and tensile strength. In particular, electrospinning enables the production of multifunctional nanofibrous materials, which not only provide functionalities such as physical protection and optimal moisture environment for the wound, but also permit a sustained release of the therapeutic agents to provide the requirements of the healing procedure [196]. These functional materials should be non-toxic and provide comfort for the patient through gas permeability, and easy removal. More advanced needs, such as microbial control can accelerate the wound healing process. A suitably designed dressing should provide appropriate warmth in cold weather and coolness in warm conditions and also permit sufficient breathability and permeability. Despite the growing research works in this area, there are still challenges and it is not yet been clearly realized how to incorporate all the criteria of an ideal wound dressing into one material to fulfill the multifunctional requirements in facilitating the healing procedure, cell growth, drug release behavior, and kinetics, as well as many other features. Additive manufacturing can be used to print biopolymer-incorporated cells, drugs, growth factors or even nanoparticles in the desired shape and high accuracy to be used as multifunctional-type dressings also in combination with electrospinning.

In making molecules available on the nanoscale, the maximum interaction of the drug molecules with cellular components will be obtained, which enhances the efficiency of the drugs.

Despite promising wound healing outcomes, nanofibrous wound dressings need to be still improved in various aspects. Bio-based nanofibrous dressings for wound healing are not commercially available yet, as the related industrial production is expensive and limited. This issue arises from the limitations of the electrospinning process such as low production rate, and complexity of adjusting the process on a large scale. However, several efforts are in place to produce electrospun nonwoven products using suitable multi-needle systems. Biocompatibility is also another concern to be considered owing to additives and the remaining solvent in the electrospun fibers. Specifically, biopolymers like chitosan, cellulose, and gelatin with low water solubility, are usually dissolved in toxic, acidic solvents for electrospinning [197].

In the case of active dressings, during the electrospinning process, high voltage and high shearing forces may dehydrate and harm some bioactive agents, even though collection of the mesh at room temperature often without post-treatments gives important advantages to this technology. Electrospinning is highly affected by numerous variables, which directly influence the physical–mechanical properties of the nanofibers. Hence, controlling the ultimate features of the electrospun wound dressings and release behavior of the entrapped biomolecules via those systems may be difficult and require a costly and time-consuming experimental campaign. In this regard, the electrospinning procedure must be extensively standardized, and other technologies, such as artificial intelligence could help to obtain optimally produced manufacturing [198].

Structurally, nanofibrous dressings should be engineered from the perspective of architecture, fiber arrangement and orientation, and network porosity [199]. In addition to an enhanced mechanical performance, the desired morphology and architecture of nanofibrous dressing should target the wound healing demands. Due to the various tensions applied to the dressing depending on its application place and time, the mechanical properties are very relevant. The relatively low mechanical strength, low pore size, and the difficulty in tailoring nanofibrous meshes have limited their applications as scaffolds for skin regeneration. Fabricating twisted yarns from electrospun fibers with improved mechanical properties will find advanced applications in this field. The nanofiber coating onto textile structures, in particular fabrics, filaments, and yarns, is another flexible method to provide required mechanical performance by virtue of the underlayer substrates. These proposed dressings can benefit from the characteristics of mechanical strength of the conventional spun bio- based yarns and outstanding features of nanofibers.

Stimuli-responsive fibrous materials with shape-memory properties have recently attracted great attention for their potential in regenerative medicine. Smart electrospun wound dressings combine the shape memory effect and nanofiber features with controlled delivery of biomolecules, which enhances their performance, thus appearing good candidates to fulfill the growing requirements of wound healing. Biological molecules derived from plants can be useful in protecting from oxidative stress. For example, olive leaf extract (OLE) has been revealed to protect endothelial cells (HUVECs) from ROS in a 3D model and can be useful for release from biopolymers for wound healing [200]. Further studies are aimed at incorporating OLE inside biopolymer fibers.

Application of suitable biosensors integrated into the nanofibrous dressings for monitoring wound conditions could also be another approach to develop smart materials [199]. These classes of dressings have been less developed and studied and should be further considered under the advanced wound dressing perspective. For some kinds of injuries, especially cavity wounds, the direct electrospinning of nanofibers as coating on the wound surface could also be a promising approach. Finally, great promise is held for developing novel cellulosic materials as superabsorbent for exudative wounds. Along with plant-derived cellulose, bacteria-derived cellulose shows an intrinsic hydrogel structure useful to drain serum from open wounds, thus supporting skin homeostasis and reparative processes [201]. Obtaining solvent-clean, easy, and industrially affordable methods to produce cellulose nanofibers represents the next step in this area.

8. Conclusions

The skin is the largest organ in the body to be damaged and any relevant wound needs prompt and effective treatment. Since wound healing is often a complex dynamic process wherein the interaction between the cells, secretory factors, and ECM molecules play an important role in the fate of the healing process, efficient wound management can result in difficulty. Biomaterial dressings should protect the wounds from environmental pollutants, provide a physiologically moist microenvironment, allow sufficient oxygen diffusion to the wound site, induce cell proliferation to facilitate tissue regeneration, while minimizing the risk of bacterial infection. The wound dressing should be soft, biocompatible, non-toxic, and non-allergenic. The healing process is greatly expected to improve by using multifunctional wound dressings. Several research works in this field have confirmed the efficiency of nanofibrous structures compared to conventional dressings. Owing to their outstanding characteristics, like small diameter, high surface to volume ratio and porous network with sufficient pore size, micro- and nano-fibers offer a great potential in the development of wound dressings. Electrospinning is a versatile and efficient method to generate fibrous structures from diverse materials with tunable compositions, features, and morphologies. Optimization of different electrospinning variables enables nanofibrous dressings to be obtained with the desired physical and biological performance. Multifunctional nanofibrous dressings possess all the requirements for effective wound healing and can be designed by blending different natural or synthetic polymers and incorporating drugs, nanoparticles, and bioactive agents through the electrospinning process. Electrospinning of natural and synthetic biopolymers from different origins, natural and renewable sources, such as cellulose, chitosan, PHAs, and PLAs have own superior features for producing dressings and tissue engineering scaffolds, thus being valuable to the current wound care industry due to their superior biocompatibility, biodegradability, bioactivity, and other specific structures. The presented review highlights the recent developments of bio-based nanofibrous wound dressings and stresses the electrospinning process as an efficient method to design an ideal wound dressing from natural and synthetic bio-based polymers in order to comply with the multifunctional requirements in facilitating the healing process, cell growth, and release behavior of drug and other bioactive agents.

Author Contributions: Conceptualization, B.A., H.M., and S.D.; methodology, B.A. and H.M.; investigation, B.A., L.Z., J.G.D.l.O., and H.M.; writing—original draft preparation, B.A., L.Z., and H.M.; writing—review and editing, J.G.D.l.O., S.L., and S.D.; visualization, B.A.; supervision, S.D.; project administration, A.L. and S.D.; funding acquisition, A.L. and S.D. All authors have read and agreed to the published version of the manuscript.

Funding: The project received funding from the Bio-Based Industries Joint Undertaking (BBI JU) under the European Union's Horizon 2020 research and innovation programme by POLYBIOSKIN (Grant agreement No. 7N5839).

Acknowledgments: J.G.D.l.O. and S.D. acknowledge the Doctorate Degree program PEGASO of the Tuscany region. B.A., L.Z., S.L and S.D. thank for 4NanoEARDRM project (EuroNanoMed 3, 2017).

Conflicts of Interest: The authors declare no conflict of interest.

References

1. Adeli, H.; Khorasani, M.T.; Parvazinia, M. Wound dressing based on electrospun PVA/chitosan/starch nanofibrous mats: Fabrication, antibacterial and cytocompatibility evaluation and in vitro healing assay. *Int. J. Biol. Macromol.* **2019**, *122*, 238–254. [CrossRef]
2. Memic, A.; Abudula, T.; Mohammed, H.S.; Joshi Navare, K.; Colombani, T.; Bencherif, S.A. Latest Progress in Electrospun Nanofibers for Wound Healing Applications. *ACS Appl. Bio. Mater.* **2019**, *2*, 952–969. [CrossRef]
3. Augustine, R.; Rehman, S.R.U.; Ahmed, R.; Zahid, A.A.; Sharifi, M.; Falahati, M.; Hasan, A. Electrospun chitosan membranes containing bioactive and therapeutic agents for enhanced wound healing. *Int. J. Biol. Macromol.* **2020**, *156*, 153–170. [CrossRef] [PubMed]
4. Zhu, Z.; Liu, Y.; Xue, Y.; Cheng, X.; Zhao, W.; Wang, J.; He, R.; Wan, Q.; Pei, X. Tazarotene Released from Aligned Electrospun Membrane Facilitates Cutaneous Wound Healing by Promoting Angiogenesis. *ACS Appl. Mater. Interfaces* **2019**, *11*, 36141–36153. [CrossRef] [PubMed]
5. Chen, Q.; Wu, J.; Liu, Y.; Li, Y.; Zhang, C.; Qi, W.; Yeung, K.W.K.; Wong, T.M.; Zhao, X.; Pan, H. Electrospun chitosan/PVA/bioglass Nanofibrous membrane with spatially designed structure for accelerating chronic wound healing. *Mater. Sci. Eng. C Mater. Biol. Appl.* **2019**, *105*, 110083. [CrossRef]
6. Fang, Y.; Zhu, X.; Wang, N.; Zhang, X.; Yang, D.; Nie, J.; Ma, G. Biodegradable core-shell electrospun nanofibers based on PLA and γ-PGA for wound healing. *Eur. Polym. J.* **2019**, *116*, 30–37. [CrossRef]
7. Perumal, G.; Pappuru, S.; Chakraborty, D.; Nandkumar, A.M.; Chand, D.K.; Doble, M. Synthesis and characterization of curcumin loaded PLA—Hyperbranched polyglycerol electrospun blend for wound dressing applications. *Mater. Sci. Eng. C* **2017**, *76*, 1196–1204. [CrossRef]
8. Khan, N.; Misra, M.; Koch, T.; Mohanty, A. Applications of electrospun nanofibers in the biomedical field. *SURG J.* **2012**, *5*, 63–73.
9. Miguel, S.P.; Sequeira, R.S.; Moreira, A.F.; Cabral, C.S.D.; Mendonca, A.G.; Ferreira, P.; Correia, I.J. An overview of electrospun membranes loaded with bioactive molecules for improving the wound healing process. *Eur. J. Pharm. Biopharm.* **2019**, *139*, 1–22. [CrossRef]
10. Zhang, S.; Ye, J.; Sun, Y.; Kang, J.; Liu, J.; Wang, Y.; Li, Y.; Zhang, L.; Ning, G. Electrospun fibrous mat based on silver (I) metal-organic frameworks-polylactic acid for bacterial killing and antibiotic-free wound dressing. *Chem. Eng. J.* **2020**, *390*, 124523. [CrossRef]
11. Dong, Y.; Zheng, Y.; Zhang, K.; Yao, Y.; Wang, L.; Li, X.; Yu, J.; Ding, B. Electrospun Nanofibrous Materials for Wound Healing. *Adv. Fiber Mater.* **2020**, *2*, 212–227. [CrossRef]
12. Ilomuanya, M.O.; Adebona, A.C.; Wang, W.; Sowemimo, A.; Eziegbo, C.L.; Silva, B.O.; Adeosun, S.O.; Joubert, E.; De Beer, D. Development and characterization of collagen-based electrospun scaffolds containing silver sulphadiazine and Aspalathus linearis extract for potential wound healing applications. *SN Appl. Sci.* **2020**, *2*, 1–13. [CrossRef]
13. Huang, W.C.; Ying, R.; Wang, W.; Guo, Y.; He, Y.; Mo, X.; Xue, C.; Mao, X. A Macroporous Hydrogel Dressing with Enhanced Antibacterial and Anti-Inflammatory Capabilities for Accelerated Wound Healing. *Adv. Funct. Mater.* **2020**, *30*, 2000644. [CrossRef]
14. Qu, J.; Zhao, X.; Liang, Y.; Xu, Y.; Ma, P.X.; Guo, B. Degradable conductive injectable hydrogels as novel antibacterial, anti-oxidant wound dressings for wound healing. *Chem. Eng. J.* **2019**, *362*, 548–560. [CrossRef]

15. Koivuniemi, R.; Hakkarainen, T.; Kiiskinen, J.; Kosonen, M.; Vuola, J.; Valtonen, J.; Luukko, K.; Kavola, H.; Yliperttula, M. Clinical study of nanofibrillar cellulose hydrogel dressing for skin graft donor site treatment. *Adv. Wound Care* **2020**, *9*, 199–210. [CrossRef]
16. Kong, D.; Zhang, Q.; You, J.; Cheng, Y.; Hong, C.; Chen, Z.; Jiang, T.; Hao, T. Adhesion loss mechanism based on carboxymethyl cellulose-filled hydrocolloid dressings in physiological wounds environment. *Carbohydr. Polym.* **2020**, *235*, 115953. [CrossRef]
17. Chin, C.-Y.; Ng, P.-Y.; Ng, S.-F. Moringa oleifera standardised aqueous leaf extract-loaded hydrocolloid film dressing: In vivo dermal safety and wound healing evaluation in STZ/HFD diabetic rat model. *Drug Deliv. Transl. Res.* **2019**, *9*, 453–468. [CrossRef]
18. Chamorro, A.M.; Thomas, M.C.V.; Mieras, A.S.; Leiva, A.; Martínez, M.P.; Yeste, M.M.S.H.; Grupo, U. Multicenter randomized controlled trial comparing the effectiveness and safety of hydrocellular and hydrocolloid dressings for treatment of category II pressure ulcers in patients at primary and long-term care institutions. *Int. J. Nurs. Stud.* **2019**, *94*, 179–185. [CrossRef]
19. Zhang, K.; Bai, X.; Yuan, Z.; Cao, X.; Jiao, X.; Li, Y.; Qin, Y.; Wen, Y.; Zhang, X. Layered nanofiber sponge with an improved capacity for promoting blood coagulation and wound healing. *Biomaterials* **2019**, *204*, 70–79. [CrossRef]
20. Khalid, A.; Naeem, N.; Khan, T.; Wahid, F. Polysaccharide Composites as a Wound-Healing Sponge. *Adv. Appl. Polysacch. Their Compos.* **2020**, *73*, 1.
21. Hao, Y.; Zhao, W.; Zhang, L.; Zeng, X.; Sun, Z.; Zhang, D.; Shen, P.; Li, Z.; Han, Y.; Li, P. Bio-multifunctional alginate/chitosan/Fucoidan sponges with enhanced angiogenesis and hair follicle regeneration for promoting full-thickness wound healing. *Mater. Des.* **2020**, *193*, 108863. [CrossRef]
22. Lazzeri, L.; Cascone, M.G.; Danti, S.; Serino, L.P.; Moscato, S.; Bernardini, N. Gelatine/PLLA sponge-like scaffolds: Morphological and biological characterization. *J. Mater. Sci. Mater. Med.* **2007**, *18*, 1399–1405. [CrossRef] [PubMed]
23. Zhao, W.Y.; Fang, Q.Q.; Wang, X.F.; Wang, X.W.; Zhang, T.; Shi, B.H.; Zheng, B.; Zhang, D.D.; Hu, Y.Y.; Ma, L. Chitosan-calcium alginate dressing promotes wound healing: A preliminary study. *Wound Repair Regen.* **2020**, *28*, 326–337. [CrossRef] [PubMed]
24. Xing, L.; Ma, Y.; Tan, H.; Yuan, G.; Li, S.; Li, J.; Jia, Y.; Zhou, T.; Niu, X.; Hu, X. Alginate membrane dressing toughened by chitosan floccule to load antibacterial drugs for wound healing. *Polym. Test.* **2019**, *79*, 106039. [CrossRef]
25. Zare-Gachi, M.; Daemi, H.; Mohammadi, J.; Baei, P.; Bazgir, F.; Hosseini-Salekdeh, S.; Baharvand, H. Improving anti-hemolytic, antibacterial and wound healing properties of alginate fibrous wound dressings by exchanging counter-cation for infected full-thickness skin wounds. *Mater. Sci. Eng. C* **2020**, *107*, 110321. [CrossRef] [PubMed]
26. Bi, H.; Feng, T.; Li, B.; Han, Y. In Vitro and In Vivo Comparison Study of Electrospun PLA and PLA/PVA/SA Fiber Membranes for Wound Healing. *Polymers* **2020**, *12*, 839. [CrossRef]
27. Alberti, T.B.; Coelho, D.S.; de Prá, M.; Maraschin, M.; Veleirinho, B. Electrospun PVA nanoscaffolds associated with propolis nanoparticles with wound healing activity. *J. Mater. Sci.* **2020**, *55*, 1–16. [CrossRef]
28. Dhivya, S.; Vijaya Padma, V.; Santhini, E. Wound dressings–a review. *Biomedicine* **2015**, *5*, 22. [CrossRef]
29. Gong, W.; Li, J.; Ren, G.; Lv, L. Wound healing and inflammation characteristics of the submicrometric mats prepared from electrospinning. *J. Bioact. Compat. Polym.* **2018**, *34*, 83–96. [CrossRef]
30. Safaee-Ardakani, M.R.; Hatamian-Zarmi, A.; Sadat, S.M.; Mokhtari-Hosseini, Z.B.; Ebrahimi-Hosseinzadeh, B.; Rashidiani, J.; Kooshki, H. Electrospun Schizophyllan/polyvinyl alcohol blend nanofibrous scaffold as potential wound healing. *Int. J. Biol. Macromol.* **2019**, *127*, 27–38. [CrossRef]
31. Moradkhannejhad, L.; Abdouss, M.; Nikfarjam, N.; Shahriari, M.H.; Heidary, V. The effect of molecular weight and content of PEG on in vitro drug release of electrospun curcumin loaded PLA/PEG nanofibers. *J. Drug Deliv. Sci. Technol.* **2020**, *56*, 101554. [CrossRef]
32. Liu, X.; Nielsen, L.H.; Qu, H.; Christensen, L.P.; Rantanen, J.; Yang, M. Stability of lysozyme incorporated into electrospun fibrous mats for wound healing. *Eur. J. Pharm. Biopharm.* **2019**, *136*, 240–249. [CrossRef]
33. Joseph, B.; Augustine, R.; Kalarikkal, N.; Thomas, S.; Seantier, B.; Grohens, Y. Recent advances in electrospun polycaprolactone based scaffolds for wound healing and skin bioengineering applications. *Mater. Today Commun.* **2019**, *19*, 319–335. [CrossRef]

34. Dodero, A.; Scarfì, S.; Pozzolini, M.; Vicini, S.; Alloisio, M.; Castellano, M. Alginate-Based Electrospun Membranes Containing ZnO Nanoparticles as Potential Wound Healing Patches: Biological, Mechanical, and Physicochemical Characterization. *ACS Appl. Mater. Interfaces* **2020**, *12*, 3371–3381. [CrossRef] [PubMed]
35. Zahedi, E.; Esmaeili, A.; Eslahi, N.; Shokrgozar, M.A.; Simchi, A. Fabrication and Characterization of Core-Shell Electrospun Fibrous Mats Containing Medicinal Herbs for Wound Healing and Skin Tissue Engineering. *Mar. Drugs* **2019**, *17*, 27. [CrossRef] [PubMed]
36. Dai, X.; Liu, J.; Zheng, H.; Wichmann, J.; Hopfner, U.; Sudhop, S.; Prein, C.; Shen, Y.; Machens, H.G.; Schilling, A.F. Nano-formulated curcumin accelerates acute wound healing through Dkk-1-mediated fibroblast mobilization and MCP-1-mediated anti-inflammation. *NPG Asia Mater.* **2017**, *9*, e368. [CrossRef]
37. Ullah, A.; Ullah, S.; Khan, M.Q.; Hashmi, M.; Nam, P.D.; Kato, Y.; Tamada, Y.; Kim, I.S. Manuka honey incorporated cellulose acetate nanofibrous mats: Fabrication and in vitro evaluation as a potential wound dressing. *Int. J. Biol. Macromol.* **2020**, *155*, 479–489. [CrossRef]
38. Miao, J.; Pangule, R.C.; Paskaleva, E.E.; Hwang, E.E.; Kane, R.S.; Linhardt, R.J.; Dordick, J.S. Lysostaphin-functionalized cellulose fibers with antistaphylococcal activity for wound healing applications. *Biomaterials* **2011**, *32*, 9557–9567. [CrossRef]
39. Mutlu, G.; Calamak, S.; Ulubayram, K.; Guven, E. Curcumin-loaded electrospun PHBV nanofibers as potential wounddressing material. *J. Drug Deliv. Sci. Technol.* **2018**, *43*, 185–193. [CrossRef]
40. Ghorbani, S.; Eyni, H.; Tiraihi, T.; Asl, L.S.; Soleimani, M.; Atashi, A.; Beiranvand, S.P.; Warkiani, M.E. Combined effects of 3D bone marrow stem cell-seeded wet-electrospun poly lactic acid scaffolds on full-thickness skin wound healing. *Int. J. Polym. Mater. Polym. Biomater.* **2018**, *67*, 905–912. [CrossRef]
41. Malgarim Cordenonsi, L.; Faccendini, A.; Rossi, S.; Bonferoni, M.C.; Malavasi, L.; Raffin, R.; Scherman Schapoval, E.E.; Del Fante, C.; Vigani, B.; Miele, D.; et al. Platelet lysate loaded electrospun scaffolds: Effect of nanofiber types on wound healing. *Eur. J. Pharm. Biopharm.* **2019**, *142*, 247–257. [CrossRef]
42. Cui, S.; Sun, X.; Li, K.; Gou, D.; Zhou, Y.; Hu, J.; Liu, Y. Polylactide nanofibers delivering doxycycline for chronic wound treatment. *Mater. Sci. Eng. C Mater. Biol. Appl.* **2019**, *104*, 109745. [CrossRef] [PubMed]
43. Rezvani Ghomi, E.; Khalili, S.; Nouri Khorasani, S.; Esmaeely Neisiany, R.; Ramakrishna, S. Wound dressings: Current advances and future directions. *J. Appl. Polym. Sci.* **2019**, *136*, 47738. [CrossRef]
44. Dabiri, G.; Damstetter, E.; Phillips, T. Choosing a Wound Dressing Based on Common Wound Characteristics. *Adv. Wound Care* **2016**, *5*, 32–41. [CrossRef] [PubMed]
45. Tao, G.; Cai, R.; Wang, Y.; Liu, L.; Zuo, H.; Zhao, P.; Umar, A.; Mao, C.; Xia, Q.; He, H. Bioinspired design of AgNPs embedded silk sericin-based sponges for efficiently combating bacteria and promoting wound healing. *Mater. Des.* **2019**, *180*, 107940. [CrossRef]
46. Fahimirad, S.; Ajalloueian, F. Naturally-derived electrospun wound dressings for target delivery of bio-active agents. *Int. J. Pharm.* **2019**, *566*, 307–328. [CrossRef] [PubMed]
47. Maleki, H.; Barani, H. Morphological and mechanical properties of drawn poly(l-lactide) electrospun twisted yarns. *Polym. Eng. Sci.* **2018**, *58*, 1091–1096. [CrossRef]
48. Azimi, B.; Milazzo, M.; Lazzeri, A.; Berrettini, S.; Uddin, M.J.; Qin, Z.; Buehler, M.J.; Danti, S. Electrospinning Piezoelectric Fibers for Biocompatible Devices. *Adv. Healthc. Mater.* **2019**, *9*, 1901287. [CrossRef]
49. Günday, C.; Anand, S.; Gencer, H.B.; Munafò, S.; Moroni, L.; Fusco, A.; Donnarumma, G.; Ricci, C.; Hatir, P.C.; Türeli, N.G.; et al. Ciprofloxacin-loaded polymeric nanoparticles incorporated electrospun fibers for drug delivery in tissue engineering applications. *Drug Deliv. Transl. Res.* **2020**, *10*, 706–720. [CrossRef]
50. Dart, A.; Bhave, M.; Kingshott, P. Antimicrobial Peptide-Based Electrospun Fibers for Wound Healing Applications. *Macromol. Biosci.* **2019**, *19*, e1800488. [CrossRef]
51. Yang, X.; Li, L.; Yang, D.; Nie, J.; Ma, G. Electrospun Core–Shell Fibrous 2D Scaffold with Biocompatible Poly (Glycerol Sebacate) and Poly-L-Lactic Acid for Wound Healing. *Adv. Fiber Mater.* **2020**, *2*, 105–117. [CrossRef]
52. .Wang, J.; Windbergs, M. Functional electrospun fibers for the treatment of human skin wounds. *Eur. J. Pharm. Biopharm.* **2017**, *119*, 283–299. [CrossRef] [PubMed]
53. Huang, W.; Zou, T.; Li, S.; Jing, J.; Xia, X.; Liu, X. Drug-Loaded Zein Nanofibers Prepared Using a Modified Coaxial Electrospinning Process. *AAPS PharmSciTech* **2013**, *14*, 675–681. [CrossRef] [PubMed]
54. Yao, Z.-C.; Zhang, C.; Ahmad, Z.; Huang, J.; Li, J.-S.; Chang, M.-W. Designer fibers from 2D to 3D–Simultaneous and controlled engineering of morphology, shape and size. *Chem. Eng. J.* **2018**, *334*, 89–98. [CrossRef]

55. Ura, D.P.; Karbowniczek, J.E.; Szewczyk, P.K.; Metwally, S.; Kopyscianski, M.; Stachewicz, U. Cell Integration with Electrospun PMMA Nanofibers, Microfibers, Ribbons, and Films: A Microscopy Study. *Bioengineering* **2019**, *6*, 41. [CrossRef] [PubMed]
56. Wang, C.; Wang, J.; Zeng, L.; Qiao, Z.; Liu, X.; Liu, H.; Zhang, J.; Ding, J. Fabrication of Electrospun Polymer Nanofibers with Diverse Morphologies. *Molecules* **2019**, *24*, 834. [CrossRef]
57. Maleki, H.; Semnani Rahbar, R.; Saadatmand, M.M.; Barani, H. Physical and morphological characterisation of poly(L-lactide) acid-based electrospun fibrous structures: Tunning solution properties. *Plast. Rubber Compos.* **2018**, *47*, 438–446. [CrossRef]
58. Maleki, H.; Gharehaghaji, A.A.; Toliyat, T.; Dijkstra, P.J. Drug release behavior of electrospun twisted yarns as implantable medical devices. *Biofabrication* **2016**, *8*, 35019. [CrossRef]
59. Barani, H.; Khorashadizadeh, M.; Haseloer, A.; Klein, A. Characterization and Release Behavior of a Thiosemicarbazone from Electrospun Polyvinyl Alcohol Core-Shell Nanofibers. *Polymers* **2020**, *12*, 1488. [CrossRef]
60. Gaharwar, A.K.; Mihaila, S.M.; Kulkarni, A.A.; Patel, A.; Di Luca, A.; Reis, R.L.; Gomes, M.E.; van Blitterswijk, C.; Moroni, L.; Khademhosseini, A. Amphiphilic beads as depots for sustained drug release integrated into fibrillar scaffolds. *J. Control. Release* **2014**, *187*, 66–73. [CrossRef]
61. Li, T.X.; Ding, X.; Sui, X.; Tian, L.L.; Zhang, Y.; Hu, J.Y.; Yang, X.D. Sustained Release of Protein Particle Encapsulated in Bead-on-String Electrospun Nanofibers. *J. Macromol. Sci. Part B* **2015**, *54*, 887–896. [CrossRef]
62. Li, T.; Ding, X.; Tian, L.; Hu, J.; Yang, X.; Ramakrishna, S. The control of beads diameter of bead-on-string electrospun nanofibers and the corresponding release behaviors of embedded drugs. *Mater. Sci. Eng. C Mater. Biol. Appl.* **2017**, *74*, 471–477. [CrossRef] [PubMed]
63. Maleki, H.; Mathur, S.; Klein, A. Antibacterial Ag containing core-shell polyvinyl alcohol-poly (lactic acid) nanofibers for biomedical applications. *Polym. Eng. Sci.* **2020**, *60*, 1221–1230. [CrossRef]
64. Hadisi, Z.; Farokhi, M.; Bakhsheshi-Rad, H.R.; Jahanshahi, M.; Hasanpour, S.; Pagan, E.; Dolatshahi-Pirouz, A.; Zhang, Y.S.; Kundu, S.C.; Akbari, M. Hyaluronic Acid (HA)-Based Silk Fibroin/Zinc Oxide Core–Shell Electrospun Dressing for Burn Wound Management. *Macromol. Biosci.* **2020**, *20*, 1900328. [CrossRef] [PubMed]
65. Maleki, H.; Gharehaghaji, A.A.; Criscenti, G.; Moroni, L.; Dijkstra, P.J. The influence of process parameters on the properties of electrospun PLLA yarns studied by the response surface methodology. *J. Appl. Polym. Sci.* **2015**, *132*. [CrossRef]
66. Maleki, H.; Gharehaghaji, A.A.; Moroni, L.; Dijkstra, P.J. Influence of the solvent type on the morphology and mechanical properties of electrospun PLLA yarns. *Biofabrication* **2013**, *5*, 035014. [CrossRef]
67. Danti, S.; D'Alessandro, D.; Mota, C.; Bruschini, L.; Berrettini, S. Applications of bioresorbable polymers in skin and eardrum. In *Bioresorbable Polymers for Biomedical Applications*; Elsevier: Amsterdam, The Netherlands, 2017; pp. 423–444.
68. Feng, X.; Li, J.; Zhang, X.; Liu, T.; Ding, J.; Chen, X. Electrospun polymer micro/nanofibers as pharmaceutical repositories for healthcare. *J. Control. Release* **2019**, *302*, 19–41. [CrossRef]
69. Alves, P.E.; Soares, B.G.; Lins, L.C.; Livi, S.; Santos, E.P. Controlled delivery of dexamethasone and betamethasone from PLA electrospun fibers: A comparative study. *Eur. Polym. J.* **2019**, *117*, 1–9. [CrossRef]
70. Li, J.; Hu, Y.; He, T.; Huang, M.; Zhang, X.; Yuan, J.; Wei, Y.; Dong, X.; Liu, W.; Ko, F.; et al. Electrospun Sandwich-Structure Composite Membranes for Wound Dressing Scaffolds with High Antioxidant and Antibacterial Activity. *Macromol. Mater. Eng.* **2017**, *303*, 1700270. [CrossRef]
71. Pour Khalili, N.; Moradi, R.; Kavehpour, P.; Islamzada, F. Boron nitride nanotube clusters and their hybrid nanofibers with polycaprolacton: Thermo-pH sensitive drug delivery functional materials. *Eur. Polym. J.* **2020**, *127*, 109585. [CrossRef]
72. Gačanin, J.; Hedrich, J.; Sieste, S.; Glaßer, G.; Lieberwirth, I.; Schilling, C.; Fischer, S.; Barth, H.; Knöll, B.; Synatschke, C.V. Autonomous Ultrafast Self-Healing Hydrogels by pH-Responsive Functional Nanofiber Gelators as Cell Matrices. *Adv. Mater.* **2019**, *31*, 1805044. [CrossRef] [PubMed]
73. He, H.; Cheng, M.; Liang, Y.; Zhu, H.; Sun, Y.; Dong, D.; Wang, S. Intelligent Cellulose Nanofibers with Excellent Biocompatibility Enable Sustained Antibacterial and Drug Release via a pH-Responsive Mechanism. *J. Agric. Food Chem.* **2020**, *68*, 3518–3527. [CrossRef]

74. Mamidi, N.; Zuníga, A.E.; Villela-Castrejón, J. Engineering and evaluation of forcespun functionalized carbon nano-onions reinforced poly (ε-caprolactone) composite nanofibers for pH-responsive drug release. *Mater. Sci. Eng. C* **2020**, *112*, 110928. [CrossRef] [PubMed]
75. Abdalkarim, S.Y.H.; Yu, H.; Wang, C.; Chen, Y.; Zou, Z.; Han, L.; Yao, J.; Tam, K.C. Thermo and light-responsive phase change nanofibers with high energy storage efficiency for energy storage and thermally regulated on–off drug release devices. *Chem. Eng. J.* **2019**, *375*, 121979. [CrossRef]
76. Yang, X.; Li, W.; Sun, Z.; Yang, C.; Tang, D. Electrospun P (NVCL-co-MAA) nanofibers and their pH/temperature dual-response drug release profiles. *Colloid Polym. Sci.* **2020**, *298*, 629–636. [CrossRef]
77. Hoque, J.; Sangaj, N.; Varghese, S. Stimuli-Responsive Supramolecular Hydrogels and Their Applications in Regenerative Medicine. *Macromol. Biosci.* **2019**, *19*, 1800259. [CrossRef]
78. Jang, H.H.; Park, S.B.; Hong, J.S.; Lee, H.L.; Song, Y.H.; Kim, J.; Jung, Y.H.; Kim, C.; Kim, D.-M.; Lee, S.E.; et al. Piperlongumine-Eluting Gastrointestinal Stent Using Reactive Oxygen Species-Sensitive Nanofiber Mats for Inhibition of Cholangiocarcinoma Cells. *Nanoscale Res. Lett.* **2019**, *14*, 58. [CrossRef]
79. Gao, S.; Tang, G.; Hua, D.; Xiong, R.; Han, J.; Jiang, S.; Zhang, Q.; Huang, C. Stimuli-responsive bio-based polymeric systems and their applications. *J. Mater. Chem. B* **2019**, *7*, 709–729. [CrossRef]
80. Liu, M.; Duan, X.P.; Li, Y.M.; Yang, D.P.; Long, Y.Z. Electrospun nanofibers for wound healing. *Mater. Sci. Eng. C Mater. Biol. Appl.* **2017**, *76*, 1413–1423. [CrossRef]
81. Nagarajan, S.; Bechelany, M.; Kalkura, N.S.; Miele, P.; Bohatier, C.P.; Balme, S. Chapter 20—Electrospun Nanofibers for Drug Delivery in Regenerative Medicine. In *Applications of Targeted Nano Drugs and Delivery Systems*; Mohapatra, S.S., Ed.; Elsevier: Amsterdam, The Netherlands, 2019; pp. 595–625.
82. Leung, C.M.; Dhand, C.; Mayandi, V.; Ramalingam, R.; Lim, F.P.; Barathi, V.A.; Dwivedi, N.; Orive, G.; Beuerman, R.W.; Ramakrishna, S.; et al. Wound healing properties of magnesium mineralized antimicrobial nanofibre dressings containing chondroitin sulphate—A comparison between blend and core–shell nanofibres. *Biomater. Sci.* **2020**, *8*, 3454–3471. [CrossRef]
83. Li, J.; Xu, W.; Chen, J.; Li, D.; Zhang, K.; Liu, T.; Ding, J.; Chen, X. Highly Bioadhesive Polymer Membrane Continuously Releases Cytostatic and Anti-Inflammatory Drugs for Peritoneal Adhesion Prevention. *ACS Biomater. Sci. Eng.* **2018**, *4*, 2026–2036. [CrossRef]
84. Sofi, H.S.; Ashraf, R.; Khan, A.H.; Beigh, M.A.; Majeed, S.; Sheikh, F.A. Reconstructing nanofibers from natural polymers using surface functionalization approaches for applications in tissue engineering, drug delivery and biosensing devices. *Mater. Sci. Eng. C* **2019**, *94*, 1102–1124. [CrossRef] [PubMed]
85. Miroshnichenko, S.; Timofeeva, V.; Permyakova, E.; Ershov, S.; Kiryukhantsev-Korneev, P.; Dvořaková, E.; Shtansky, D.V.; Zajíčková, L.; Solovieva, A.; Manakhov, A. Plasma-coated polycaprolactone nanofibers with covalently bonded platelet-rich plasma enhance adhesion and growth of human fibroblasts. *Nanomaterials* **2019**, *9*, 637. [CrossRef] [PubMed]
86. Maleki, H.; Gharehaghaji, A.A.; Dijkstra, P.J. A novel honey-based nanofibrous scaffold for wound dressing application. *J. Appl. Polym. Sci.* **2013**, *127*, 4086–4092. [CrossRef]
87. Alippilakkotte, S.; Kumar, S.; Sreejith, L. Fabrication of PLA/Ag nanofibers by green synthesis method using Momordica charantia fruit extract for wound dressing applications. *Colloids Surf. A Physicochem. Eng. Asp.* **2017**, *529*, 771–782. [CrossRef]
88. Milazzo, M.; Gallone, G.; Marcello, E.; Mariniello, M.D.; Bruschini, L.; Roy, I.; Danti, S. Biodegradable polymeric micro/nano-structures with intrinsic antifouling/antimicrobial properties: Relevance in damaged skin and other biomedical applications. *J. Funct. Biomater.* **2020**, *11*, 60. [CrossRef] [PubMed]
89. Zhao, J.; Han, F.; Zhang, W.; Yang, Y.; You, D.; Li, L. Toward improved wound dressings: Effects of polydopamine-decorated poly (lactic-co-glycolic acid) electrospinning incorporating basic fibroblast growth factor and ponericin G1. *RSC Adv.* **2019**, *9*, 33038–33051. [CrossRef]
90. Dwivedi, C.; Pandey, H.; Pandey, A.C.; Patil, S.; Ramteke, P.W.; Laux, P.; Luch, A.; Singh, A.V. In vivo biocompatibility of electrospun biodegradable dual carrier (antibiotic+ growth factor) in a mouse model—Implications for rapid wound healing. *Pharmaceutics* **2019**, *11*, 180. [CrossRef] [PubMed]
91. Seonwoo, H.; Shin, B.; Jang, K.-J.; Lee, M.; Choo, O.-S.; Park, S.-B.; Kim, Y.C.; Choi, M.-J.; Kim, J.; Garg, P.; et al. Epidermal Growth Factor–Releasing Radially Aligned Electrospun Nanofibrous Patches for the Regeneration of Chronic Tympanic Membrane Perforations. *Adv. Healthc. Mater.* **2019**, *8*, 1801160. [CrossRef]

92. Balakrishnan, S.B.; Thambusamy, S. Preparation of silver nanoparticles and riboflavin embedded electrospun polymer nanofibrous scaffolds for in vivo wound dressing application. *Process Biochem.* **2020**, *88*, 148–158. [CrossRef]
93. Luef, K.P.; Stelzer, F.; Wiesbrock, F. Poly (hydroxy alkanoate) s in medical applications. *Chem. Biochem. Eng. Q.* **2015**, *29*, 287–297. [CrossRef] [PubMed]
94. Kakoria, A.; Sinha-Ray, S. A review on biopolymer-based fibers via electrospinning and solution blowing and their applications. *Fibers* **2018**, *6*, 45. [CrossRef]
95. Hamdan, S.; Pastar, I.; Drakulich, S.; Dikici, E.; Tomic-Canic, M.; Deo, S.; Daunert, S. Nanotechnology-driven therapeutic interventions in wound healing: Potential uses and applications. *ACS Cent. Sci.* **2017**, *3*, 163–175. [CrossRef] [PubMed]
96. Thakur, V.K.; Thakur, M.K.; Kessler, M.R. *Handbook of Composites from Renewable Materials, Biodegradable Materials*; John Wiley & Sons: Hoboken, NJ, USA, 2017; Volume 5.
97. Lin, N.; Dufresne, A. Nanocellulose in biomedicine: Current status and future prospect. *Eur. Polym. J.* **2014**, *59*, 302–325. [CrossRef]
98. Prasanth, R.; Nageswaran, S.; Thakur, V.K.; Ahn, J.-H. Electrospinning of cellulose: Process and applications. In *Nanocellulose Polymer Nanocomposites: Fundamentals and Applications*; Thakur, V.K., Ed.; John Wiley & Sons, Inc.: Hoboken, NJ, USA, 2014; pp. 311–340.
99. Kucińska-Lipka, J.; Gubanska, I.; Janik, H. Bacterial cellulose in the field of wound healing and regenerative medicine of skin: Recent trends and future prospectives. *Polym. Bull.* **2015**, *72*, 2399–2419. [CrossRef]
100. Hakkarainen, T.; Koivuniemi, R.; Kosonen, M.; Escobedo-Lucea, C.; Sanz-Garcia, A.; Vuola, J.; Valtonen, J.; Tammela, P.; Mäkitie, A.; Luukko, K. Nanofibrillar cellulose wound dressing in skin graft donor site treatment. *J. Control. Release* **2016**, *244*, 292–301. [CrossRef]
101. Bodin, A.; Concaro, S.; Brittberg, M.; Gatenholm, P. Bacterial cellulose as a potential meniscus implant. *J. Tissue Eng. Regen. Med.* **2007**, *1*, 406–408. [CrossRef]
102. Czaja, W.; Krystynowicz, A.; Bielecki, S.; Brown, R.M., Jr. Microbial cellulose—The natural power to heal wounds. *Biomaterials* **2006**, *27*, 145–151. [CrossRef]
103. Muangman, P.; Opasanon, S.; Suwanchot, S.; Thangthed, O. Efficiency of microbial cellulose dressing in partial-thickness burn wounds. *J. Am. Coll. Certif. Wound Spec.* **2011**, *3*, 16–19.
104. Sahana, T.; Rekha, P. Biopolymers: Applications in wound healing and skin tissue engineering. *Mol. Biol. Rep.* **2018**, *45*, 2857–2867. [CrossRef]
105. Vatankhah, E.; Prabhakaran, M.P.; Jin, G.; Mobarakeh, L.G.; Ramakrishna, S. Development of nanofibrous cellulose acetate/gelatin skin substitutes for variety wound treatment applications. *J. Biomater. Appl.* **2014**, *28*, 909–921. [CrossRef] [PubMed]
106. Frenot, A.; Henriksson, M.W.; Walkenström, P. Electrospinning of cellulose-based nanofibers. *J. Appl. Polym. Sci.* **2007**, *103*, 1473–1482. [CrossRef]
107. He, X.; Cheng, L.; Zhang, X.; Xiao, Q.; Zhang, W.; Lu, C. Tissue engineering scaffolds electrospun from cotton cellulose. *Carbohydr. Polym.* **2015**, *115*, 485–493. [CrossRef] [PubMed]
108. Lee, K.Y.; Jeong, L.; Kang, Y.O.; Lee, S.J.; Park, W.H. Electrospinning of polysaccharides for regenerative medicine. *Adv. Drug Deliv. Rev.* **2009**, *61*, 1020–1032. [CrossRef]
109. Frey, M.W. Electrospinning cellulose and cellulose derivatives. *Polym. Rev.* **2008**, *48*, 378–391. [CrossRef]
110. Liu, X.; Lin, T.; Gao, Y.; Xu, Z.; Huang, C.; Yao, G.; Jiang, L.; Tang, Y.; Wang, X. Antimicrobial electrospun nanofibers of cellulose acetate and polyester urethane composite for wound dressing. *J. Biomed. Mater. Res. Part B Appl. Biomater.* **2012**, *100*, 1556–1565. [CrossRef]
111. Park, T.-J.; Jung, Y.J.; Choi, S.-W.; Park, H.; Kim, H.; Kim, E.; Lee, S.H.; Kim, J.H. Native chitosan/cellulose composite fibers from an ionic liquid via electrospinning. *Macromol. Res.* **2011**, *19*, 213–215. [CrossRef]
112. Rao, S.S.; Jeyapal, S.G.; Rajiv, S. Biodegradable electrospun nanocomposite fibers based on Poly (2-hydroxy ethyl methacrylate) and bamboo cellulose. *Compos. Part B Eng.* **2014**, *60*, 43–48. [CrossRef]
113. Anitha, S.; Brabu, B.; Thiruvadigal, D.J.; Gopalakrishnan, C.; Natarajan, T. Optical, bactericidal and water repellent properties of electrospun nano-composite membranes of cellulose acetate and ZnO. *Carbohydr. Polym.* **2012**, *87*, 1065–1072. [CrossRef]
114. Song, J.; Birbach, N.L.; Hinestroza, J.P. Deposition of silver nanoparticles on cellulosic fibers via stabilization of carboxymethyl groups. *Cellulose* **2012**, *19*, 411–424. [CrossRef]

115. Min, B.-M.; Lee, S.W.; Lim, J.N.; You, Y.; Lee, T.S.; Kang, P.H.; Park, W.H. Chitin and chitosan nanofibers: Electrospinning of chitin and deacetylation of chitin nanofibers. *Polymer* **2004**, *45*, 7137–7142. [CrossRef]
116. Jayakumar, R.; Prabaharan, M.; Kumar, P.S.; Nair, S.; Tamura, H. Biomaterials based on chitin and chitosan in wound dressing applications. *Biotechnol. Adv.* **2011**, *29*, 322–337. [CrossRef] [PubMed]
117. Teng, S.H.; Wang, P.; Kim, H.J. Blend fibers of chitosan–agarose by electrospinning. *Mater. Lett.* **2009**, *63*, 2510–2512.
118. Naseri, N.; Algan, C.; Jacobs, V.; John, M.; Oksman, K.; Mathew, A.P. Electrospun chitosan-based nanocomposite mats reinforced with chitin nanocrystals for wound dressing. *Carbohydr. Polym.* **2014**, *109*, 7–15. [CrossRef] [PubMed]
119. Yousefi, I.; Pakravan, M.; Rahimi, H.; Bahador, A.; Farshadzadeh, Z.; Haririan, I. An investigation of electrospun Henna leaves extract-loaded chitosan based nanofibrous mats for skin tissue engineering. *Mater. Sci. Eng. C* **2017**, *75*, 433–444. [CrossRef] [PubMed]
120. Zhao, R.; Li, X.; Sun, B.; Zhang, Y.; Zhang, D.; Tang, Z.; Chen, X.; Wang, C. Electrospun chitosan/sericin composite nanofibers with antibacterial property as potential wound dressings. *Int. J. Biol. Macromol.* **2014**, *68*, 92–97. [CrossRef]
121. Cai, Z.-X.; Mo, X.-M.; Zhang, K.-H.; Fan, L.-P.; Yin, A.-L.; He, C.-L.; Wang, H.-S. Fabrication of chitosan/silk fibroin composite nanofibers for wound-dressing applications. *Int. J. Mol. Sci.* **2010**, *11*, 3529–3539. [CrossRef]
122. Ardila, N.; Medina, N.; Arkoun, M.; Heuzey, M.-C.; Ajji, A.; Panchal, C.J. Chitosan–bacterial nanocellulose nanofibrous structures for potential wound dressing applications. *Cellulose* **2016**, *23*, 3089–3104. [CrossRef]
123. Datta, S.; Rameshbabu, A.P.; Bankoti, K.; Maity, P.P.; Das, D.; Pal, S.; Roy, S.; Sen, R.; Dhara, S. Oleoyl-chitosan-based nanofiber mats impregnated with amniotic membrane derived stem cells for accelerated full-thickness excisional wound healing. *ACS Biomater. Sci. Eng.* **2017**, *3*, 1738–1749. [CrossRef]
124. Kalantari, K.; Afifi, A.M.; Jahangirian, H.; Webster, T.J. Biomedical applications of chitosan electrospun nanofibers as a green polymer–Review. *Carbohydr. Polym.* **2019**, *207*, 588–600. [CrossRef]
125. Patel, S.K.S.; Sandeep, K.; Singh, M.; Singh, G.P.; Lee, J.-K.; Bhatia, S.K.; Kalia, V.C.; de Souza, L.; Shivakumar, S. Biotechnological Application of Polyhydroxyalkanoates and Their Composites as Anti-microbials Agents. In *Biotechnological Applications of Polyhydroxyalkanoates*; Kalia, V.C., Ed.; Springer: Singapore, 2019; pp. 227–290.
126. Osswald, T.A.; Garcia-Rodriguez, S.; Teramoto, N.; Rai, R.; Roy, I.; Visakh, P.; Thomas, S.; Pothan, L.A.; Fukushima, K.; Tabuani, D. *A Handbook of Applied Biopolymer Technology: Synthesis, Degradation and Applications*; Royal Society of Chemistry: London, UK, 2011.
127. Wu, Q.; Wang, Y.; Chen, G.Q. Medical application of microbial biopolyesters polyhydroxyalkanoates. *Artif. Cells Blood Substit. Immobil. Biotechnol.* **2009**, *37*, 1–12. [CrossRef] [PubMed]
128. Junyu, Z.; Ekaterina, I.S.; Tatiana, G.V.; Luiziana, F.D.S.; Guo-Qiang, C. Polyhydroxyalkanoates (PHA) for therapeutic applications. *Mater. Sci. Eng.* **2018**, *86*, 144–150.
129. Chen, G.-Q.; Zhang, J. Microbial polyhydroxyalkanoates as medical implant biomaterials. *Artif. Cells Nanomed. Biotechnol.* **2018**, *46*, 1–18. [CrossRef] [PubMed]
130. Cheng, S.; Wu, Q.; Yang, F.; Xu, M.; Leski, M.; Chen, G.-Q. Influence of dl-β-Hydroxybutyric Acid on Cell Proliferation and Calcium Influx. *Biomacromolecules* **2005**, *6*, 593–597. [CrossRef]
131. Yang, X.-D.; Zou, X.-H.; Dai, Z.-W.; Luo, R.-C.; Wei, C.-J.; Chen, G.-Q. Effects of Oligo(3-hydroxyalkanoates) on the Viability and Insulin Secretion of Murine Beta Cells. *J. Biomater. Sci. Polym. Ed.* **2009**, *20*, 1729–1746. [CrossRef] [PubMed]
132. Sudesh, K.; Abe, H.; Doi, Y. Synthesis, structure and properties of polyhydroxyalkanoates: Biological polyesters. *Prog. Polym. Sci.* **2000**, *25*, 1503–1555. [CrossRef]
133. Sanhueza, C.; Acevedo, F.; Rocha, S.; Villegas, P.; Seeger, M.; Navia, R. Polyhydroxyalkanoates as biomaterial for electrospun scaffolds. *Int. J. Biol. Macromol.* **2019**, *124*, 102–110. [CrossRef]
134. Volova, T.; Goncharov, D.; Sukovatyi, A.; Shabanov, A.; Nikolaeva, E.; Shishatskaya, E. Electrospinning of polyhydroxyalkanoate fibrous scaffolds: Effects on electrospinning parameters on structure and properties. *J. Biomater. Sci. Polym. Ed.* **2014**, *25*, 370–393. [CrossRef]
135. Li, W.; Cicek, N.; Levin, D.B.; Liu, S. Enabling electrospinning of medium-chain length polyhydroxyalkanoates (PHAs) by blending with short-chain length PHAs. *Int. J. Polym. Mater. Polym. Biomater.* **2019**, *68*, 499–509. [CrossRef]
136. Zhu, M.; Zuo, W.; Yu, H.; Yang, W.; Chen, Y. Superhydrophobic surface directly created by electrospinning based on hydrophilic material. *J. Mater. Sci.* **2006**, *41*, 3793–3797. [CrossRef]

137. Shishatskaya, E.I.; Nikolaeva, E.D.; Vinogradova, O.N.; Volova, T.G. Experimental wound dressings of degradable PHA for skin defect repair. *J. Mater. Sci. Mater. Med.* **2016**, *27*, 165. [CrossRef] [PubMed]
138. Suwantong, O.; Waleetorncheepsawat, S.; Sanchavanakit, N.; Pavasant, P.; Cheepsunthorn, P.; Bunaprasert, T.; Supaphol, P. In vitro biocompatibility of electrospun poly(3-hydroxybutyrate) and poly(3-hydroxybutyrate-co-3-hydroxyvalerate) fiber mats. *Int. J. Biol. Macromol.* **2007**, *40*, 217–223. [CrossRef] [PubMed]
139. Han, I.; Shim, K.J.; Kim, J.Y.; Im, S.U.; Sung, Y.K.; Kim, M.; Kang, I.-K.; Kim, J.C. Effect of Poly(3-hydroxybutyrate-co-3-hydroxyvalerate) Nanofiber Matrices Cocultured With Hair Follicular Epithelial and Dermal Cells for Biological Wound Dressing. *Artif. Organs* **2007**, *31*, 801–808. [CrossRef] [PubMed]
140. Kuppan, P.; Vasanthan, K.S.; Sundaramurthi, D.; Krishnan, U.M.; Sethuraman, S. Development of Poly(3-hydroxybutyrate-co-3-hydroxyvalerate) Fibers for Skin Tissue Engineering: Effects of Topography, Mechanical, and Chemical Stimuli. *Biomacromolecules* **2011**, *12*, 3156–3165. [CrossRef]
141. Abdul, M.; Kasturi, M.; Sivakumar, M.; Kumar, S.; Syed, S.; Rahman, S.; Noor, A.; Nanthini, S. Fabrication and Characterization of an Electrospun PHA/Graphene Silver Nanocomposite Scaffold for Antibacterial Applications. *Materials* **2018**, *11*, 15.
142. Yuan, J.; Geng, J.; Xing, Z.; Shim, K.J.; Han, I.; Kim, J.C.; Kang, I.K.; Shen, J. Novel wound dressing based on nanofibrous PHBV–keratin mats. *J. Tissue Eng. Regen. Med.* **2015**, *9*, 1027–1035. [CrossRef]
143. Salvatore, L.; Carofiglio, V.E.; Stufano, P.; Bonfrate, V.; Calò, E.; Scarlino, S.; Nitti, P.; Centrone, D.; Cascione, M.; Leporatti, S. Potential of electrospun poly (3-hydroxybutyrate)/collagen blends for tissue engineering applications. *J. Healthc. Eng.* **2018**, *2018*, 6573947. [CrossRef]
144. Meng, W.; Kim, S.-Y.; Yuan, J.; Kim, J.C.; Kwon, O.H.; Kawazoe, N.; Chen, G.; Ito, Y.; Kang, I.-K. Electrospun PHBV/collagen composite nanofibrous scaffolds for tissue engineering. *J. Biomater. Sci. Polym. Ed.* **2007**, *18*, 81–94. [CrossRef]
145. Kandhasamy, S.; Perumal, S.; Madhan, B.; Umamaheswari, N.; Banday, J.A.; Perumal, P.T.; Santhanakrishnan, V.P. Synthesis and Fabrication of Collagen-Coated Ostholamide Electrospun Nanofiber Scaffold for Wound Healing. *ACS Appl. Mater. Interfaces* **2017**, *9*, 8556–8568. [CrossRef]
146. Azimi, B.; Thomas, L.; Fusco, A.; Kalaoglu Altan, O.; Basnett, P.; Cinelli, P.; De Clerck, K.; Roy, I.; Donnarumma, G.; Coltelli, M.B.; et al. Electrosprayed Chitin Nanofibril/Electrospun Polyhydroxyalkanoate Fiber Mesh as Functional Nonwoven for Skin Applications. *J. Funct. Biomater* **2020**, *11*, 62. [CrossRef]
147. Danti, S.; Trombi, L.; Fusco, A.; Azimi, B.; Lazzeri, A.; Morganti, P.; Coltelli, M.-B.; Donnarumma, G. Chitin nanofibrils and nanolignin as functional agents in skin regeneration. *Int. J. Mol. Sci.* **2019**, *20*, 2669. [CrossRef] [PubMed]
148. Ditaranto, N.; Basoli, F.; Trombetta, M.; Cioffi, N.; Rainer, A. Electrospun nanomaterials implementing antibacterial inorganic nanophases. *Appl. Sci.* **2018**, *8*, 1643. [CrossRef]
149. Melendez-Rodriguez, B.; Figueroa-Lopez, K.J.; Bernardos, A.; Martínez-Máñez, R.; Cabedo, L.; Torres-Giner, S.M.; Lagaron, J. Electrospun antimicrobial films of poly (3-hydroxybutyrate-co-3-hydroxyvalerate) containing eugenol essential oil encapsulated in mesoporous silica nanoparticles. *Nanomaterials* **2019**, *9*, 227. [CrossRef] [PubMed]
150. Azizi, S.; Ahmad, M.; Mahdavi, M.; Abdolmohammadi, S. Preparation, characterization, and antimicrobial activities of ZnO nanoparticles/cellulose nanocrystal nanocomposites. *BioResources* **2013**, *8*, 1841–1851. [CrossRef]
151. Gold, K.; Slay, B.; Knackstedt, M.; Gaharwar, A.K. Antimicrobial Activity of Metal and Metal-Oxide Based Nanoparticles. *Adv. Ther.* **2018**, *1*, 1700033. [CrossRef]
152. Figueroa-Lopez, K.J.; Torres-Giner, S.; Enescu, D.; Cabedo, L.; Cerqueira, M.A.; Pastrana, L.M.; Lagaron, J.M. Electrospun active biopapers of food waste derived poly (3-hydroxybutyrate-co-3-hydroxyvalerate) with short-term and long-term antimicrobial performance. *Nanomaterials* **2020**, *10*, 506. [CrossRef]
153. Khezerlou, A.; Alizadeh-Sani, M.; Azizi-Lalabadi, M.; Ehsani, A. Nanoparticles and their antimicrobial properties against pathogens including bacteria, fungi, parasites and viruses. *Microb. Pathog.* **2018**, *123*, 505–526. [CrossRef]
154. Abdalkarim, S.Y.H.; Yu, H.-Y.; Wang, D.; Yao, J. Electrospun poly(3-hydroxybutyrate-co-3-hydroxyvalerate)/cellulose reinforced nanofibrous membranes with ZnO nanocrystals for antibacterial wound dressings. *Cellulose* **2017**, *24*, 2925–2938. [CrossRef]
155. Siddiqi, K.S.; ur Rahman, A.; Husen, A. Properties of Zinc Oxide Nanoparticles and Their Activity Against Microbes. *Nanoscale Res. Lett.* **2018**, *13*, 141. [CrossRef]

156. Sen, C.K.; Khanna, S.; Gordillo, G.; Bagchi, D.; Bagchi, M.; Roy, S. Oxygen, oxidants, and antioxidants in wound healing: An emerging paradigm. *Ann. NY Acad. Sci.* **2002**, *957*, 239–249. [CrossRef]
157. Baynes, J.W. Role of oxidative stress in development of complications in diabetes. *Diabetes* **1991**, *40*, 405–412. [CrossRef] [PubMed]
158. Augustine, R.; Hasan, A.; Patan, N.K.; Dalvi, Y.B.; Varghese, R.; Antony, A.; Unni, R.N.; Sandhyarani, N.; Moustafa, A.E.A. Cerium Oxide Nanoparticle Incorporated Electrospun Poly(3-hydroxybutyrate-co-3-hydroxyvalerate) Membranes for Diabetic Wound Healing Applications. *ACS Biomater.* **2020**, *6*, 58–70. [CrossRef]
159. Cristallini, C.; Danti, S.; Azimi, B.; Tempesti, V.; Ricci, C.; Ventrelli, L.; Cinelli, P.; Barbani, N.; Lazzeri, A. Multifunctional Coatings for Robotic Implanted Device. *Int. J. Mol. Sci.* **2019**, *20*, 5126. [CrossRef] [PubMed]
160. Li, W.; Cicek, N.; Levin, D.B.; Logsetty, S.; Liu, S. Bacteria-triggered release of a potent biocide from core-shell polyhydroxyalkanoate (PHA)-based nanofibers for wound dressing applications. *J. Biomater. Sci. Polym. Ed.* **2020**, *31*, 394–406. [CrossRef]
161. Fattahi, F.-S.; Khoddami, A.; Avinc, O. Poly(lactic acid) (PLA) Nanofibers for Bone Tissue Engineering. *J. Text. Polym.* **2019**, *7*, 47.
162. Kian, L.; Saba, N.; Jawaid, M.; Sultan, M. A review on processing techniques of bast fibers nanocellulose and its polylactic acid (PLA) nanocomposites. *Int. J. Biol. Macromol.* **2019**, *121*, 1314–1328. [CrossRef]
163. Li, W.; Fan, X.; Wang, X.; Shang, X.; Wang, Q.; Lin, J.; Hu, Z.; Li, Z. Stereocomplexed micelle formation through enantiomeric PLA-based Y-shaped copolymer for targeted drug delivery. *Mater. Sci. Eng. C Mater. Biol. Appl.* **2018**, *91*, 688–695. [CrossRef]
164. Maleki, H.; Barani, H. Stereocomplex electrospun fibers from high molecular weight of poly (L-lactic acid) and poly (D-lactic acid). *J. Polym. Eng.* **2020**, *40*, 136–142. [CrossRef]
165. Augustine, R.; Zahid, A.A.; Hasan, A.; Wang, M.; Webster, T.J. CTGF loaded electrospun dual porous core-shell membrane for diabetic wound healing. *Int. J. Nanomed.* **2019**, *14*, 8573. [CrossRef]
166. Pankongadisak, P.; Sangklin, S.; Chuysinuan, P.; Suwantong, O.; Supaphol, P. The use of electrospun curcumin-loaded poly(L-lactic acid) fiber mats as wound dressing materials. *J. Drug Deliv. Sci. Technol.* **2019**, *53*, 101121. [CrossRef]
167. Zou, F.; Sun, X.; Wang, X. Elastic, hydrophilic and biodegradable poly (1, 8-octanediol-co-citric acid)/polylactic acid nanofibrous membranes for potential wound dressing applications. *Polym. Degrad. Stab.* **2019**, *166*, 163–173. [CrossRef]
168. Yang, C.; Yan, Z.; Lian, Y.; Wang, J.; Zhang, K. Graphene oxide coated shell-core structured chitosan/PLLA nanofibrous scaffolds for wound dressing. *J. Biomater. Sci. Polym. Ed.* **2020**, *31*, 622–641. [CrossRef] [PubMed]
169. Majchrowicz, A.; Roguska, A.; Krawczyńska, A.; Lewandowska, M.; Martí-Muñoz, J.; Engel, E.; Castano, O. In vitro evaluation of degradable electrospun polylactic acid/bioactive calcium phosphate ormoglass scaffolds. *Arch. Civ. Mech. Eng.* **2020**, *20*, 50. [CrossRef]
170. Azimi, B.; Nourpanah, P.; Rabiee, M.; Arbab, S.; Cascone, M.; Baldassare, A.; Lazzeri, L. Application of the dry-spinning method to produce poly (ε-caprolactone) fibers containing bovine serum albumin laden gelatin nanoparticles. *J. Appl. Polym. Sci.* **2016**, *133*, 44233. [CrossRef]
171. Szentivanyi, A.; Chakradeo, T.; Zernetsch, H.; Glasmacher, B. Electrospun cellular microenvironments: Understanding controlled release and scaffold structure. *Adv. Drug Deliv. Rev.* **2011**, *63*, 209–220. [CrossRef] [PubMed]
172. Khalf, A.; Madihally, S.V. Recent advances in multiaxial electrospinning for drug delivery. *Eur. J. Pharm. Biopharm.* **2016**, *112*, 1–17. [CrossRef]
173. Spolenak, R.; Gorb, S.; Arzt, E. Adhesion design maps for bio-inspired attachment systems. *Acta Biomater.* **2005**, *1*, 5–13. [CrossRef]
174. Bolgen, N.; Vargel, I.; Korkusuz, P.; Menceloglu, Y.Z.; Piskin, E. In vivo performance of antibiotic embedded electrospun PCL membranes for prevention of abdominal adhesions. *J. Biomed. Mater. Res. B Appl. Biomater.* **2007**, *81B*, 530–543. [CrossRef]
175. Chen, C.; Lv, G.; Pan, C.; Song, M.; Wu, C.H.; Guo, D.D.; Wang, X.M.; Chen, B.A.; Gu, Z.Z. Poly(lactic acid) (PLA) based nanocomposites—A novel way of drug-releasing. *Biomed. Mater.* **2007**, *2*, L1–L4. [CrossRef]
176. Conway, J.; Whettam, J. Adverse reactions to wound dressings. *Nurs. Stand.* **2002**, *16*, 52–60. [CrossRef]
177. Stoddard, R.J.; Steger, A.L.; Blakney, A.K.; Woodrow, K.A. In pursuit of functional electrospun materials for clinical applications in humans. *Ther. Deliv.* **2016**, *7*, 387–409. [CrossRef] [PubMed]

178. Uppal, R.; Ramaswamy, G.N.; Arnold, C.; Goodband, R.; Wang, Y. Hyaluronic acid nanofiber wound dressing—Production, characterization, and in vivo behavior. *J. Biomed. Mater. Res.* **2011**, *97B*, 20–29. [CrossRef]
179. Mulholland, E.J. Electrospun Biomaterials in the Treatment and Prevention of Scars in Skin Wound Healing. *Front. Bioeng. Biotech.* **2020**, *8*, 1–15. [CrossRef] [PubMed]
180. Wijeyaratne, S.M.; Kannangara, L. Safety and Efficacy of Electrospun Polycarbonate-Urethane Vascular Graft for Early Hemodialysis Access: First Clinical Results in Man. *JAVA* **2011**, *12*, 28–35. [CrossRef] [PubMed]
181. Floyd, C.T.; Rothwell, S.W.; Risdahl, J.; Martin, R.; Olson, C.; Rose, N. Salmon thrombin-fibrinogen dressing allows greater survival and preserves distal blood flow compared with standard kaolin gauze in coagulopathic swine with a standardized lethal femoral artery injury. *J. Spec. Oper. Med.* **2012**, *12*, 16–26.
182. Silva, S.Y.; Rueda, L.C.; López, M.; Vélez, I.D.; Rueda-Clausen, C.F.; Smith, D.J.; Muñoz, G.; Mosquera, H.; Silva, F.A.; Buitrago, A.; et al. Double blind, randomized controlled trial, to evaluate the effectiveness of a controlled nitric oxide releasing patch versus meglumine antimoniate in the treatment of cutaneous leishmaniasis [NCT00317629]. *Trials* **2006**, *7*, 14. [CrossRef]
183. Silva, S.Y.; Rueda, L.C.; Márquez, G.A.; López, M.; Smith, D.J.; Calderón, C.A.; Castillo, J.C.; Matute, J.; Rueda-Clausen, C.F.; Orduz, A.; et al. Double blind, randomized, placebo controlled clinical trial for the treatment of diabetic foot ulcers, using a nitric oxide releasing patch: Pathon. *Trials* **2007**, *8*, 26. [CrossRef]
184. Arenbergerova, M.; Arenberger, P.; Bednar, M.; Kubat, P.; Mosinger, J. Light-activated nanofibre textiles exert antibacterial effects in the setting of chronic wound healing. *Exp. Dermatol.* **2012**, *21*, 619–624. [CrossRef]
185. Safety and Performance Study of a Novel Fibrin Dressing for Cancellous Bone Bleeding. Available online: https://stteresamedical.com/PDFS/SRS_2016_Abstr.pdf (accessed on 7 September 2020).
186. Balain, B.; Craig, N.; Gnanalingham, K.; Madan, S.; Sharma, H.; Wynne-Jones, G.; Hoseth, J.; Øsytein, L.; Punsvik, V.; Sura, S.; et al. Safety and Efficacy of a Novel Fibrin Dressing on Bleeding Cancellous Bone. *J Clin. Exp. Orthop.* **2018**, *4*, 50. [CrossRef]
187. Floyd, C.T.; Padua, R.A.; Olson, C.E. Hemostasis and Safety of a Novel Fibrin Dressing Versus Standard Gauze in Bleeding Cancellous Bone in a Caprine Spine Surgery Model. *Spine Deform.* **2017**, *5*, 310–313. [CrossRef]
188. Performance of a Dextran-only DressingVersus SurgiClot®to Achieve Hemostasis in a Swine Injury ModelofCancellous BoneBleeding. Available online: https://stteresamedical.com/PDFS/170-19-ORS_2016_FINAL-copy.pdf (accessed on 7 September 2020).
189. Spincare™. Available online: https://nanomedic.com/ (accessed on 7 September 2020).
190. Case Report 1: Using Spincare™ for Donor Site Wounds. Available online: https://nanomedic.com/research/donor-site-wounds/ (accessed on 7 September 2020).
191. Case Report 2: Using Spincare™ for Partial Thickness Wounds. Available online: https://nanomedic.com/research/partial-thickness-burns/ (accessed on 7 September 2020).
192. Clinical Trials Using Electrospinning. Available online: https://clinicaltrials.gov/ct2/results?cond=&term=electrospinning&cntry=&state=&city=&dist= (accessed on 7 September 2020).
193. Persano, L.; Camposeo, A.; Tekmen, C.; Pisignano, D. Industrial Upscaling of Electrospinning and Applications of Polymer Nanofibers: A Review. *Macromol. Mater. Eng.* **2013**, *298*, 504–520. [CrossRef]
194. Bosworth, L.A. Travelling along the Clinical Roadmap: Developing Electrospun Scaffolds for Tendon Repair. *Conf. Pap. Sci.* **2014**, 304974. [CrossRef]
195. Paul, R. High Performance Technical Textiles: An Overview. In *High Performance Technical Textiles: 2019*; John Wiley & Sons Ltd.: Hoboken, NJ, USA, 2019; pp. 1–10.
196. Gizaw, M.T.J.; Faglie, A.; Lee, S.; Neuenschwander, P.; Chou, S. Electrospun Fibers as a Dressing Material for Drug and Biological Agent Delivery in Wound Healing Applications. *Bioengineering* **2018**, *5*, 9. [CrossRef] [PubMed]
197. Rieger, K.A.; Birch, N.P.; Schiffman, J.D. Designing electrospun nanofiber mats to promote wound healing–a review. *J. Mater. Chem. B* **2013**, *1*, 4531–4541. [CrossRef] [PubMed]
198. Nasouri, K. Novel estimation of morphological behavior of electrospun nanofibers with artificial intelligence system (AIS). *Polym. Test.* **2018**, *69*, 499–507. [CrossRef]

199. Homaeigohar, S.; Boccaccini, A.R. Antibacterial biohybrid nanofibers for wound dressings. *Acta Biomater.* **2020**, *107*, 25–49. [CrossRef] [PubMed]
200. De la Ossa, J.G.; Felice, F.; Azimi, B.; Salsano, J.E.; Digiacomo, M.; Macchia, M.; Danti, S.; Di Stefano, R. Waste autochthonous tuscan olive leaves (Olea europaea var. olivastra seggianese) as antioxidant source for biomedicine. *Int. J. Mol. Sci.* **2019**, *20*, 5918. [CrossRef]
201. Orlando, I.; Roy, I. Cellulose-Based Hydrogels for Wound Healing. In *Cellulose-Based Superabsorbent Hydrogels. Polymers and Polymeric Composites: A Reference Series*; Mondal, M., Ed.; Springer: Cham, Switzerland, 2019; pp. 1–18. [CrossRef]

© 2020 by the authors. Licensee MDPI, Basel, Switzerland. This article is an open access article distributed under the terms and conditions of the Creative Commons Attribution (CC BY) license (http://creativecommons.org/licenses/by/4.0/).

MDPI
St. Alban-Anlage 66
4052 Basel
Switzerland
Tel. +41 61 683 77 34
Fax +41 61 302 89 18
www.mdpi.com

Journal of Functional Biomaterials Editorial Office
E-mail: jfb@mdpi.com
www.mdpi.com/journal/jfb

www.ingramcontent.com/pod-product-compliance
Lightning Source LLC
LaVergne TN
LVHW070407100526
838202LV00014B/1405